Springer Proceedings in Physics

Volume 170

More information about this series at http://www.springer.com/series/361

Piero Nicolini · Matthias Kaminski
Jonas Mureika · Marcus Bleicher
Editors

1st Karl Schwarzschild Meeting on Gravitational Physics

Springer

Editors
Piero Nicolini
Frankfurt Institute for Advanced Studies
Frankfurt am Main
Germany

Matthias Kaminski
Department of Physics and Astronomy
University of Alabama
Tuscaloosa, AL
USA

Jonas Mureika
Department of Physics
Loyola Marymount University
Los Angeles, CA
USA

Marcus Bleicher
Frankfurt Institute for Advanced Studies
Frankfurt am Main
Germany

ISSN 0930-8989 ISSN 1867-4941 (electronic)
Springer Proceedings in Physics
ISBN 978-3-319-20045-3 ISBN 978-3-319-20046-0 (eBook)
DOI 10.1007/978-3-319-20046-0

Library of Congress Control Number: 2015945121

Springer Cham Heidelberg New York Dordrecht London

Printed on acid-free paper

Springer International Publishing AG Switzerland is part of Springer Science+Business Media
(www.springer.com)

Preface

The 2013 Karl Schwarzschild Meeting on Gravitational Physics (KSM) was a top international event involving the worldwide highest qualified scientific personalities in the field of black hole physics, general relativity, and related topics. It featured the participation of 91 scientists from 15 countries over 4 continents. These attendees included undergraduate and graduate students, postdoctoral researchers, as well as junior and senior faculty. We envisioned the foundational spirit of the conference to be: "by acknowledging the past we open a route to the future." Here "the past" refers to the pioneering black hole studies of Karl Schwarzschild, a native of Frankfurt am Main, who published his first two papers while attending the Frankfurt-Gymnasium (now the Lessing-Gymnasium) in Fürstenbergerstra*β*e 166 in the late 1880s.

The year 2013 marked the 140th anniversary of Schwarzschild's birth. Although inspired by the key historical work of Karl Schwarzschild, this meeting served to highlight its various repercussions in a variety of aspects of frontier theoretical physics. Black holes are no longer pedagogical curiosities of mathematics physics. Instead, these objects constitute one of the primary testbeds of the ultimate theory of nature: quantum gravity. Their presence in many distinct branches of physics is striking. Historically, this is the result of a long process started with the discovery of the Hawking effect (also known as black hole evaporation), which can be regarded as the first attempt to reconcile gravity and quantum mechanics. The Hawking effect has since contributed to the collapse of barriers between gravitation and particle physics, thermodynamics, and even condensed matter physics.

Over the past decade, black holes have become a central feature of attempts to address the hierarchy problem through the introduction of extra spatial dimensions. In such frameworks, it has become commonplace in high energy physics to suppose that microscopic black holes could be produced in current and future accelerator experiments. Black holes have also gained a more solid reception in the field of observational astronomy due to the improved technology of modern radio-telescopes and are a common topic to astronomers and theoretical particle physicists. In contrast to the various international meetings on gravitation, the KSM sought to offer a complementary program: rather than quantity, we aimed for quality.

A smaller group of eminent scientists came to review groundbreaking frontier results in gravity research by means of comprehensive plenary lectures.

As a completely new feature, we provided a worldwide platform specifically dedicated to young scientists. To foster the synergy and the collaborative spirit among senior academics and forthcoming scientific leaders, a major component of the KSM featured plenary sessions highlighting the work of the "next generation" of gravitational physicists. This included a variety of levels of researchers: young research group leaders, assistant professors, postdoctoral researchers, and even doctoral candidates. Selected senior scientists served as mentors to the group of young participants, starting off the week with a "Meet Your Mentor" session to discuss their research and career paths. A competition for "Best Student Talk" and "Best Junior Scientist Talk" was sponsored by Springer. The winners were honored at an evening gala onward the end of the week. The prize for Best Student Talk was won by Daniel Siegel (MPI, Potsdam), with Honorable Mentions going to Maximiliano Isi (Loyola Marymount University, Los Angeles), Stefan Janiszewski (University of Washington, Seattle), and Benjamin Niedner (Oxon). The prize for Best Junior Scientist Talk was awarded to Daniele Malafarina (Fudan University, Shanghai), with Honorable Mentions to Shohreh Abdolrahimi (Oldenburg University), Michele Fontanini (Sao Paolo University), Benjamin Koch (PUC, Santiago).

In a time of online-journals and electronic communication, the KSM was envisioned as not merely a venue for exchange of information, but rather as a place where new ideas are developed through complementary knowledge and encouraged interactions of the participants. We feel the conference more than met these goals. In this volume, we share the fruits of this labor.

Frankfurt am Main

Piero Nicolini
Matthias Kaminski
Jonas Mureika
Marcus Bleicher

The participants and organizers of the Karl Schwarzschild Meeting 2013 in the FIAS Lecture Hall

Keynote Speakers:
Alfio Bonanno (Catania Observatory)
Silke Britzen (MPI Bonn)
Xavier Calmet (Sussex)
Bernard Carr (QMU London)
Georgi Dvali (LMU / MPI Munich / NYU)
Steven Gubser (Princeton)
Andreas Karch (UW Seattle)
Frans Klinkhamer (KIT)
Claus Laemmerzahl (Bremen / ZARM)
Greg Landsberg (Brown / CERN)
Frank Linhard (Frankfurt)
Reinhard Meinel (Jena)
Robert Mann (Waterloo / PI)
John Moffat (Waterloo / PI)
Leonardo Modesto (Fudan)
Thanu Padmanabhan (IUCAA)
Luciano Rezzolla (MPI Potsdam)
Kellog Stelle (Imperial)
Leonard Susskind (Stanford)
Herman Verlinde (Princeton)
Robert Wald (Chicago / EFI)
Clifford Will (Florida / WUSTL)
DFG
HIC for FAIR
Helmholtz International Center
FREUNDE DER UNIVERSITÄT
Springer
FIAS Frankfurt Institute for Advanced Studies
Karl Schwarzschild Meeting 2013
Frankfurt am Main
22-26 July 2013
Organizers:
Piero Nicolini (FIAS)
Marcus Bleicher (FIAS)
Jonas Mureika (LMU - LA)
Matthias Kaminski (UW Seattle)
Scientific Advisory Board:
Jacob Bekenstein (Jerusalem)
Ted Jacobson (Maryland)
Joseph Polchinski (UCSB \ KITP)
Martin Reuter (Mainz)
Carlo Rovelli (Aix-Marseille / CPT)
Dam T. Son (Chicago)
KSM 2013

Acknowledgments

The organizers of the 2013 Karl Schwarzschild Meeting on Gravitational Physics are grateful for the generosity of the Frankfurt Institute for Advanced Studies for providing the spectacular venue at which the meeting took place. We thank the Deutsche Forschungsgemeinschaft (DFG), the Helmholtz International Center for FAIR, and the Freunde und Förderer der Goethe-Universität Frankfurt for their financial support, as well as the Physikalischer Verein Frankfurt for hosting a visit to the old Physics Institute where the Stern-Gerlach Experiment was performed in 1922. Special thanks goes to Springer for providing the prizes for Best Student and Best Junior Scientist presentations, the basis for an integral part of our philosophy.

For their help and support of our efforts, we extend a special acknowledgment to our Advisory Board: Jacob Bekenstein (Hebrew University of Jerusalem), Ted Jacobson (University of Maryland, College Park), Joseph Polchinski (University of California and Kavli Institute for Theoretical Physics, Santa Barbara), Martin Reuter (Johannes-Gutenberg University, Mainz), Carlo Rovelli (Aix-Marseille University and Centre de Physique Théorique, Marseille), and Dam T. Son (The University of Chicago).

Finally, we are indebted to all participants who helped to make the KSM a success, and hopefully the first of many such meetings to come.

Contents

Contributors

Shohreh Abdolrahimi Institut für Physik, Universität Oldenburg, Oldenburg, Germany

Giovanni Acquaviva Dipartimento di Fisica, Università degli Studi di Trento, Trento, Italy

Victor E. Ambruş Consortium for Fundamental Physics, School of Mathematics and Statistics, The University of Sheffield, Sheffield, UK

Martin Ammon Theoretisch-Physikalisches Institut, Friedrich-Schiller-Universität Jena, Jena, Germany

Ivan Arraut Theory Center, Institute of Particle and Nuclear Studies, The High Energy Accelerator Research Organization, Tsukuba, Ibaraki, Japan

Max R. Atkin Fakultät für Physik, Universität Bielefeld, Bielefeld, Germany; Université Catholique de Louvain, Louvain-la-Neuve, Belgium

Shan Bai Theoretisch-Physikalisches Institut, Friedrich-Schiller-Universität Jena, Jena, Germany

Luca Baiotti Institute of Laser Engineering, Osaka University, Suita, Osaka, Japan

Lorenzo Bianchi Institut für Physik und Integrative Research Institute for the Sciences IRIS Adlershof, Humboldt-Universität zu Berlin, Berlin, Germany

Thomas Boller Max-Planck-Institut für Extraterrestrische Physik, Garching, Germany

Alfio Bonanno Istituto Nazionale di Astrofisica, Osservatorio Astrofisico di Catania, Catania, Italy; Istituto Nazionale di Fisica Nucleare, Sezione di Catania, Catania, Italy

Silke Britzen Max-Planck-Institut für Radioastronomie, Bonn, Germany

Xavier Calmet Physics and Astronomy, University of Sussex, Falmer, Brighton, UK

Bernard J. Carr Department of Physics and Astronomy, Queen Mary University of London, London, UK

Roberto Casadio Dipartimento di Fisica e Astronomia, Università di Bologna, Bologna, Italy; Istituto Nazionale di Fisica Nucleare, Sezione di Bologna, Bologna, Italy

Gunther Caspar Frankfurt Institute for Advanced Studies, Frankfurt am Main, Germany

Riccardo Ciolfi Max-Planck-Institut für Gravitationsphysik (Albert-Einstein-Institut), Potsdam, Germany

Alessandro Codello CP3–Origins and Danish Institute for Advanced Study, University of Southern Denmark, Odense, Denmark; Scuola Internazionale Superiore di Studi Avanzati, Trieste, Italy

Carlos Contreras Departamento de Física, Universidad Técnica Federico Santa María, Valparaíso, Chile

Giulio D'Odorico Institute for Mathematics, Astrophysics and Particle Physics, Radboud University, Nijmegen, The Netherlands

Jiří Daněk Faculty of Mathematics and Physics, Institute of Theoretical Physics, Charles University in Prague, Prague, Czech Republic

Dennis D. Dietrich Institut für Theoretische Physik, Johann Wolfgang Goethe-Universität Frankfurt am Main, Frankfurt am Main, Germany; Arnold-Sommerfeld-Zentrum für Theoretische Physik, Ludwig-Maximilians-Universität München, Munich, Germany

Johanna Erdmenger Max-Planck-Institut für Physik (Werner-Heisenberg-Institut), Munich, Germany

Carlo Ewerz Institut für Theoretische Physik, Ruprecht-Karls-Universität Heidelberg, Heidelberg, Germany; ExtreMe Matter Institute, GSI Helmholtzzentrum für Schwerionenforschung, Darmstadt, Germany

Michele Fontanini Instituto de Física, Universidade de São Paulo, São Paulo, SP, Brazil

Valentina Forini Institut für Physik und Integrative Research Institute for the Sciences IRIS Adlershof, Humboldt-Universität zu Berlin, Berlin, Germany

Antonia M. Frassino Frankfurt Institute for Advanced Studies, Frankfurt am Main, Germany; Institut für Theoretische Physik, Johann Wolfgang Goethe-Universität Frankfurt am Main, Frankfurt am Main, Germany

Sijie Gao Department of Physics, Beijing Normal University, Beijing, China

Remo Garattini Dipartimento di Ingegneria, Università degli Studi di Bergamo, Dalmine, Italy; Istituto Nazionale di Fisica Nucleare, Sezione di Milano, Milan, Italy

Amir M. Ghezelbash Department of Physics and Engineering Physics, University of Saskatchewan, Saskatoon, SK, Canada

Walter Greiner Frankfurt Institute for Advanced Studies, Frankfurt am Main, Germany

Daniel C. Guariento Instituto de Física, Universidade de São Paulo, São Paulo, SP, Brazil; Perimeter Institute for Theoretical Physics, Waterloo, ON, Canada

Norman Gürlebeck Zentrum für angewandte Raumfahrttechnologie und Mikrogravitation, Universität Bremen, Bremen, Germany

Eva Hackmann Zentrum für angewandte Raumfahrttechnologie und Mikrogravitation, Universität Bremen, Bremen, Germany

Peter O. Hess Instituto de Ciencias Nucleares, Universidad Nacional Autónoma de México, Circuito Exterior, Mexico, DF, Mexico

Ben Hoare Institut für Physik und Integrative Research Institute for the Sciences IRIS Adlershof, Humboldt-Universität zu Berlin, Berlin, Germany

Mei Huang Institute of High Energy Physics, Chinese Academy of Sciences, Beijing, China

Giorgio Immirzi Montopoli di Sabina, Italy

Lorenzo Iorio Ministero dell'Istruzione, dell'Università e della Ricerca, Bari, Italy

Maximiliano Isi Department of Physics, Loyola Marymount University, Los Angeles, CA, USA

Stefan Janiszewski Department of Physics, University of Washington, Seattle, WA, USA

Andreas Karch Department of Physics, University of Washington, Seattle, WA, USA

Frans R. Klinkhamer Institut für Theoretische Physik, Karlsruher Institut für Technologie, Karlsruhe, Germany

Benjamin Koch Instituto de Fisica, Pontificia Universidad Católica de Chile, Santiago, Chile

Claus Lämmerzahl Zentrum für angewandte Raumfahrttechnologie und Mikrogravitation, Universität Bremen, Bremen, Germany

Danning Li Institute of High Energy Physics, Chinese Academy of Sciences, Beijing, China

Frank Linhard Physikalischer Verein, Frankfurt am Main, Germany

Francisco S.N. Lobo Instituto de Astrofísica e Ciências do Espaço, Faculdade de Ciências da Universidade de Lisboa, Lisbon, Portugal

Daniele Malafarina Center for Field Theory and Particle Physics & Department of Physics, Fudan University, Shanghai, China; Department of Physics, SST Nazarbayev University, Astana, Kazakhstan

Robert B. Mann Department of Physics and Astronomy, University of Waterloo, Waterloo, ON, Canada

Dimitri Marinelli Dipartimento di Fisica, Università degli Studi di Pavia, Pavia, Italy

Reinhard Meinel Theoretisch-Physikalisches Institut, Friedrich-Schiller-Universität Jena, Jena, Germany

John W. Moffat Perimeter Institute for Theoretical Physics, Waterloo, ON, Canada; Department of Physics and Astronomy, University of Waterloo, Waterloo, ON, Canada

Andreas Müller Exzellenzcluster Origin and Structure of the Universe, Technische Universität München, Garching, Germany

Jonas Mureika Department of Physics, Loyola Marymount University, Los Angeles, CA, USA

Piero Nicolini Frankfurt Institute for Advanced Studies, Frankfurt am Main, Germany; Institut für Theoretische Physik, Johann Wolfgang Goethe-Universität Frankfurt am Main, Frankfurt am Main, Germany

Benjamin Niedner Rudolf Peierls Centre for Theoretical Physics, Oxford, UK

Niall Ó Murchadha Physics Department, University College Cork, Cork, Ireland

Gonzalo J. Olmo Departamento de Física Teórica and Instituto de Física Corpuscular, Centro Mixto Universidad de Valencia - Consejo Superior de Investigaciones Científicas. Universidad de Valencia, Burjassot, Valencia, Spain

Alessio Orlandi Dipartimento di Fisica e Astronomia, Università di Bologna, Bologna, Italy

Tanu Padmanabhan Inter-University Centre for Astronomy and Astrophysics, Pune University Campus, Pune, India

Carlo Pagani Scuola Internazionale Superiore di Studi Avanzati, Trieste, Italy; Istituto Nazionale di Fisica Nucleare, Trieste, Italy

Da-Wei Pang Max-Planck-Institut für Physik (Werner-Heisenberg-Institut), Munich, Germany

Roberto Percacci Scuola Internazionale Superiore di Studi Avanzati, Trieste, Italy

Leslaw Rachwal Abdus Salam International Centre for Theoretical Physics, Trieste, Italy; Scuola Internazionale Superiore di Studi Avanzati, Trieste, Italy

Luciano Rezzolla Max-Planck-Institut für Gravitationsphysik (Albert-Einstein-Institut), Potsdam, Germany

Paola Rioseco Instituto de Fisica, Pontificia Universidad Católica de Chile, Santiago, Chile

Diego Rubiera-Garcia Departamento de Física, Universidade Federal da Paraíba, João Pessoa, PB, Brazil

Andreas Samberg Institut für Theoretische Physik, Ruprecht-Karls-Universität Heidelberg, Heidelberg, Germany; ExtreMe Matter Institute, GSI Helmholtzzentrum für Schwerionenforschung, Darmstadt, Germany

Frank Saueressig Institute for Mathematics, Astrophysics and Particle Physics, Radboud University, Nijmegen, The Netherlands

Mirko Schäfer Frankfurt Institute for Advanced Studies, Frankfurt am Main, Germany

Thomas Schönenbach Frankfurt Institute for Advanced Studies, Frankfurt am Main, Germany

Daniel M. Siegel Max-Planck-Institut für Gravitationsphysik (Albert-Einstein-Institut), Potsdam, Germany

Kellogg S. Stelle The Blackett Laboratory, Imperial College London, London, UK

Kentaro Takami Max-Planck-Institut für Gravitationsphysik (Albert-Einstein-Institut), Potsdam, Germany

John F. Wheater Rudolf Peierls Centre for Theoretical Physics, Oxford, UK

Elizabeth Winstanley Consortium for Fundamental Physics, School of Mathematics and Statistics, The University of Sheffield, Sheffield, UK

Hansjörg Zeller Max-Planck-Institut für Physik (Werner-Heisenberg-Institut), Munich, Germany

Part I
The Life and Work of Karl Schwarzschild

Chapter 1
Karl Schwarzschild and Frankfurt

Frank Linhard

Abstract This paper will give a short account of the life of Karl Schwarzschild and his relation to Frankfurt. Focus will be on the early papers he published, while he was still at school. And his last papers, that made him famous in the community of astrophysicists focusing on black holes will be mentioned in biographical context as well.

1.1 Short Overview of Karl Schwarzschild's Life

1.1.1 Karl Schwarzschild and Jewish Frankfurt

Karl Schwarzschild was a Frankfurt born Astronomer of Jewish descent. The name *Schwarzschild* does not only sound similar to the name *Rothschild*. Actually both names have their origin in the Frankfurt ghetto. Jews were forced into a ghetto since 1462. On one hand this was done to separate them from the others, but the argument mentioned, was that they should receive protection by the Kaiser. Therefore ghettos were established in the so called free cities of the "Reich". Of course the protection had a price: Taxes were raised by the church—in Frankfurt represented by the Archbishop of Mainz—by the city, i.e. the City Council, and by the Kaiser himself. In 1806 the French appointed Grand Duke of Frankfurt Karl von Dalberg ordered that equal rights be granted to all religious creeds. Despite von Dalberg's efforts, Frankfurt issued a new set of Jewish regulations in 1807 that attempted to reestablish the ghetto. Finally in 1811 Dalberg's Highest Regulation, for the equality of civil right of the Jewish Municipality eliminated the requirements to live in the ghetto and abolished all special Jewish taxes.[1] Karl Schwarzschild was born after this, but the former situation had lasted for centuries and thus formed the social environment. Karl was born on the 9th of October 1873 and died at the age of 42 on

[1] This description is based on the documentation at the *Museum Judengasse*, Frankfurt am Main.

F. Linhard (✉)
Physikalischer Verein, Frankfurt Am Main, Germany
e-mail: frank@linhard.com

P. Nicolini et al. (eds.), *1st Karl Schwarzschild Meeting on Gravitational Physics*, Springer Proceedings in Physics 170, DOI 10.1007/978-3-319-20046-0_1

May 16th 1916. There is still no scientific biography of Karl Schwarzschild, but a scetch of 25 pages Biography in the Collected Works by Voigt, Göttingen (see [1, 14], p. 1). Furthermore there is an entry in the Dictionary of Scientific Biography by Dieke (6 pages) [2]. For his early years within his Frankfurt family we have a short account in the memoirs of his brother Alfred—this covers up to 1907 (see [14], p. 2). The main steps in his life[2] were Karls early years in Frankfurt (1873–1891), his studies in Strasbourg (1891–1893) and Munich (1893–1896). A position at the Observatory in Vienna (1896–1899), were he did scientific work to do his *Habilitation* in Munich (habil. 1899–1901 Privatdozent). From there he was appointed Professor and Director of the Observatory at Göttingen (1901–1909). Finally he changed to Potsdam for the most prestigious position an astronomer could have at this time in Germany (1909–1913). Karl was born the eldest of 7 children, the first in his family to become a scientist. His father Moses Martin was a successful stock broker in Frankfurt and the family was comparably wealthy [2]. Alfred Schwarzschild described his brother in his memoirs like this:

> He built a telescope around an objective lens, using a tube of newspaper blackened inside, and showed his brothers and sisters the rings of Saturn. He had piano lessons, but he was just as interested in the mechanism inside the piano and the theory of sound vibrations as he was in music itself (see [14], p. 1).

Karl first attended the Jewish Community School and later the municipal Gymnasium in Frankfurt (now called Lessing Gymnasium). Apart from pure mathematics, astronomy soon became a major interest. His fathers friend J. Epstein was a mathematician with a private observatory (see [14], p. 1). In his last year at school, Karl was fully occupied with studies, preparation for the Abitur,[3] dance lessons, writing poems, etc. On September 3rd 1891, he came top in the Abitur and therefore had to give the farewell address in Latin (see [14], p. 4).

1.1.2 Schwarzschild as a Student and Postdoc

The same year he began to study astronomy in Strasburg with Ernst Becker the Director of the Observatory in Strasburg, which was German since 1871.

In 1893 he did his *Einjähriges*—the German military service—at Munich but continued his studies in 1894 under Hugo von Seeliger at the Ludwig Maximilians University. His doctorate he gained in 1896 with "summa cum laude" on "The Poincare Theory of the Equilibrium of a Homogeneous, Rotating, Fluid Body".

After his doctorate he became Assistant at the von Kuffner Observatory near Vienna, where he did scientific work on photometry.

---

[2]For this chronology I will combine the outlines from the biographies mentioned [6], p. 1–28, [2, 6].

[3]Final exams that pupils take at the end of their secondary education in Germany.

Contributions on the Photographic Photometry of Stars (see [14], Vol. 2 [6.2]) was accepted as *Habilitationsschrift* at Munich and in 1899 he became *Privatdozent* at the Ludwig Maximilians University.

1.1.3 The Göttingen Years

The Göttingen position was very prestigious and Berlin hesitated to appoint a young scholar like Karl to Director of the Observatory and full Professor. But other candidates, like his mentor Seeliger did not want to come to Göttingen. Karl sent his parents a telegram:

> Extraordinarius and Director. Arrive Monday—Karl. (see [14], p. 10)

October the 19th 1901 his appointment as assistant Professor was announced. Karl was to spend the happiest years of his life in Göttingen. By May 24th 1902 he was appointed full Professor. In 1904, together with Klein, Hilbert and Minkowski he organized a seminar on mathematical physics. His main contributions to science during the Göttingen years were: The so called *Göttingen Actinometry* and the confirmation of the existence of Red Giants and the explanation of the statistical distribution of luminosities found by Hertzsprung. Furthermore he gained material for stellar Statistics.

In 1905 Karl published three fundamental papers on geometrical optics, of which Max Born later wrote, these were the backbone for his own "Optics", the most famous book in the field for decades. Also in 1905 he did work on solar physics and followed an Expedition to Algiers for the eclipse on August 30th 1905. He did spectroscopic and photographic research and documented the entire expedition in a large paper [8]. In 1906 he did theoretical work on the equilibrium of the Sun's atmosphere, and in 1907 he worked on *proper motion of the stars*, a proposal of the ellipsodial velocity-distribution. In 1909 he married Else Rosenbach, but that was in his last year at Göttingen.

1.1.4 Schwarzschild and the Astrophysical Observatory at Potsdam and World War I

The Astrophysical Observatory at Potsdam was the greatest research institution for Astronomy in Germany (see [14], p. 18). And while Karl already had problems to get the appointment at Göttingen, because of his age, the situation was even worth with Potsdam. It was the most prominent position for an Astronomer in Germany and Karl was aged only 36. Again he gained the appointement and reached the peak of his career the young.

What he had to do in Potsdam more than in Göttingen was Administration, as the research institution was so large, there was a lot to organise and supervise.

But he also did work on stellar statistics and stellar dynamics, and on the spectral classification of stars. Because of the value of his scientific work, together with his successful fulfillment of the prestigious Potsdam position, in 1912 Prussian Academy of Sciences Berlin awarded him with the Membership, while a Honorary Professorship at Berlin followed only in 1916. But this was already during the First World War.

In 1914: Schwarzschild immediately volunteered for service, although he would not have been drafted, because of his age and position. Else Schwarzschild later explained to their children: He felt obliged to take this step, precisely because he was a German Jew (see [14], p. 23).

From the many contributions Schwarzschild made to Astronomy and Astrophysics some were named after him, as

- the Schwarzschild exponent in photography
- the Schuster–Schwarzschild model
- the Schwarzschild criterion for convection in stellar atmospheres
- the Schwarzschild equation in stellar statistics
- the Schwarzschild distribution of stellar velocities
- and—of course—the famous Schwarzschild radius

As already mentioned his main projects were:

- Photographic stellar Photometry (law of density of photographic films)
- Göttingen Actinometry
- Ellipsoidal hypothesis
- Examination of Stellar-statistical Problems

Karl Schwarzschild published 120 papers in total. He also worked on the construction and theory of instruments, and on geometrical optics: From the theorem of the shortest light-path a theory of optical instruments was developped (Eikonal 1905).

When Schwarzschild died it was mentioned by 15 addresses and 20 short notices. Some appeared in the most eminent journals by the greatest scientists of the time, as in e.g.:

- Astronomische Nachrichten 1916
- The Observatory 1916
- Astrophysical Journal 1917 (Hertzsprung) [5]
- MNRAS 1917 (Eddington) [4]
- Einstein's memorial address June 29th, 1916 Berlin Academy (see [14], pp. 34–35)

> Karl Schwarzschild was born at Frankfurt 1873 October 9. He was educated successively at the public school at Frankfurt and at Strasbourg and Munich universities. He gave early evidence of remarkable mathematical ability, and when barely sixteen contributed two excellent papers to the Astronomische Nachrichten on the problem of determining an orbit from three observations [5].

This is the beginning of the obituary by Hertzsprung (1873–1967), who was a close friend of Karl since 1902 in Göttingen. He worked with him in Potsdam since 1909 as well.

1.2 Young Schwarzschild: The Frankfurt Papers

When Karl was a 16 years old pupil at the Gymnasium (now Lessing Gymnasium) he was already acquainted with his fathers friend Epstein and his son Paul, who later became Professor of Mathematics at Strasburg. It is recorded, that Karl forgot his logarithm tables at school, and that he was able, to calculate the required values himself using a series expansion (see [14], p. 2).

Almost every Obituary or account of Karl's life mentions his early papers. His brother Alfred remembered, what happened when he did send his work on orbital determination to the *Astronomische Nachrichten* and it was accepted.

Karl gave his father a printed copy at his birthday. The family was very proud and everybody read through the paper several times. As Alfred says—*not one of us understood a word of it* (see [14], p. 2). Karl wrote papers on Celestial Mechanics during the period between 1890 and 1903 in Frankfurt, Munich and at the Von Kuffner Observatory, Vienna. The first papers were focused on Methods for the determination of orbits, later papers on the Capture of Comets by Jupiter, Periodic Orbits and General Celestial Mechanics.

The Franfurt papers developed a method by Bruns for determining an orbit from three positions further: Heinrich Bruns (1848–1919) was the director of Leipzig Observatory.

Karl Schwarzschild puts the finishing touches to the method by taking Bruns' elements as a first approximation for the computation of further approximations. This works until satisfactory accuracy is reached. In the paper he applies differentiation of the equations by Bruns. After that he neglects terms of higher order. With his results he applies an iteration [3, 9, 10].

Zur Bahnbestimmung nach Bruns.

Von *K. Schwarzschild.*

Die Bahnbestimmungs-Methode, die Herr Prof. Bruns in Nr. 2824 der Astr. Nachr. angegeben hat, löst die Aufgabe nicht vollständig, eine Bahn an drei gegebene Beobachtungen eines Himmelskörpers anzuschliessen. Will man dies aber erreichen, so kann man folgende Methode benutzen, die die Bruns'sche Methode als erste Annäherung giebt und zeigt, wie weitere Annäherungen zu finden sind.

Nach den in der erwähnten Abhandlung angegebenen Bezeichnungen hat man die Gleichungen:

$$\xi'' + \frac{k^2\xi}{r^3} + k^2 XP = 0; \quad \eta'' + \frac{k^2\eta}{r^3} + k^2 YP = 0; \quad \zeta'' + \frac{k^2\zeta}{r^3} + k^2 ZP = 0$$

$$\xi = \varrho\cdot f; \quad \eta = \varrho\cdot g; \quad \zeta = \varrho\cdot h; \quad 1 = f^2 + g^2 + h^2; \quad P = \frac{\mu}{R^3} - \frac{1}{r^3}$$

demnach:

$$\left.\begin{aligned} f\varrho'' + 2f'\varrho' + f''\varrho + k^2\frac{\varrho\cdot f}{r^3} + k^2 XP &= 0 \\ g\varrho'' + 2g'\varrho' + g''\varrho + k^2\frac{\varrho\cdot g}{r^3} + k^2 YP &= 0 \\ h\varrho'' + 2h'\varrho' + h''\varrho + k^2\frac{\varrho\cdot h}{r^3} + k^2 ZP &= 0 \end{aligned}\right\} \quad (1)$$

Karl gets 6 equations from 3 observations. Likewise Karl Schwarzschild applies the same method to the determination of the orbital elements of visual binaries. Unfortunately most scientists who worked in the field ignored Schwarzschild's papers. At least Aitken refers to Karl Schwarzschild's second paper in his 1935 *The Binary Stars* [1]. In Newcomb-Engelmann's *Populäre Astronomie* Karl Schwarzschild himself admits, that there has not much changed in the method of determining orbits, since the days of Gauss and Olbers. He speaks of ño substantial improvements during the last centuries.[4]

So kann man durch Differenziren beliebig viele Gleichungen aufstellen. Die Grössen X, Y, Z sind stets bekannt. Für die Grössen f, g, h und deren Differentialquotienten werden durch drei Beobachtungen folgende Gleichungen geliefert, wenn man die Grössen, die zu der als Ausgangspunkt genommenen Zeit t gehören, ohne Index lässt, dagegen diejenigen, die zu den Zeiten t_1, t_2, t_3 gehören, durch die Indices 1, 2, 3 kennzeichnet:

$$\left.\begin{aligned} f_1 &= f + (t_1 - t)f' + \frac{(t_1 - t)^2}{1.2} f'' + \frac{(t_1 - t)^3}{1.2.3} f''' + \cdots \\ f_2 &= f + (t_2 - t)f' + \frac{(t_2 - t)^2}{1.2} f'' + \frac{(t_2 - t)^3}{1.2.3} f''' + \cdots \\ f_3 &= f + (t_3 - t)f' + \frac{(t_3 - t)^2}{1.2} f'' + \frac{(t_3 - t)^3}{1.2.3} f''' + \cdots \end{aligned}\right\} \quad (6)$$

Entsprechende Gleichungen hat man für g und h.

1.3 World War I and the Final Papers

At the outbreak of World War I Karl immediately volunteered and he was appointed "Offiziersstellvertreter" at a Weather Station in Namur. Later he had to stay with the Artillery in France and Russia, where he caught an infection at the Eastern front in late 1915. Eddington gives an account of this final episode in Schwarzschild's life at the end of his Eloge for the Astronomical Society:

> Soon after the outbreak of war Schwarzschild was placed in charge of a meteorological station for the German army at Namur. He was afterwards attached to the artillery staff, at first in France and later on the Eastern front. He received the order of Knight of the Iron Cross. As might be expected, the German army organisation took care to give scope to his scientific ability, and in 1915 November he contributed to the Berlin Academy a paper on "The Effect of Wind and Air-density on the Path of a Projectile": publication has, of course, been postponed. On military service he contracted a disease which ultimately proved fatal. After a long illness, attended with much suffering, he died on 1916 May 11. His lively and attractive character and his readiness to co-operate with others brought him into cordial relationship with astronomers in many parts of the world, and his early death is felt as a deep personal loss. He was elected an Associate on 1909 June 11. A.S.E. [4].

His final papers made him most famous: He solved the equations of Einstein's new General Theory of Relativity for the metric around a point mass. Then Schwarzschild solved for the metric inside an incompressible fluid sphere, a solution fundamental to gravitational collapse and black hole formation.

His solutions exhibited the Schwarzschild radius characterizing Black Holes, but it also permits the calculation of the precession of the perihelion of elliptical planetary orbits characteristic of General Relativity. The papers were published as:

> K. Schwarzschild: Über das Gravitationsfeld eines Massenpunktes nach der Einsteinschen Theorie, Sitzungsberichte der Mathematisch Physikalischen Klasse der Deutschen Akademie der Wissenschaften Berlin 16, 189 (1916) and Ibid. 424 [11, 12].

[4]The impact of the early work was—as already mentioned—very modest: Methods for determining a preliminary orbit are programmed today, while the determination of definitive orbits for comets and asteroids still requires laborious, detailed work [7].

His very last paper was *Zur Quantenhypothese*, written during his illness, bears on the application of the quantum theory to spectral series [13].

Hertzsprung accounted on his last impression of Karl Schwarzschild in his obituary:

> The ardent wish for scientific occupation did not leave him until the last. It seemed incredible to the visitor at his sickbed that he was a man so near death. At such an occasion one feels the mastery of a strong spirit over earthly misery [5].

Schwarzschild died a week after publication of the last paper. What killed him was 'pemphigus' (a form of acute necrotizing ulcerative gingivitis), a disease that is still quite dreadful and can be extremely resistant even to modern antibiotics [6].

He and Else had three children, Agathe, Martin, and Alfred. Martin Schwarzschild (1912–1997) fled Nazi Germany in 1935 for an outstanding career in Astrophysics at Princeton University (see [14], p. 25). Alfred stayed in Nazi Germany and was murdered in the Holocaust.

I will close this short account of Karl Schwarzschilds life and work with his own words from his Admission speech before the Berlin Academy of Science 1913, translated by Eddington:

> Mathematics, physics, chemistry, astronomy, march in one front. Whichever lags behind is drawn after. Whichever hastens ahead helps on the others. The closest solidarity exists between astronomy and the whole circle of exact science. [...] From this aspect I may count it well that my interest has never been limited to the things beyond the moon, but has followed the threads which spin themselves from there to our sublunar knowledge; I have often been untrue to the heavens (cited from [4]).

References

1. R.G. Aitken, *The Binary Stars* (McGraw-Hill, New York, 1935)
2. S.H. Dieke, Schwarzschild, Karl, in *Dictionary of Scientific Biography*, vol. 12 (Charles Scribner's Sons, New York, 1975), pp. 247–253
3. R. Dvorak, Celestial mechanics, commentary, in *Karl Schwarzschild: Gesammelte Werke: Collected Works*, vol. 1–3 (Springer, Berlin, 1992), pp. 45–49
4. A. Eddington, Karl Schwarzschild. Mon. Not. R. Astron. Soc. **77**, 314–319 (1917)
5. E. Hertzsprung, Karl Schwarzschild. Astrophys. J. **45**, 285–292 (1917)
6. J.J. O'Connor, E.F. Robertson, Karl Schwarzschild, in *MacTutor History of Mathematics Archive* (University of St Andrews, 2003). http://www-history.mcs.st-andrews.ac.uk/Biographies/Schwarzschild.html. Cited 21 Mar 2015
7. K. Schwarzschild, Bahnbestimmung (Determination of orbits), in *Newcomb-Engelmann's Populäre Astronomie*, 5th edn., ed. by P. Kempf (Leipzig, 1914), pp. 89–93
8. K. Schwarzschild, Über die totale Sonnenfinsternis vom 30. August 1905. Abhandl. Kgl. Ges. d. Wiss. zu Gött. NF **5**(2), 3–73 (1907)
9. K. Schwarzschild, Methode zur Bahnbestimmung der Doppelsterne. Astron. Nachr. **124**, 216–218 (1890)
10. K. Schwarzschild, Zur Bahnbestimmung nach Bruns. Astron. Nachr. **124**, 211–216 (1890)
11. K. Schwarzschild, Über das Gravitationsfeld eines Massenpunktes nach der Einsteinschen Theorie. Sitzungsberichte der Mathematisch Physikalischen Klasse der Deutschen Akademie der Wissenschaften Berlin **16**, 189 (1916)

12. K. Schwarzschild, Über das Gravitationsfeld einer Kugel aus inkompressibler Flüssigkeit nach der Einsteinschen Theorie. Sitzungsberichte der Mathematisch Physikalischen Klasse der Deutschen Akademie der Wissenschaften Berlin **16**, 424 (1916)
13. K. Schwarzschild, Zur Quantenhypothese. Sitzungsberichte der Mathematisch Physikalischen Klasse der Deutschen Akademie der Wissenschaften Berlin **16**, 548 (1916)
14. H.H. Voigt (ed.), in *Karl Schwarzschild: Gesammelte Werke: Collected Works*, vol. 1–3 (Springer, Berlin, 1992)

Part II
Black Holes in Classical General Relativity, Numerical Relativity, Astrophysics, Cosmology and Alternative Theories of Gravity

Chapter 2
Black Hole Observations—Towards the Event Horizon

Silke Britzen

Abstract Black Holes are probably the most elusive solutions of Einstein's theory of General Relativity. Despite numerous observations of the direct galactic environment and indirect influence of astrophysical black holes (e.g. jets, variable emission across the wavelength spectrum, feedback processes, etc.)—a direct proof of their existence is still lacking. This article highlights some aspects deduced from many observations and concentrates on the experimental results with regard to black holes with masses from millions to billions of solar masses. The focus will be on the challenges and remaining questions. The Event Horizon Telescopce (EHT) project to image the photon sphere of Sgr A* and its potential is briefly sketched. This instrumental approach shall lead to highest resolution observations of the supermassive black hole at the center of the Milky Way (Sgr A*).

2.1 Active Galactic Nuclei (AGN)

Black Holes (hereafter BH) as physical objects play an important role in many astrophysical environments and processes (e.g. [13]). Whether as the endproduct and final stages of massive stars or as the central machines at the cores of actively radiating galaxies as Active Galactic Nuclei (AGN). Many questions related to AGN are still open. Assuming that the standard big bang scenario describes correctly the early phases of the Universe (for alternative models see e.g. [57]), then the first question to ask is: How could the earliest supermassive BHs (with masses in excess of 10^9 $M_\odot$), observed as luminous AGN at redshifts $z > 6$, be already in place when the Universe was less than a billion years old (e.g. [19, 20, 43, 58, 59])? How were supermassive BHs formed in the early phases of the Universe? Some of the competing theories suggest they formed as the remnants of the first generation of Pop III stars (e.g. [30, 42]) or through the 'direct collapse' of atomic-cooling gas (e.g. [4, 11]). Did the BHs grow through mergers of their host galaxies with other galaxies (e.g. [52])? Is accretion the more important process as some investigations (e.g. [1])

S. Britzen (✉)
Max-Planck-Institut für Radioastronomie, Auf dem Huegel 69, 53121 Bonn, Germany
e-mail: sbritzen@mpifr.de

P. Nicolini et al. (eds.), *1st Karl Schwarzschild Meeting on Gravitational Physics*,
Springer Proceedings in Physics 170, DOI 10.1007/978-3-319-20046-0_2

suggest? Are collisions the major and common cause of growth and activity? Major mergers of galaxies are suspected to be associated with quasar activity (e.g. [2, 3, 38, 50]). Is activity a phase in the life of a BH and how is this activity triggered? Observations at lower redshifts ($z \leq 1.5$) show that only a small fraction of supermassive BHs are undergoing quasar (the most luminous type of AGN) episodes, which are estimated to last for 1–100 Myr (e.g. [31, 55]). This would argue against the expectation of uninterruped accretion. How many life cycles will an average AGN undergo and how can this be tested? A most influential result was the discovery of a tight correlation between BH mass and the velocity dispersion of the bulge component of the host galaxy (e.g. [21, 24]). This and similar correlations lead to the assumption of a coevolution of black holes and galactic bulges. However, recent investigations showed, that supermassive BHs correlate differently with different galaxy components (e.g. [39]) and the details remain to be studied. AGN interact with their astrophysical environment in a variety of ways and can induce feedback processes, which in turn may influence the evolution of AGN themselves or members of galaxy clusters within which they lie. Clusters of galaxies constitute an excellent laboratory for the study of the feedback between the accretion and star formation processes. It is still unclear whether the AGN activity can account for the energy budget in clusters of galaxies and the Intracluster medium.

2.2 Jets

Zooming in on the central BH: Our knowledge regarding supermassive Black Holes has to be deduced from what is observable and can be extrapolated into the unobservable regime. AGN produce up to 5 % of the energy content of the Universe. The energy output can be more easily measured and observed compared to the feeding process (e.g. [7]). It seems extremely difficult to decipher the details of the accretion process. A careful monitoring of the complex details of feedback processes on all observable scales around the central object seems to be required (in the case of Cen A: e.g. [36]; NUGA-NUclei of GAlaxies—sample: e.g. [29]). Direct imaging of the accretion disk is beyond the current observational capabilities. Though, by help of gravitational lensing, a size measurement and spectral data of an AGN accretion disk could be obtained with HST-measurements [44].

The most obvious and probably best studied observational signatures of supermassive Black Holes are their jets (e.g. [33]), which seem to be a natural consequence of accretion disks with magnetic fields under certain conditions (e.g. [41]). These plasma-streams extend from the most central region of a galxy much beyond the optically visible galaxy into intergalactic space. While the jets are observable from large scales down to micro-arcsecond scales, the jet base as well as the jet-launching mechanisms are still not amenable to observations and have to be inferred indirectly [7]. Several theoretical jet launching scenarios have been proposed. The most prominent and often discussed are the Blandford and Znajek [5] and the Blandford and Payne [6] mechanism. The former model describes jet production via electromagnetic extraction of energy from Kerr BHs. The latter model explores the possibility that angular momentum could be removed magnetically from an accretion disk by field lines that leave the disk surface, and could eventually be carried off in a jet moving perpendicular to the disk. Acceleration and collimation of jets, the jet composition, the dominant dissipation and radiation processes remain to be studied and clarified (e.g. [49]). AGN jets have been studied in great detail in many Very Long Baseline Interferometry (VLBI) radio observations for several decades. The observational work has been accompanied by theoretical modelling and simulations (e.g. [32, 51]). Jets, although observed with all available radio telescope arrays at all available frequencies in the last decades remain enigmatic (e.g. [8, 28]). Many questions are related to the jet "components" which reveal apparent superluminal motions. Are these features shocks or instabilities in the plasma flow or do both phenomena co-exist in one jet? Helicity and geometrical effects seem to play a more important role than previously thought and can be used to explain the observed patterns [10]. These findings quite naturally lead to a questioning of the standard paradigm inevitably causally connecting broad-band flaring and jet component ejection (e.g. [9]).

2.2.1 Supermassive Binary Black Holes

The collision of galaxies and subsequent merging of the black holes at the centres of these galaxies is most likely one of the two relevant processes in the growth of the supermassive black hole mass and in the evolution of its spin. Theoretical and computational studies of the evolution of merging binaries taking into account conservative dynamics (post-Newtonian relativistic and finite size couplings, like spinorbit, spinspin, mass quadrupole mass monopole) or dissipative elements (gravitational radiation, dynamical friction and interaction with accretion disks) are of importance (e.g. [12]). Of equal importance is the observational evidence for supermassiv binary BHs through high-resolution radio imaging and optical spectroscopy as well as modelling of periodicities observed in total flux density and structural variations (e.g. [40, 48]).

2.2.2 Broadband—Flaring

AGN reveal variability over the entire electromagnetic spectrum on different timescales. Assuming that variable emission carries information about the structure and energy changes which occur at the centre of the AGN, non-imaging techniques (such as time series analysis) can be helpful in studying variable emission. Blazars—those AGN that are most likely viewed under a small viewing angle, pointing their jet directly at the observer—are known for significant and rapid variability at all wavelengths observed. Luminosities can appear to be (quasi)-periodic and might reveal important insight into the physics of the BH-system. As an example, [56] reconfirm based on a detailed analysis of the periodicities in 1156+295 that global p-mode oscillations of the accretion disc are coupled to the jet. They suspect that the osciallation of the disc is driven by Kelvin-Helmholtz instabilites in the inner edge of the accretion disc. High energy emission can probe the inner region around the BH. In particular, the Fermi Gamma-ray space telescope currently detects and observes many AGN producing so far unprecedented lightcurves in these wavelength regimes (e.g. [47]). However, it remains unclear where the ultra-relativistic particles are produced either close to the core or far outside (e.g. [14, 37]).

2.3 The Galactic Center Black Hole Sgr A*

The closest supermassive BH is the source Sgr A* at the centre of the Milky Way with a mass of about 4 million solar masses inferred from stellar orbits (e.g. [17, 25]). Whether this close-by supermassive BH can serve as prototype for the more active but also more distant AGN is so far unclear. Ideally, Sgr A* could be studied as a less active and lightweight version of the typical BH in AGN. Sgr A* is the most promising candidate and best case for a direct "image-proof" of the photon sphere around the dark object, which may be detectable by mm- and/or submm-VLBI. Imaging of the shadow of the BH might probe directly the existence of BHs (e.g. [15, 18]). It is a technically challenging project and still requires more time and improvement in resolution (e.g. [15]). 1-mm observations suggest that there might be structure on sizes below the size of the event horizon [22]. It is not yet clear, whether this indicates the need for an event horizon dependent on BH rotation (BH spin), hot spots orbiting in the accretion disk, or relativistic distortions of the inner accretion disk near the last stable orbit. Future (sub-) mm interferometric observations with about 10 micro-arcseconds angular resolution corresponding to a spatial scale of about the size of the BH in the Galactic Centre (1 Schwarzschild radius) are required to solve this question. The comparison of the shape and spectrum of the emission around the BH with BH-simulations will provide information on the observable GR effects and MHD physics in the accretion disk. Sgr A* displays extremely rapidly variable outbursts a few times a day in the Near-infrared and X-rays (e.g. [35, 53, 60]). These flares probe regions in the accretion flow as compact as or even smaller than the Schwarzschild radius. The interpretation of this flaring and the question whether this is periodic or not bears potential for a better understanding of the accretion process. Ever larger numbers of young, massive stars are being found throughout the Nuclear star cluster around Sgr A (e.g. [25]). Another challenge is to understand how star formation in this region works and whether the existing evidence for an initial stellar mass function different from the rest of the Galaxy will be confirmed. Further questions related to Sgr A* deal with the amount of the current gas inflow rate into the central parsec around the BH and whether or not a detectable collimated outflow from Sgr A* exists. Several scientific teams aim at proving the strength of gravity closer and closer to the assumed event horizon. Up-coming large observation facilities—

interferometric beam combiners at the VLTI (GRAVITY, e.g. [27]), instrumentation for the E-ELT (METIS), and mm/submm-VLBI (Event Horizon Telescope, e.g. [15]) will play an important role in probing the inner regions or providing the first direct evidence of a BH.

2.3.1 Event Horizon or Apparent Horizon

The observational evidence in favor of an event horizon of Sgr A* seems compelling [45]. However, recent publications (e.g. [54]) discuss the point that event horizons are (generically) not physically observable. In contrast, apparent horizons (and the closely related trapping horizons) seem to be generically physically observable—in the sense that they can be detected by observers working in finite-size regions of spacetime. Event horizons thus seem inappropriate tools for defining astrophysical black holes, or indeed for defining any notion of evolving black hole, (evolving either due to accretion or Hawking radiation). According to [54] the only situation in which an event horizon becomes physically observable is for the very highly idealized stationary or static black holes. This topic was recently discussed by [34] by claiming that a gravitational collapse produces apparent horizons but no event horizons behind which information is lost. Most important will be to test theoretical predictions with regard to the physics of BH versus the observationl results. In particular it will be as important as challengig to search for possible quantum effects of the BH (e.g. [16, 26]).

Despite the relevance of BHs in the astrophysical context, the most important questions related to their pure existence or proof of existence and their physics are still open. With technological progress, several scientific teams aim at proving the strength of gravity closer and closer to the assumed event horizon. Many questions with regard to BHs are still open. Yet they all touch important scientific questions.

Acknowledgments I would like to thank the organizers of the Karl Schwarzschild Meeting 2013 for an excellent scientific conference and awesome hospitality. The here sketched BH-related questions were subject of projects and collaborations within the scientific theme of the COST Action MP0905. This work has been supported by the COST Action MP0905 Black Holes in a Violent Universe.

References

1. M.A. Abramowicz, P.C. Fragile, Foundations of black hole accretion disk theory. Living Rev. Rel. **16**, 1 (2013)
2. V. Allevato, A. Finoguenov, N. Cappelluti, T. Miyaji et al., Astrophys. J. **736**, 99 (2011)
3. J.N. Bahcall, S. Kirhakos, D.H. Saxe, D.P. Schneider, Hubbel space telescope images of a sample of 20 nearby luminous quasars. Astrophys. J. **479**, 642 (1997)
4. J.E. Barnes, L.E. Hernquist, Fueling starburst galaxies with gas-rich mergers. Astrophys. J. **370**, L65–L68 (1991)

5. M.C. Begelman, M. Volonteri, M.J. Rees, Formation of supermassive black holes by direct collapse in pre-galactic haloes. Mon. Not. R. Astron. Soc. **370**, 289–298 (2006)
6. R.D. Blandford, R.L. Znajek, Electromagnetic extraction of energy from Kerr black holes. Mon. Not. R. Astron. Soc. **179**, 433 (1977)
7. R.D. Blandford, D.G. Payne, Hydromagnetic flows from accretion disks and the production of radio jets. Mon. Not. R. Astron. Soc. **199**, 883 (1982)
8. R.D. Blandford, **199** (2003)
9. S. Britzen, N.A. Kudryavtseva, A. Witzel, R.M. Campbell, et al., The kinematics in the pc-scale jets of AGN. The case of S5 1803+784. Astron. Astrophys. **511** 57 (2010)
10. S. Britzen, A. Witzel, B.P. Gong, J.W. Zhang, et al., Understanding BL Lacertae objects. Structural and kinematic mode changes in the BL Lac object PKS 0735+178. Astron. Astrophys. **515**, 105 (2010)
11. V. Bromm, A. Loeb, Formation of the first supermassive black holes. Astrophys. J. **596**, 34–46 (2003)
12. M. Colpi, *Massive Binary Black Holes in Galactic Nuclei and Their Path to Coalescence*. Space Science Reviews, Online First (2014)
13. C. DeWitt, *Black Holes (Les Houches Lectures)* (Harwood Academic, 1972). ISBN 13:978-0677156101
14. C.D. Dermer, in *Jets at all Scales, Proceedings of the International Astronomical Union, IAU Symposium*, vol. 275 (2010), p. 111
15. S. Doeleman et al., *Astro 2010: The Astronomy and Astrophysics Decadal Survey*, Science White Papers, No. 68 (2009)
16. G. Dvali, C. Gomez, *Black Hole Macro-Quantumness*. arXiv:1212.0765
17. A. Eckart, The Galactic black hole. Lectures on general relativity and astrophysics, in *Series in High Energy Physics, Cosmology and Gravitation*, eds. by H. Falcke, F.W. Hehl (IoP, Institute of Physics Publishing, Bristol, 2003), p. 229
18. H. Falcke, F. Melia, E. Agol, Astrophys. J. **528**(1), 13–16 (2000)
19. X. Fan, V.K. Narayanan, R.H. Lupton, M.A. Strauss et al., A survey of $z > 5.8$ quasars in the sloan digital sky survey. I. Discovery of three new quasars and the spatial density ofluminous quasars at $z > 6$. Astrophys. J. **122**, 2833 (2001)
20. X. Fan, M.A. Strauss, D.P. Schneider, R.H., Becker et al.: A survey of $z > 5.7$ quasars in the sloan digital sky survey. II. Discovery of three additional quasars at $z > 6$. Astrophys. J. **125**, 1649 (2003)
21. F. Ferrarese, D. Merritt, A fundamental relation between supermassive black holes and their host galaxies. Astrophys. J. **539**, L9–L12 (2000)
22. V.L. Fish, S. Doeleman, C. Beaudoin et al., 1.3 mm wavelength VLBI of Sagittarius A*: detection of time-variable emission on event horizon scales. Astrophys. J. **727**(2), 36 (2011)
23. S. Doeleman, E. Agol, D. Backer, F. Baganoff et al., Imaging an event horizon: submm-VLBI of a super massive black hole, in *Astro2010: The Astronomy and Astrophysics Decadal Survey*, Science White Papers, no. 68 (2009)
24. K. Gebhardt et al., A relationship between nuclear black hole mass and galaxy velocity dispersion. Astrophys. J. **539**, L13–L16 (2000)
25. R. Genzel, E. Eisenhauer, S. Gillessen, Rev. Mod. Phys. **82**(4), 3121 (2010)
26. S. Giddings, Possible observational windows for quantum effects from black holes. arXiv:1406.7001 (2014)
27. S. Gillessen, F. Eisenhauer, G. Perrin, W. Brandner et al., GRAVITY: a four-telescope beam combiner instrument for the VLTI. Proc. SPIE **7734**, 77340Y (2010)
28. Gómez et al., (2004)
29. S. Haan, E. Schinnerer, E. Emsellem, S. Garca-Burillo et al., Dynamical evolution of AGN host galaxies-gas in/out-flow rates in seven NUGA galaxies. Astrophys. J. **692**(2), 1623–1661 (2009)
30. Z. Haiman, A. Loeb, What is the highest plausible redshift of luminous quasars? Astrophys. J. **552**, 459 (2001)

31. Z. Haiman, L. Ciotti, J.P. Ostriker, Reasoning from fossils: learning from the local black hole population about the evolution of quasars. Astrophys. J. **606**, 763–773 (2004)
32. P.E. Hardee, AGN jets: a review of stability and structure. Relativistic jets: the common physics of AGN, microquasars, and gamma-ray bursts. AIP Conf. Proc. **856**, 57–77 (2014)
33. D. Homan, Physical properties of jets in AGN. Int. J. Mod. Phys. Conf. Ser. **08**, 163 (2012)
34. Hawking, S.: *Information Preservation and Weather Forecasting for Black Holes*. arXiv:1401.5761 (2014)
35. J.L. Hora, G. Witzel, M.L.N. Ashby, E.E. Becklin et al., *Spitzer/IRAC Observations of the Variability of Sgr A* and the Object G2 at 4.5 Microns*. eprint arXiv:1408.1951 (2014)
36. F.P. Israel, R. Güsten, R. Meijerink, A.F., Loenen et al., The molecular circumnuclear disk (CND) in Centaurus A. A multi-transition CO and [CI] survey with Herschel, APEX, JCMT, and SEST. Astron. Astrophys. **562**, 96 (2014)
37. A.P. Marscher, S.G. Jorstad, *Rapid Variability of Gamma-ray Emission from Sites near the 43 GHz Cores of Blazar Jets* (2010). arXiv:1005.5551
38. G. Kauffmann, M. Haehnelt, A unified model for the evolution of galaxies and quasars. Mon. Not. R. Astron. Soc. **311**, 576–588 (2000)
39. J. Kormendy, L.C. Ho, Coevolution (or not) of supermassive black holes and host galaxies. Ann. Rev. Astron. Astrophys. **51**, 511–653 (2013)
40. E. Kun, K.É. Gabányi, M. Karouzos, S. Britzen, L.Á. Gergely, A spinning supermassive black hole binary revealed by VLBI data on the jet of S5 1928+738. arXiv:1402.2644
41. D. Lynden-Bell, On why discs generate magnetic towers and collimate jets. Mon. Not. R. Astron. Soc. **341**, 1360–1372 (2003)
42. P. Madau, M.J. Rees, Massive black holes as population III remnants. Astrophys. J. **551**, L27–L30 (2001)
43. D.J. Mortlock, S.J. Warren, B.B.P. Venemans, M. Patel et al., A luminous quasar at a redshift of z = 7.085. Nature **474**, 616–619 (2011)
44. J.A. Muoz, E. Mediavilla, C.S. Kochanek, E.E. Falco, A.M. Mosquera, A study of gravitational lens chromaticity with the hubble space telescope. Astrophys. J. **742**(2), 67 (2011)
45. R. Narayan, J.E. McClintock, Advection-dominated accretion and the black hole event horizon. New Astron. Rev. **51**(10–12), 733–751 (2008)
46. R. Narayan, J.E. McClintock, *Observational Evidence for Black Holes*. arXiv:1312.6698 (2013)
47. P.L. Nolan, Fermi large area telescope second source catalog. Astrophys. J. Suppl. **199**(2), article id. 31, 46 (2012)
48. S.-J. Qian, S. Britzen, A. Witzel, T.P. Krichbaum, et al., A possible precessing nozzle and the Lense-Thirring effect in blazar 3C 454.3. Res. Astron. Astrophys. **14**(3), 249–274 (2014)
49. G.E. Romero, R.A. Sunyaev, T. Belloni, et al., Jets at all scales, in *Proceedings of IAU Symposium*, vol. 275 eds. by G.E. Romero, R.A. Sunyaev, T. Belloni (Cambridge University Press, Cambridge, 2011)
50. D.B. Sanders, B.T. Soifer, J.H. Elias et al., Ultraluminous infrared galaxies and the origin of quasars. Astrophys. J. **325**, 74–91 (1988)
51. M. Sikora, Radiation processes in blazars. in the fourth compton symposium. AIP Conf. Proc. **410**, 494–505 (1997)
52. T.L. Tanaka, Driving the growth of the earliest supermassive black holes with major mergers of host galaxies. arXiv:1406.3023
53. G. Trap, A. Goldwurm, K. Dodds-Eden et al., Concurrent X-ray, near-infrared, sub-millimeter, and GeV gamma-ray observations of Sagittarius A*. Astron. Astrophys. **528**, 140 (2011)
54. M. Visser, Physical observability of horizons. arXiv:1407.7295 (2014)
55. J.-M. Wang, Y.-M. Chen, F. Zhang, Cosmological evolution of the duty cycle of quasars. Astrophys. J. **647**, L17–L20 (2006)
56. J.-Y. Wang, T. An, W.A. Baan, X.-L. Lu, Periodic radio variabilities of the blazar 1156 + 295: harmonic oscillations. Mon. Not. R. Astron. Soc. **443**(1), 58–66 (2014)
57. C. Wetterich, External Universe. Phys. Rev. D **90**, 3520 (2014)
58. C.J. Willott, R.J. McLure, M.J. Jarvis, 3×10^9 black hole in the quasar SDSS J1148 + 5251 at z = 6.41. Astrophys. J. **587**, 15 (2003)

59. C.J. Willott, P. Delorme, C. Reylé, L. Albert et al., The Canada-France high-z quasar survey: nine new quasars and the luminosity funtion at redshift 6. Astrophys. J. **139**, 906–918 (2010)
60. G. Witzel, A., Eckart, M., Bremer, M., Zamaninasab et al., Source-intrinsic Near-infrared properties of Sgr A*: total intensity measurements. Astrophys. J. Suppl. **203**(2), article id. 18, 36 (2012)

Chapter 3
Primordial Black Holes and Quantum Effects

Bernard J. Carr

Abstract Primordial black holes are of special interest because of the crucial role of quantum effects in their formation and evaporation. This means that they provide a unique probe of the early universe, high-energy physics and quantum gravity. We highlight some recent developments in the subject, including improved limits on the fraction of the Universe going into evaporating PBHs in the mass range 10^9–10^{17} g and the possibility of using PBHs to probe a cosmological bounce.

3.1 Introduction

A comparison of the cosmological density at a time t after the big bang with the density associated with a black hole of mass M shows that primordial black holes (PBHs) should have of order the particle horizon mass, $M_H(t) \approx 10^{15}(t/10^{-23})$ g, at formation. PBHs could thus span an enormous mass range: from 10^{-5} g for those formed at 10^{-43} s to $10^5 M_\odot$ for those formed at 1 s. By contrast, black holes forming at the present epoch could never be smaller than about $1 M_\odot$. However, the high density of the early Universe is not a *sufficient* condition for PBH formation. One either needs large-amplitude density fluctuations, possibly of inflationary origin, so that overdense regions can eventually stop expanding and recollapse, or some sort of cosmological phase transition at which PBHs can form spontaneously (e.g. via the collapse of cosmic loops or the collisions of bubbles of broken symmetry). All these formation mechanisms depend in some sense on quantum effects and they are discussed in detail in [1] (henceforth CKSY).

The realization that PBHs might be small prompted Hawking to study their quantum consequences. This led to his famous discovery [2] that black holes radiate thermally with a temperature $T \approx 10^{-7}(M/M_\odot)^{-1}$ K and evaporate on a timescale $\tau(M) \approx 10^{64}(M/M_\odot)^3$y. Only black holes smaller than 10^{15} g would have evaporated by the present epoch and 10^{15} g ones would be exploding today. Since the

B.J. Carr (✉)
Department of Physics and Astronomy, Queen Mary University of London,
Mile End Rd, London E1 4NS, UK
e-mail: B.J.Carr@qmul.ac.uk

P. Nicolini et al. (eds.), *1st Karl Schwarzschild Meeting on Gravitational Physics*,
Springer Proceedings in Physics 170, DOI 10.1007/978-3-319-20046-0_3

latter would be producing photons with energy of order 100 MeV, the observational limits on the γ-ray background intensity imply that their density could not exceed 10^{-8} times the critical density. Nevertheless, this does not preclude PBHs playing other important cosmological roles. Indeed, their study provides a unique probe of four areas of physics: gravitational collapse for $M > 10^{15}$ g, high energy physics for $M \sim 10^{15}$ g, the early Universe for $M < 10^{15}$ g and quantum gravity for $M \sim 10^{-5}$ g.

Since both their formation and evaporation are a consequence of quantum effects, PBHs may offer the only astrophysical realization of what might be termed "quantum black holes" (i.e. holes for which quantum effects are important) [3]. This article will focus on their evaporation rather than their formation. In particular, it will discuss the upper limit on the fraction of the Universe going into PBHs as a function of mass because this provides important constraints on models (such as inflation) predicting their formation. The fraction of the Universe collapsing into PBHs at time t is related to their current density parameter $\Omega_{\rm PBH}$ by

$$\beta \approx 10^{-6}\Omega_{\rm PBH}(t/s)^{1/2} \approx 10^{-18}\Omega_{\rm PBH}(M/10^{15}g)^{1/2}, \tag{3.1}$$

where the t dependence reflects the decreasing ratio of the PBH and radiation densities at early times. Any limit on $\Omega_{\rm PBH}$ therefore places a constraint on $\beta(M)$. The constraints on $\beta(M)$ have been studied by numerous authors but the most recent and comprehensive discussion is that of [1]. The limits cover the mass range 10^9–10^{17} g and are shown in Fig. 3.1. The important point is that the value of $\beta(M)$ must be tiny throughout this mass range, so any cosmological model which entails an appreciable fraction of the Universe going into PBHs is immediately excluded. The most stringent limits—associated with big bang nucleosynthesis (BBN), the extragalactic

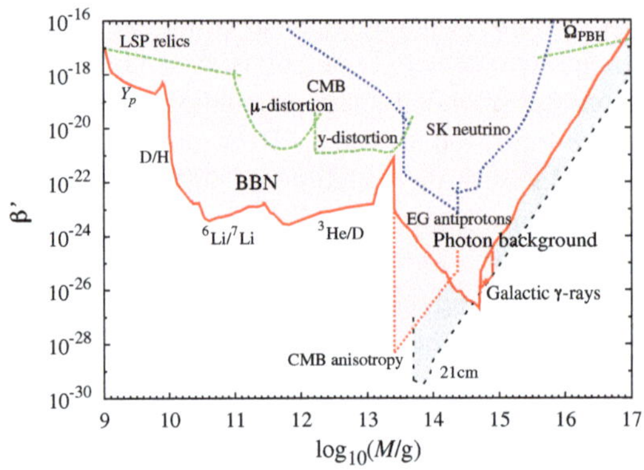

Fig. 3.1 Combined BBN and EGB limits (*solid*), compared to other constraints on evaporating PBHs from LSP relics and CMB distortions (*short-dashed*), extragalactic antiprotons and neutrinos (*dotted*), the Galactic γ-ray background (*long-dashed*), CMB anisotropies (*dash-dotted*) and the density limit from the smallest unevaporated black holes (*dashed*). From [1]

γ-ray background (EGB) and observation of anisotropies in the cosmic microwave background (CMB)—are discussed below. Positive evidence for PBHs might come from cosmic rays or short-period gamma-ray bursts but this is not covered below since the status of the observations is still ambiguous.

3.2 PBH Evaporations

A black hole with mass $M \equiv M_{10} \times 10^{10}$ g emits thermal radiation with temperature

$$T_{\mathrm{BH}} = \frac{1}{8\pi\, G\, M} \approx 1.06\, M_{10}^{-1}\, \mathrm{TeV}\,. \tag{3.2}$$

The average energy of the emitted particles is (4–6)kT_{BH}, depending on their spin. Charge and angular momentum are neglected because these will be lost through quantum emission on a shorter timescale. The mass loss rate can be expressed as

$$\frac{\mathrm{d}M_{10}}{\mathrm{d}t} = -5.34 \times 10^{-5}\, f(M)\, M_{10}^{-2}\, \mathrm{s}^{-1}\,. \tag{3.3}$$

Here $f(M)$ is a measure of the number of emitted particle species, normalised to unity for a black hole with $M \gg 10^{17}$ g, this emitting only particles which are (effectively) massless: photons, neutrinos and gravitons. Holes with $10^{15}\,\mathrm{g} < M < 10^{17}\,\mathrm{g}$ emit electrons, while those with $10^{14}\,\mathrm{g} < M < 10^{15}\,\mathrm{g}$ also emit muons, which subsequently decay into electrons and neutrinos.

Once M falls to around 10^{14} g, a black hole can also begin to emit hadrons. However, hadrons are composite particles, made up of quarks held together by gluons, so for temperatures exceeding $\Lambda_{\mathrm{QCD}} = 250–300$ MeV, one would expect the emission of quark and gluon jets rather than hadrons [5]. The jets would subsequently fragment into hadrons but only after travelling a distance $\Lambda_{\mathrm{QCD}}^{-1} \sim 10^{-13}$ cm, which is much larger than the size of the hole. The QCD fragmentation has been calculated using the PYTHIA [1] and HERWIG codes [6] but with similar results. Since there are many quark and gluon degrees of freedom, the value of f should roughly quadruple once the QCD temperature is reached. If we sum up the contributions from all particles in the Standard Model up to 1 TeV, this gives $f(M) = 15.35$ and a lifetime

$$\tau \approx 407 \left(\frac{f(M)}{15.35}\right)^{-1} M_{10}^{3}\, \mathrm{s}\,. \tag{3.4}$$

The critical mass for which τ equals the age of the Universe ($t_0 \approx 13.7$ Gyr) is $M_* \approx 5.1 \times 10^{14}$ g, corresponding to $f_* = 1.9$ and $T_{\mathrm{BH}}(M_*) = 21$ MeV.

The direct Hawking emission is termed the *primary* component, while the jet fragmentation emission is termed the *secondary* component. The spectrum of secondary photons is dominated by the 2γ-decay of soft neutral pions and peaks around

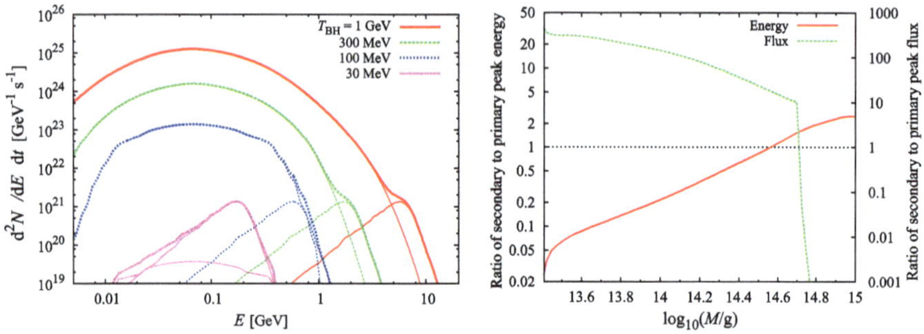

Fig. 3.2 *Left* Instantaneous emission rate of photons for four typical black hole temperatures, For each temperature, the curve with the peak to the right (*left*) represents the primary (secondary) component and the *thick curve* denotes their sum. *Right* Ratios of secondary to primary peak energies (*solid*) and fluxes (*dashed*). From [1]

$E_\gamma \simeq m_{\pi^0}/2 \approx 68\,\text{MeV}$. The emission rates of primary and secondary photons for four typical temperatures are shown in Fig. 3.2. Although QCD effects are initially small for PBHs with $M = M_*$, only contributing a few percent, they become important once M falls to $M_q \approx 0.4 M_* \approx 2 \times 10^{14}$ g since the peak energy becomes comparable to Λ_{QCD} then. This means that an appreciable fraction of the time-integrated emission from the PBHs evaporating at the present epoch goes into quark and gluon jet products. However, a PBH with somewhat larger initial mass, $M = (1+\mu)\, M_*$ will today have a mass $M(t_0) \approx (3\,\mu)^{1/3} M_*$ for $\mu \ll 1$. Since this falls below M_q for $\mu < 0.02$, the fraction of the black hole mass going into secondaries falls off sharply above M_*. The ratio of the secondary to primary peak energies and time-integrated fluxes are shown as functions of M in Fig. 3.2.

There has been some dispute about the interactions between emitted particles beyond the QCD scale. The usual assumption that there is no interaction has been refuted by Heckler [7], who claims that QED interactions could produce an optically thick photosphere once the black hole temperature exceeds $T_{\text{BH}} = 45\,\text{GeV}$. He has proposed that a similar effect may operate at an even lower temperature, $T_{\text{BH}} \approx 200\,\text{MeV}$, due to QCD effects [8]. Variants of these models and their astrophysical implications have been studied by various authors. However, MacGibbon et al. [9] have reviewed all these models and identified a number of effects which invalidate them. They conclude that emitted particles do not interact sufficiently to form a QED photosphere and that the conditions for QCD photosphere formation could only be temporarily satisfied (if at all) when the black hole temperature is of order Λ_{QCD}.

3.3 Constraints on $\beta(M)$ Imposed by BBN, EGB and CMB

PBHs with $M \sim 10^{10}$ g and $T_{\text{BH}} \sim 1\,\text{TeV}$ have a lifetime $\tau \sim 10^3$ s and therefore evaporate at the epoch of big bang nucleosynthesis (BBN). The effect of these evaporations on BBN has been a subject of long-standing interest and jet-produced hadrons are particularly important. Long-lived ones remain long enough in the ambient medium to leave an observable signature on BBN. These effects were first discussed for the relatively low mass PBHs evaporating in the early stages of BBN [10] but the analysis has now been extended by CKSY to incorporate the effects of heavier PBHs evaporating after BBN.

High energy particles emitted by PBHs modify the standard BBN scenario in three different ways: (1) high energy mesons and antinucleons induce extra interconversion between background protons and neutrons even after the weak interaction has frozen out in the background Universe; (2) high energy hadrons dissociate light elements synthesised in BBN, thereby reducing ^{4}He and increasing D, T, ^{3}He, ^{6}Li and ^{7}Li; (3) high energy photons generated in the cascade further dissociate ^{4}He. The PBH constraints depend on the initial baryon-to-photon ratio (allowing for PBH entropy production) and the ratio of the PBH number density to the entropy density, $Y_{\rm PBH} \equiv n_{\rm PBH}/s$, which is related to the initial mass fraction by $\beta = 5.4 \times 10^{21}\,(\tau/1\,{\rm s})^{1/2}\,Y_{\rm PBH}$.

The results of these calculations are summarized in Fig. 3.3 (left). PBHs with $M < 10^9$ g ($\tau < 10^{-2}$ s) are free from BBN constraints because they evaporate before weak freeze-out. PBHs with $M = 10^9$–10^{10} g ($\tau = 10^{-2}$–10^2 s) are constrained by process (1), those with $M = 10^{10}$–10^{12} g ($\tau = 10^2$–10^7 s) by process (2) and those with $M > 10^{12}$–10^{13} g ($\tau = 10^7$–10^{12} s) by process (3). We also show as a broken line the limits obtained earlier [10]. The helium limit is weaker because the primordial abundance is now known to be smaller, while the deuterium limit is stronger because of its extra production by hadrodissociation of helium.

It has been known for 40 years that observations of the diffuse extragalactic γ-ray background (EGB) constrain $\Omega_{\rm PBH}(M_*)$ to be less than around 10^{-8} [11]. This limit has subsequently been refined by numerous authors and most recently by CKSY. In

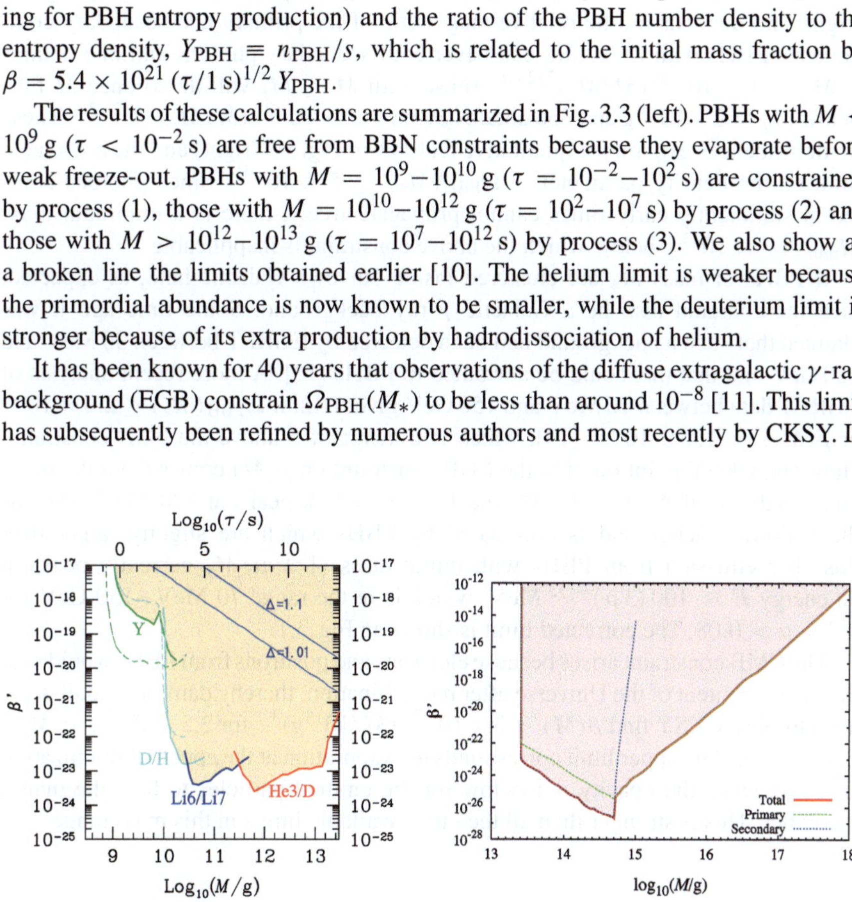

Fig. 3.3 *Left* Upper bounds on $\beta(M)$ from BBN, with broken line giving earlier limit. *Right* Upper bounds on $\beta(M)$ from the extragalactic photon background, with no other contributors to the background having been subtracted. From [1]

order to determine the present background spectrum of photons generated by PBH evaporations, one must integrate over the lifetime of the black holes, allowing for the fact that particles generated in earlier cosmological epochs will be redshifted in energy by now. The highest energy photons are associated with PBHs of mass M_*. Those from PBHs with $M > M_*$ are at lower energies because they are cooler, while those from PBHs with $M < M_*$ (although initially hotter) are at lower energies because they are redshifted.

The most recent X-ray and γ-ray observations are summarized by CKSY and correspond to an intensity $I^{\rm obs} \propto E_{\gamma 0}^{-(1+\varepsilon)}$ where ε lies between 0.1 and 0.4. The origin of these backgrounds is thought to be primarily distant astrophysical sources, such as blazars, and in principle one should remove these contributions before calculating the PBH constraints [12]. CKSY do not attempt such a subtraction, so their constraints may be overly conservative. The limits on $\beta(M)$ are shown in Fig. 3.3 (right) and depends on the relative magnitude of the primary and secondary components. PBHs with $M > M_*$ can never emit secondary photons and one obtains $\beta(M) \leq 4 \times 10^{-26}\,(M/M_*)^{7/2+\varepsilon}$. Those with $M \leq M_*$ will do so once M falls below $M_q \approx 2 \times 10^{14}$ g and one obtains $\beta(M) \leq 3 \times 10^{-27}\,(M/M_*)^{-5/2-2\,\varepsilon}$. These M-dependences explain the qualitative features of Fig. 3.3 (right) and the associated limit on the density parameter is $\Omega_{\rm PBH}(M_*) \leq 5 \times 10^{-10}$. Since photons emitted at sufficiently early times cannot propagate freely, there is a minimum mass $M_{\rm min} \approx 3 \times 10^{13}$ g below which the above constraint is inapplicable.

If PBHs of mass M_* are clustered inside our own Galactic halo, as expected, then there should also be a Galactic γ-ray background. Some time ago it was claimed that such a background had been detected by EGRET between 30 MeV and 120 GeV and that this could be attributed to PBHs [13]. A more recent analysis of EGRET data between 70 MeV and 150 GeV gives a limit $\Omega_{\rm PBH}(M_*) \leq 2.6 \times 10^{-9}$ or $\beta(M_*) < 1.4 \times 10^{-26}$ [14], which is a factor of 5 above the EGB constraint. However, CKSY point out that the EGB constraint on $\beta(M)$ comes from the time-integrated contribution of the M_* black holes, which peaks at 120 MeV, whereas the Galactic background is dominated by PBHs which are slightly larger than this. The emission from PBHs with initial mass $(1 + \mu)\,M_*$ currently peaks at an energy $E \approx 100\,(3\,\mu)^{-1/3}$ MeV, which is in the range 70 MeV–150 GeV for $0.7 > \mu > 0.08$. The corrected limit is shown in Fig. 3.1.

The CMB constraint arises because electrons and positrons from PBHs would heat the matter content of the Universe after recombination, thereby damping small-scale anisotropies. CKSY find $\beta(M) < 3 \times 10^{-30}\,(M/10^{13}{\rm g})^{3.1}$ for $2.5 \times 10^{13}\,{\rm g} < M < 2.4 \times 10^{14}$ g. The upper limit corresponds to evaporation at the epoch of reionization ($z = 6$), since the opacity is too low for the emitted particles to heat the matter thereafter. This is stronger than all the other available limits in this mass range.

3.4 PBHs and Dark Matter

Roughly 30 % of the total density of the Universe is now thought to be in the form of "cold dark matter". There has been a lot of interest in whether PBHs could provide this, since those larger than 10^{15} g would not have evaporated yet and would certainly be massive enough to be dynamically cold. One possibility is that PBHs with a mass of around $1 M_{\odot}$ could have formed efficiently at the quark-hadron phase transition at 10^{-5} s because of a temporary reduction in pressure [15]. At one stage there seemed to be evidence for this from microlensing observations. The data no longer support this but there are no constraints excluding PBHs in the sublunar range 10^{20} g $< M <$ 10^{26} g [16] or intermediate mass range $10^2 M_{\odot} < M < 10^4 M_{\odot}$ [17] from having an appreciable density.

Some people have speculated that black hole evaporation could cease once the hole gets close to the Planck mass (M_P) due to the influence of extra dimensions, higher order corrections to the gravitational Lagrangian, string effects, the generalized uncertainty principle etc. The resulting stable relics would then be natural candidates for the dark matter [18]. In an inflationary scenario, if the relics have a mass $\kappa M_{\rm Pl}$ and reheating occurs at a temperature $T_{\rm R}$ (when the PBHs form), then the requirement that the relic density be less than the dark matter density implies $\beta(M) < 2 \times 10^{-28} \kappa^{-1} (M/M_{\rm Pl})^{3/2}$ for $(T_{\rm R}/T_{\rm Pl})^{-2} < M/M_{\rm Pl} < 10^{11} \kappa^{2/5}$ [19]. The lower mass limit arises because PBHs generated before reheating are diluted exponentially. (If there is no inflationary period, the constraint extends all the way down to the Planck mass.) The upper mass limit arises because PBHs larger than this dominate the total density before they evaporate, in which case the current cosmological photon-to-baryon ratio is determined by the baryon asymmetry associated with their emission.

3.5 PBHs as a Probe of a Cosmological Bounce

In some cosmological scenarios, the Universe is expected to eventually recollapse to a big crunch and then bounce into a new expansion phase. Such a bounce may arise through either classical or quantum gravitational effects. Even if the Universe is destined to expand forever, it may have been preceded by an earlier collapsing phase. Both past and future bounces would arise in cyclic models, as reviewed in [20]. It is therefore interesting to ask whether black holes could either be generated by a big crunch or survive it if they were formed earlier [20]. We refer to these as "big-crunch black holes" (BCBHs) and "pre-crunch black holes" (PCBHs), respectively. If such black holes were detectable today, they would provide a unique probe of the last cosmological bounce, although this raises the question of whether one could differentiate between black holes formed just before and just after the bounce.

Let us assume that the universe bounces at some density ρ_B, Since the density associated with a black hole of mass M is $\rho_{BH} = (3M/4\pi R_S^3)$, this corresponds to a

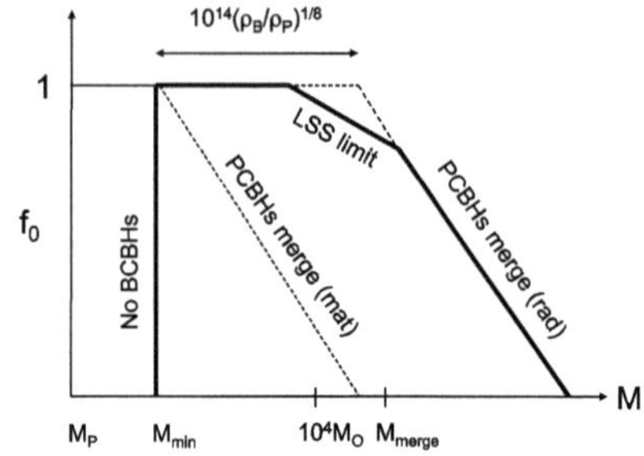

Fig. 3.4 This shows the domain in which black holes of mass M containing a fraction f_0 of the present density can form in a big crunch or avoid merging if they exist before it. From [20]

lower limit on the BCBH mass $M_{\rm min} \sim (\rho_P/\rho_B)^{1/2} M_P$. There is also a mass range in which pre-existing PCBHs lose their individual identity by merging with each other prior to the bounce. If the fraction of the cosmological density in these black holes at the bounce epoch is f_B, then the average separation between them is less than their size (i.e. the black holes merge) for $M > f_B^{-1/2} M_{\rm min}$. The important point, as indicated in Fig. 3.4, is that there is a always range of masses in which BCBHs may form and PCBHs do not merge. However, one must distinguish between f_B and the *present* fraction f_0 of the Universe's mass in black holes. Since the ratio of the black hole to radiation density scales as the cosmic scale factor, the fraction of the universe in black holes at a radiation-dominated bounce is $f_B \approx f_0 \left(\rho_{\rm eq}/\rho_B\right)^{1/4}$ where $\rho_{\rm eq} \sim 10^{12}\rho_0 \sim 10^{-17}\,{\rm g\,cm^{-3}}$. The merger condition therefore becomes $f_0 > 10^{28}\,(\rho_{\rm B}/\rho_{\rm P})^{-3/4}\,(M/M_P)^{-2}$, as indicated by the line on the right of Fig. 3.4.

There are various dynamical constraints on the form of the function $f_0(M)$ for non-evaporating PCBHs. They must have $f_0 < 1$ in order not to exceed the observed cosmological density and this gives a minimum value for the merger mass, $M_{\rm merge} \sim 10^9 (t_B/t_P)^{3/4}$ g, where t_B is the time of the bounce as measured from the notional time of infinite density. This is around 10^{15} g for $t_B \sim 10^{-35}$ s but as large as $10^4 M_\odot$ for $t_B \sim 10^{-5}$ s, so the observational consequences would be very significant. Another important constraint, deriving from Poisson fluctuations in the black hole number density, is associated with large-scale structure formation [21]. This gives a limit $f_0 < (M/10^4 M_\odot)^{-1}$, as shown by the line at the top right of Fig. 3.4.

References

1. B.J. Carr, K. Kohri, Y. Sendouda, J. Yokoyama, Phys. Rev. D. **81**, 104019 (2010)
2. S.W. Hawking, Nature **248**, 30–31 (1974)
3. X. Calmet, B.J. Carr, E. Winstanley, *Quantum Black Holes* (Springer, Berlin, 2013)
4. B.J. Carr, Astrophys. J. **201**, 1–19 (1975)
5. J.H. MacGibbon, Phys. Rev. D. **44**, 376–392 (1991)
6. J.H. MacGibbon, B.R. Webber, Phys. Rev. D. **41**, 3052–3079 (1990)
7. A. Heckler, Phys. Rev. D. **55**, 480–488 (1997)
8. A. Heckler, Phys. Rev. Lett. **78**, 3430–3433 (1997)
9. J.H. MacGibbon, B.J. Carr, D.N. Page, Phys. Rev. D. **78**, 064043 (2008). arXiv:0709.2380

10. K. Kohri, J. Yokoyama, Phys. Rev. D. **61**, 023501 (1999)
11. D.N. Page, S.W. Hawking, Astrophys. J. **206**, 1–7 (1976)
12. A. Barrau, D. Blais, G. Boudoul, D. Polarski, Phys. Lett. B. **551**, 218–225 (2003)
13. E.L. Wright, Astrophys. J. **459**, 487–490 (1996)
14. R. Lehoucq, M. Casse, J. Casandjan, I. Grnier, Astron. Astrophys. **502**, 37–43 (2009)
15. K. Jedamzik, Phys. Rev. D. **55**, 5871–5875 (1997)
16. D. Blais, T. Bringmann, C. Kiefer, D. Polarski, Phys. Rev. D. **67**, 024024 (2003)
17. R. Saito, J. Yokoyama, Phys. Rev. Lett. **102**, 161101 (2009)
18. J.H. MacGibbon, Nature **329**, 308–309 (1987)
19. B.J. Carr, J.H. Gilbert, J.E. Lidsey, Phys. Rev. D. **50**, 4853–4867 (1994)
20. B.J. Carr, A.A. Coley, Int. J. Mod. Phys. D. **20**, 2733–2738 (2011)
21. N. Afshordi, P. McDonald, D.N. Spergel, Astrophys. J. Lett. **594**, L71–74 (2003)

Chapter 4
There are No Black Holes—Pseudo-Complex General Relativity

Walter Greiner, Peter O. Hess, Mirko Schäfer, Thomas Schönenbach and Gunther Caspar

Abstract After a short review on attempts to extend General Relativity, pseudo-complex variables are introduced and their main properties are restated. A modified variational principle has to be introduced in order to obtain a new theory. This allows the appearance of an additional contribution, whose origin is a repulsive, dark energy. After the presentation of the general formalism, as examples the Schwarzschild and the Kerr solutions are discussed. It is shown that a collapsing mass increasingly accumulates dark energy until the collapse is stopped. Rather than a black hole, a gray star is formed. We discuss a possible experimental verification, investigating the orbital frequency of a particle in a circular orbit.

4.1 Introduction

General Relativity (GR) is a well accepted theory which has been verified by many experimental measurements. One prediction of this theory is the existence of *black holes*, which are formed once a very large mass suffers a gravitational collapse. Astronomical observations seem to confirm this prediction, finding large mass concentrations in the center of most galaxies. These masses vary from several million solar masses to up to several billion solar masses. However, a black hole implies the appearance of an event horizon, which an external observer cannot penetrate, thus, excluding a part of space from observation. A black hole also implies a singularity at its center. Both consequences from GR are, from a philosophical point of view, unacceptable and the task is to find a possibility to avoid them. We postulate: A physical theory should not have any singularities! A black hole is an extreme object and one would not be surprised that GR has to be modified for such situations. Eventually,

W. Greiner (✉) · M. Schäfer · T. Schonönenbach · G. Caspar
Frankfurt Institute for Advanced Studies, Ruth-Moufangstraße 1,
60438 Frankfurt am Main, Germany
e-mail: greiner@fias.uni-frankfurt.de

P.O. Hess
Instituto de Ciencias Nucleares, Universidad Nacional Autónoma de México,
Circuito Exterior, C.U., A.P. 70-543, 04510 Mexico, DF, Mexico

P. Nicolini et al. (eds.), *1st Karl Schwarzschild Meeting on Gravitational Physics*,
Springer Proceedings in Physics 170, DOI 10.1007/978-3-319-20046-0_4

the singularity can be avoided considering a quantized GR. This is not clear up to now.

There have been several attempts to generalize GR. Einstein [1, 2] introduced complex variables in order to unify GR with Electrodynamics. Later on, other groups continued this research calling the new theory *complexified GR* (see for example [3, 4] and references therein). The real component of the complex variable is given by x^μ while the imaginary component is given by $l\frac{p^\mu}{m}$, where p^μ is the momentum of a particle and m is its mass. As a by-product, for dimensional reasons a minimal length parameter l appears. One of the motivations to continue the investigation of the complex GR is *Born's equivalence principle*. Born noted [5, 6] that in GR there is an asymmetry between the coordinates and momenta, while in Quantum Mechanics they occur in a symmetric manner. In order to restore the symmetry he proposed a modified length element, adding to ds^2, the length square element, an additional term $l^2 g_{\mu\nu} du^\mu du^\nu$, with u^μ as the four velocity and $g_{\mu\nu}$ the metric (Born used instead of u^μ the notation p^μ/m). Again the minimal length parameter appears due to dimensional reasons. In [7] it was recognized that the new length element is related to a *maximal acceleration*, $a \leq 1/l$. Many other groups joined this investigation [8–14] and we will show that it is automatically contained in the proposed pseudo-complex extension of GR (which we will call from here on pc-GR). In [15, 16] a non-symmetric metric is considered and we will also show that it is contained within a pseudo-complex (pc) description.

In Sect. 4.2 we will introduce pc-variables and mention some important properties. In the same section the formulation of the pc-GR is resumed. In Sect. 4.3 we present the results of the pc-Schwarzschild and pc-Kerr solution. It will be shown that in the pc-GR dark energy accumulates around a large mass concentration, which will finally stop the gravitational collapse, forming rather a gray star than a black hole. There will be no event horizon, thus in principle allowing an external observer to access all regions of space. As an application of the theory, in the same section the circular motion of a particle around a gray star is considered, with a possible experimental verification. In Sect. 4.4 the conclusions will be drawn.

4.2 Formulation of the Pseudo-complex General Relativity

First we resume some basic properties of pc-variables: A pseudo-complex variable is given by $X = X_R + IX_I$, with X_R as the pseudo-real and X_I the pseudo-imaginary component. It is of great advantage to write it in terms of the *zero divisor basis* (the notation becomes obvious further below) $X = X_+\sigma_+ + X_-\sigma_-$, with $\sigma_\pm = \frac{1}{2}(1 \pm I)$. The expressions $\sigma_\pm$ obey the relations $\sigma_\pm^2 = \sigma_\pm$ and $\sigma_+\sigma_- = 0$. The last property is the definition of a zero divisor. When one defines as the pseudo-complex conjugate $X^* = X_R - IX_I$, which implies $\sigma_\pm^* = \sigma_\mp$, then for elements in the zero divisor basis ($X = \lambda\sigma_\pm$) the norm squared $\mid X \mid^2 = XX^*$ is zero. One can interpret this expression as a "generalized" zero. Calculations in the zero divisor basis are particularly simple. For example, products and quotients of functions can be done independently in each zero divisor component. Also differentiation and integration of pseudo-complex functions can be defined, similar to complex analysis (with some slight changes). For more details, consult [17, 18]. In the literature there exist several names for the pc-variables. Sometimes they are called hyper-complex, hyperbolic or semi-complex.

The consequences of using pc-variables for the Lorentz transformation are as follows: A finite Lorentz transformation is given by

$$\begin{aligned} e^{i\omega_{\mu\nu}\Lambda_{\mu\nu}} &= e^{i\omega^{+}_{\mu\nu}\Lambda^{+}_{\mu\nu}}\sigma_{+} + e^{i\omega^{-}_{\mu\nu}\Lambda^{-}_{\mu\nu}}\sigma_{-} \\ \Lambda_{\mu\nu} &= X_{\mu}P_{\nu} - X_{\nu}P_{\mu} \\ \Lambda^{\pm}_{\mu\nu} &= X^{\pm}_{\mu}P^{\pm}_{\nu} - X^{\pm}_{\nu}P^{\pm}_{\mu} \\ \omega_{\mu\nu} &= \omega^{+}_{\mu\nu}\sigma_{+} + \omega^{-}_{\mu\nu}\sigma_{-}. \end{aligned} \tag{4.1}$$

It divides into a Lorentz transformation in each zero-divisor component. The generators look the same, except now the variables are pseudo-complex. In the zero-divisor component the coordinates are given by $X^{\pm}_{\mu}$ and the momenta by $P^{\pm}_{\nu}$. Because $\sigma_{+}\sigma_{-} = 0$, the two Lorentz transformations commute, thus we have

$$SO_{+}(3,1) \otimes SO_{-}(3,1) \supset SO(3,1). \tag{4.2}$$

The standard Lorentz group is contained in the direct product and is reached by projecting the pseudo-complex parameters, coordinates and momenta to their real parts, i.e.,

$$\begin{aligned} \omega_{\mu\nu} \rightarrow \omega^{R}_{\mu\nu} &= \frac{1}{2}\left(\omega^{+}_{\mu\nu} + \omega^{-}_{\mu\nu}\right) \\ X_{\mu} &\rightarrow x_{\mu} \\ P_{\nu} &\rightarrow p_{\nu}. \end{aligned} \tag{4.3}$$

This projection method has to be applied also to the metric components.

That pseudo-complex variables also proved to be very useful was demonstrated in [19]: As shown in [19], the field equation for a scalar boson field is obtained from the Lagrangian density $\frac{1}{2}\left(D_{\mu}\Phi D^{\mu}\Phi - M^{2}\Phi^{2}\right)$, where Φ is the pc-boson field, $M = M_{+}\sigma_{+} + M_{-}\sigma_{-}$ is a pc-mass and D_{μ} a pc-derivative. The propagators of this theory are the ones of Pauli-Villars, which already are regularized. One obtains the same propagator in the standard theory, with a non-pc scalar field, using the Lagrange density $-\frac{1}{(M_{+}^{2}-M_{-}^{2})}\phi\left(\partial_{\mu}\partial^{\mu} + M_{+}^{2}\right)\left(\partial_{\mu}\partial^{\mu} + M_{-}^{2}\right)\phi$, where ϕ is now a real valued function, M_{+} is identified with the physical mass m and $M_{-} \gg M_{+}$ is the regularizing mass. Note, that this theory is highly non-linear while the pc-description is linear. This indicates that a pc-description can substantially simplify the structure of the theory and we can expect something similar in the pc-formulation of GR.

Let us now return to the pc-GR: The pc-extension of GR is quite direct within the zero divisor components. The first attempts are published in [20, 21] and in a more recent article [22] which includes modifications. Here we will present a short review. We introduce the pc-metric via

$$g_{\mu\nu}(X, A) = g^{+}_{\mu\nu}(X_+, A_+)\sigma_+ + g^{-}_{\mu\nu}(X_-, A_-)\sigma_-, \tag{4.4}$$

were the metric is assumed to be symmetric (in Moffat's theory of a non-symmetric metric [15, 16], the σ_+ component is the metric $g_{\mu\nu}$, while the σ_- component is its transposed, so in principle Moffat's theory is contained in our theory, if we skip the restriction to a symmetric metric). The metric components depend on the variables $X^\mu_\pm$ and parameters, denoted shortly as $A_\pm$. In each zero-divisor component a GR is constructed in the same manner as in standard GR. The pc-coordinates have the structure

$$X^\mu = x^\mu + Ilu^\mu. \tag{4.5}$$

Again, due to dimensional reasons, a minimal length parameter has to be introduced. Because it is just a parameter, it is not affected by any relativistic transformation, contrary to the believe that a minimal length is related to the breaking of Lorentz symmetry. The error made is to relate a minimal length to a physical length, which is affected by a Lorentz transformation. Here, the minimal length is a parameter and thus cannot be affected by such a transformation. The consequences are very important. For example, in [19] a pc-Field Theory was developed, demonstrating that a minimal length parameter does not affect the known symmetries: the calculations of Feynman diagrams remain very simple and the propagators of the theory are automatically regularized.

In mathematical terms we can explain the pc-extension of GR in terms of the following chain

$$G_+ \otimes G_+ \supset G. \tag{4.6}$$

In each component a standard GR is formulated. The base manifold is given by $X^\mu_\pm$ and the tangent spaces are given by $U^\mu_\pm$. Note, that U^μ includes the acceleration. Excluding the acceleration leads to G.

The pc-length square element is given by

$$d\omega^2 = g_{\mu\nu}(X, A)DX^\mu DX^\nu, \tag{4.7}$$

where D refers to a pc-differential [19, 20].

One may ask, what are the corrections due to the minimal length l? This will lead to the conclusion that all other theories, mentioned in the introduction, are a consequence of a pc-description. An expansion up to lu^μ is given by

$$g_{\mu\nu}(X) \approx g_{\mu\nu}(x) + lu_\lambda F^\lambda_{\mu\nu}(x). \tag{4.8}$$

The norm of the four-velocity can not be larger than 1. Assuming that the minimal length is probably very small (Planck length), one can safely take into account only the first term. Thus, in the metric tensor $g_{\mu\nu}(X)$ the pc-coordinates X^μ are substituted by x^μ. With this and expressing the pc-coordinates explicitly in terms of their pseudo-real and pseudo-imaginary components, the $d\omega^2$ acquires the form

$$d\omega^2 \approx g_{\mu\nu}(x)\left(dx^\mu dx^\nu + l^2 du^\mu du^\nu\right) + 2Ilg_{\mu\nu}(x)dx^\mu du^\nu. \qquad (4.9)$$

The terms in du^μ can not be neglected when effects near the maximal accelerations are considered. The du^μ are *differentials* of velocities, thus accelerations, and can reach values of the order of $1/l$. When the motion of a particle is considered, the $d\omega^2$ has to be real. This provides the condition

$$g_{\mu\nu}(x)dx^\mu du^\nu = 0, \qquad (4.10)$$

which is nothing but the dispersion relation. With (4.10) the length square element acquires the form as used in the theories mentioned in the introduction. There, the dispersion relation is introduced by hand while here it appears as a logical consequence.

When maximal acceleration effects are of no importance, one can also neglect the terms proportional to l and l^2 in (4.9).

All properties of tensors, four derivatives, Christoffel symbols, etc. can be directly extended from standard GR, defining them in each zero-divisor component as done in standard GR [20, 22, 23]. The only concept which has to be modified is the variational principle. If one uses (S denotes the action) $\delta S = \delta S_+\sigma_+ + \delta S_-\sigma_- = 0$, then we would obtain $\delta S_\pm = 0$. which correspond to two separated theories. In order to get a new theory, in [24, 25] a modified variational principle was proposed, namely that the variation has to be within the zero divisor (it can be interpreted as a "generalized zero"). This leads to field equations which on their right hand side are not zero but proportional to an element in the zero divisor. Our convention is to set it proportional to σ_-. Thus the Einstein equations read ($c = 1$)

$$G_{\mu\nu} = R_{\mu\nu} - \frac{1}{2}g_{\mu\nu}R \;=\; -8\pi\kappa T_{\mu\nu}\sigma_-. \qquad (4.11)$$

The $R_{\mu\nu}$ are the components of the pc-Ricci tensor, while R is the Ricci scalar. On the right hand side appears an energy-momentum tensor which describes the presence of an additional field which is always there in a pc-description. This field will turn out to have the properties of a dark energy and it will introduce a repulsion against gravitational collapse.

4.3 pc-Schwarzschild and pc-Kerr Solution

In [22] we presented the pc-Schwarzschild and pc-Kerr solutions. Of interest here is the g_{00} component, namely

$$g_{00} = \left(1 - \frac{2m}{r} + \frac{\Omega(r)}{r}\right). \tag{4.12}$$

(Here, we neglect for the moment a possible factor e^f [22], which we set to 1). We already restricted to the first term in the expansion in lu^μ. The $\Omega(r)$ is a not yet known function of the radial distance r. We model it by $\Omega = \frac{B}{2r^2}$. This leads to a correction in the metric of $\frac{B}{2r^3}$. The correction to the metric components have to depend at least on $1/r^3$, because a dependence on $1/r^2$ with a large B is excluded by experiments in the solar system [26].

One may speculate about the origin of the dark energy. One possibility are the vaccum fluctuations (Casimir effect): In [27] the Casimir effect in a gravitational backgound is investigated, within the Hartle-Hawking vacuum. No recoupling of the vacuum fluctuations with the gravitational field is considered. Thus, there is still the Schwarzschild metric present with an event horizon at the Schwarzschild radius. As a result, the expectation value of the trace of the energy-momentum tensor, due to the vacuum fluctuations, falls off proportional to $1/r^6$. This would mean that the mass, represented by the energy density, falls of proportional to $1/r^3$. Because no recoupling with the gravitational field is considered, the calculation has to stop at the Schwarzschild radius. Below that, no time can be defined in the same way as outside. In the pc-GR the recoupling of the dark energy energy-momentum tensor with the gravitational field is automatically included in (4.11). This leads to the correction in (4.12). Using the result in [27] literally, would imply a correction to the metric proportional to $1/r^4$. We will assume that the correction to the metric falls off like $1/r^3$ instead. This is the minimal correction which can be implemented not yet in conflict with current astronomical observations [26]. We expect to change the r-dependence, when the recoupling to the gravitational field is included in the calculation of the Casimir effect. Therefore, the model assumption that the corrections to the metric behave as $1/r^3$ is a rather good one. Higher-order corrections proportional to $1/r^n$ with $n \geq 4$ do not change our results significantly.

After this consideration, we return to the discussion of the pc-GR: In order to have the same interpretation of time in all regions of space, the g_{00} component has to be larger than zero. This introduces a minimal value of B [22].

Note, that the $\sqrt{g_{00}}$ component is proportional to an effective potential, with angular momentum zero [28]. With this, the effective potential is proportional to

$$\sqrt{g_{00}} = \sqrt{1 - \frac{2m}{r} + \frac{B}{2r^3}}. \tag{4.13}$$

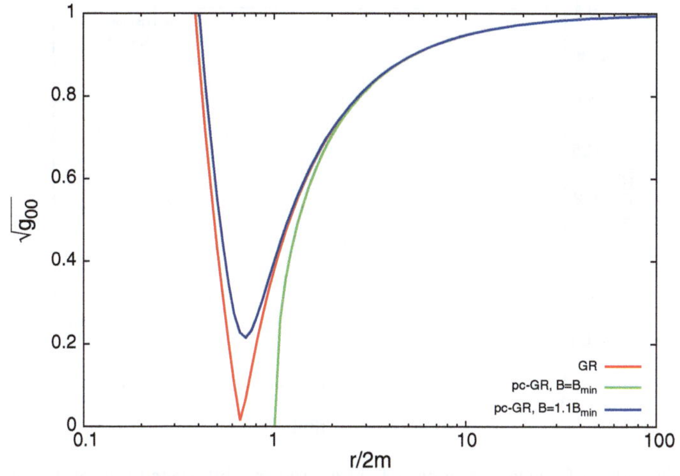

Fig. 4.1 The $\sqrt{g_{00}}$ component, which is proportional to an effective potential with angular momentum zero [28], as a function of rescaled radius $r/2m$. The *green line* indicates the classical theory with $B = 0$, whereas the *red line* and the *blue line* show $\sqrt{g_{00}}$ for pc-GR with $B = B_{min} = \frac{64}{27}m^3$ and $B = 1.1B_{min}$, respectively. We expect an analogous behavior for cosmological models in pc-GR, leading to an oscillating universe as a possible solution [21]

In Fig. 4.1 we display $\sqrt{g_{00}}$ for the classical theory and for pc-GR with the parameter B equal to the minimal value B_{min} or slightly larger.

For large distance, the potential is similar to the standard Schwarzschild solution. The differences start to appear near the Schwarzschild radius. The event horizon vanishes, because g_{00} never becomes zero. At smaller radial distances, the potential becomes repulsive, which is the consequence of the accumulation of dark energy. This changes the picture of a gravitational collapse: When a large mass is contracted due its gravitational influence, dark energy starts to accumulate and increases when the collapse advances. The collapse is finally stopped when enough dark energy accumulates and acts against the gravitational attraction. Thus, instead of a black hole the result is rather a *gray star*, though the gray star resembles pretty much a black hole seen from far apart. Therefore, from now on we will always refer to a gray star.

Today we know that the gray stars in the centers of galaxies rotate nearly at maximum speed. Thus, instead of the Schwarzschild solution one has to take the Kerr solution, which describes stars in rotation. The pc-Kerr solution was obtained in [22].

In order to relate the theory to experiment, we investigated the motion of a particle in a circular geodesic orbit around a gray star. This may be related to the possible observation of a plasma cloud orbiting such a star [29]. In Fig. 4.2 the orbital frequency is plotted versus the radial distance. As can be seen, the orbital frequency differs little from the standard Kerr solution until r is of the order of the Schwarzschild radius. Towards smaller radial distances, the orbital frequency is smaller in the

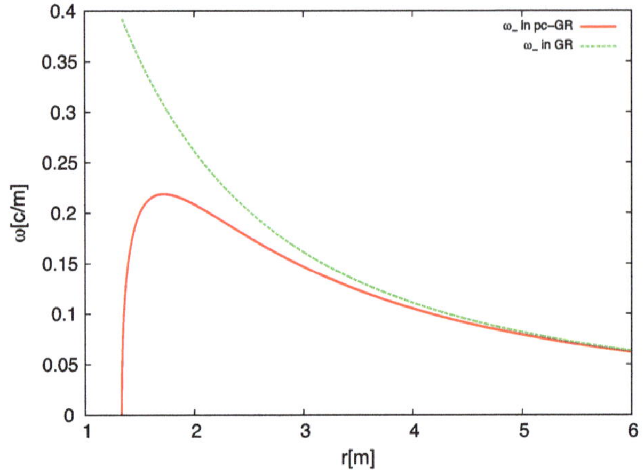

Fig. 4.2 The orbital frequency of a particle in a circular orbit around a *gray* star, as a function on the radial distance r. The units of ω are in $\frac{m}{c}$ while the radial distance is in units of half the Schwarzschild radius. $r = 2$ corresponds to the Schwarzschild radius and $\omega = 0.22$ is equivalent to about 0.11/min (For this computation we took the mass of Sagittarius A, the center of our galaxy, which is of the order $3 \times 10^7 M_{\text{sun}}$). The standard Kerr solution is given by the *upper line*, while the pc-solution is given by the *lower line*

pc-description, showing a maximum value, after which it diminishes. The maximum is a result of the structure of g_{00} which has a global minimum at about two-thirds of the Schwarzschild radius. For radii below that value the expression for the orbital frequency gets imaginary and we do not expect to observe circular geodesic orbits anymore. The curve for the pc-Kerr solution stops at this value. The curve for the standard GR stops at the point of the last stable orbit.

The result was obtained assuming that $\Omega = \frac{B}{2r^2}$. If it decreases with a larger power in r, the pc-solution approaches the standard Kerr solution, but will always show a maximum and the last stable orbit will be further out, i.e., the basic results will be the same. This result has important consequences in the experimental verification of pc-GR and we refer to [29, 30].

4.4 Conclusions

In this contribution we reviewed some essentials of the pseudo-complex General Relativity. The extension of the standard GR to pc-variables is direct due to the property that the zero-divisor components commute. In each component a standard GR is constructed. In order to obtain a new theory, the variational principle had to be changed. The variation of the action has now to be within the zero-divisor, i.e., it has to be a “generalized zero”. This introduces a new energy-momentum tensor in the Einstein equations, describing a dark energy field.

As a consequence of this dark energy-field, the gravitational collapse of a large mass is halted as soon as enough dark energy has accumulated. Due to this, no event horizon is formed and no singularity either. Instead of a black hole rather a *gray star* is formed. A possible experimental verification is proposed, determining the orbital frequency of a particle around a gray star.

Another important application of pseudo-complex General Relativity is a modified Robertson-Walker metric, implying consequences for a possible model for the evolution of the universe [21]. It seems that periodic solutions are possible. That means that the universe may oscillate! No big bangs are existing. This indicates that a new "Weltbild" emerges!

Acknowledgments Financial support from the Frankfurt Institute for Advanced Studies (FIAS), "Stiftung Polytechnische Gesellschaft Frankfurt am Main" (SPTG) and from CONACyT are acknowledged.

References

1. A. Einstein, Ann. Math. **578**, 46 (1945)
2. A. Einstein, Rev. Mod. Phys. **35**, 20 (1948)
3. C. Mantz, T. Prokopec. arXiv:gr-qc/0804.0213v1 (2008)
4. D. Lovelook, Annali di Matematica Pura ed Applicata **83**(1), 43 (1969)
5. M. Born, Proc. Roy. Soc. A **165**, 291 (1938)
6. M. Born, Rev. Mod. Phys. **21**, 463 (1949)
7. E.R. Caianiello, Lett. Nuo. Cim. **32**, 65 (1981)
8. H.E. Brandt, Found. Phys. Lett. **2**, 39 (1989)
9. H.E. Brandt, Found. Phys. Lett. **4**, 523 (1989)
10. H.E. Brandt, Found. Phys. Lett. **6**, 245 (1993)
11. R.G. Beil, Found. Phys. **33**, 1107 (2003)
12. R.G. Beil, Int. J. Theor. Phys. **26**, 189 (1987)
13. R.G. Beil, Int. J. Theor. Phys. **28**, 659 (1989)
14. R.G. Beil, Int. J. Theor. Phys. **31**, 1025 (1992)
15. J.W. Moffat, Phys. Rev. D **19**, 3554 (1979)
16. G. Kunstatter, J.W. Moffat, J. Malzan, J. Math. Phys. **24**, 886 (1983)
17. F. Antonuccio, Semi-Complex Analysis and Mathematical Physics. arXiv:gr-qc/9311032
18. I.L. Kantor, A.S. Solodovnikov, *Hypercomplex Numbers, An Elementary Introduction to Algebra* (Springer, Heidelberg, 1989)
19. P.O. Hess, W. Greiner, Int. J. Mod. Phys. E **16**, 1643 (2007)
20. P.O. Hess, W. Greiner, Int. J. Mod. Phys. E **18**, 51 (2009)
21. P.O. Hess, L. Maghlaoui, W. Greiner, Int. J. Mod. Phys. E **19**, 1315 (2010)
22. G. Caspar, T. Schönenbach, P.O. Hess, M. Schäfer, W. Greiner, Int. J. Mod. Phys. E **21**, 1250015 (2012)
23. R. Adler, M. Bazin, M. Schiffer, *Introduction to General Relativity* (McGraw Hill, New York, 1975)
24. F. P. Schuller, Dirac-Born-Infeld kinematics, maximal acceleration and almost product manifolds, Ph.D. thesis, Cambridge (2003)
25. F.P. Schuller, M.N.R. Wohlfarth, T.W. Grimm, Class. Quantum Grav. **20**(1)
26. C.M. Will, Living Rev. Relat. **9**, 3 (2006)
27. M. Visser, Phys. Rev. D **54**, 5103 (1996)

28. C.W. Misner, K.S. Thorne, J.A. Wheeler, *Gravitation* (W. H. Freeman Company, San Francisco, 1973)
29. T. Schönenbach, G. Caspar, P.O. Hess, T. Boller, A. Mueller, M. Schäfer, W. Greiner, Month. Notic. Roy. Astr. Soc. **430**, 2999 (2013)
30. T. Schönenbach, G. Caspar, P.O. Hess, T. Boller, A. Mueller, M. Schäfer, W. Greiner, see contribution within this volume

Chapter 5
Analytical Solutions for Geodesic Equation in Black Hole Spacetimes

Claus Lämmerzahl and Eva Hackmann

Abstract We review the analytical solution methods for the geodesic equation in Kerr-Newman-Taub-NUT-de Sitter spacetimes and its subclasses in terms of elliptic and hyperelliptic functions. A short guide to corresponding literature for general timelike and lightlike geodesic motion is also presented.

5.1 Introduction

Black holes belong to the most fascinating objects in astrophysics and are well suited to explore the regime of strong gravity. We consider here black holes which are described by the six parameter family of Kerr-Newman-Taub-NUT-de Sitter spacetimes including mass, rotation, electric and magnetic charge, gravitomagnetic mass (or NUT charge), and the cosmological constant. The only way to explore the gravitational field of such objects is through the observation of the motion of small massive particles and light, which is described by the geodesic equation. The complete set of orbits can best be analyzed using analytical methods. Already in 1931, Hagihara [20] used Weierstrass elliptic functions to analytically solve the geodesic equation in Schwarzschild spacetimes. Later, Darwin [5, 6] solved the geodesic equations in Schwarzschild spacetime in terms of Jacobian elliptic functions. These methods and their generalization to hyperelliptic functions can be used to solve the geodesic equation in the six parameter spacetime under consideration. Although this requires only mathematics from the 19th century, surprisingly the geodesic equations in Schwarzschild-de Sitter spacetime were analytically solved only in 2008 [15, 16]. Based on this development, the geodesic equation for many other spacetimes could be solved. Here we will review these solution methods and provide a short guide to the literature.

C. Lämmerzahl (✉) · E. Hackmann
Zentrum für angewandte Raumfahrttechnologie und Mikrogravitation,
Universität Bremen, Am Fallturm, 28359 Bremen, Germany
e-mail: claus.laemmerzahl@zarm.uni-bremen.de

E. Hackmann
e-mail: eva.hackmann@zarm.uni-bremen.de

P. Nicolini et al. (eds.), *1st Karl Schwarzschild Meeting on Gravitational Physics*,
Springer Proceedings in Physics 170, DOI 10.1007/978-3-319-20046-0_5

5.2 The Kerr-Newman-Taub-NUT-de Sitter Space-Time

We consider here spherically or axially symmetric spacetimes with up to six parameters, which are part of the family of type D Plebański-Demiański spacetimes. The metric is given by [10, 25]

$$\begin{aligned} ds^2 = &\frac{\Delta_r}{\rho^2}\left(dt - (a\sin^2\theta + 2n\cos\theta)d\varphi\right)^2 - \frac{\rho^2}{\Delta_r}dr^2 \\ &- \frac{\Delta_\theta}{\rho^2}\sin^2\theta(adt - (r^2+a^2+n^2)d\varphi)^2 - \frac{\rho^2}{\Delta_\theta}d\theta^2 \end{aligned} \tag{5.1}$$

with

$$\begin{aligned} \rho^2 &= r^2 + (n - a\cos\theta)^2\,, \\ \Delta_\theta &= 1 + \tfrac{1}{3}a^2\Lambda\cos^2\theta - \tfrac{4}{3}\Lambda an\cos\theta, \\ \Delta_r &= r^2 - 2Mr + a^2 - n^2 + Q_{\rm e}^2 + Q_{\rm m}^2 - \tfrac{1}{3}\Lambda\left(r^4 + (6n^2+a^2)r^2 + 3(a^2-n^2)n^2\right), \end{aligned} \tag{5.2}$$

where M is the mass, $a = J/M$ the specific angular momentum, Λ the cosmological constant, n the gravitomagnetic mass (NUT charge), $Q_{\rm e}$ the electric, and $Q_{\rm m}$ the magnetic charge of a gravitating source.

5.3 Analytical Solution Methods

The motion of test particles in spacetimes given by the metric (5.1) is given by the geodesic equation

$$\frac{d^2x^\mu}{ds^2} + \Gamma^\mu{}_{\nu\rho}\frac{dx^\nu}{ds}\frac{dx^\rho}{ds} = 0 \tag{5.3}$$

where $\Gamma^\mu{}_{\nu\rho}$ are the Christoffel symbols and s is an affine parameter. As the metric (5.1) is axially symmetric there exists the two constants of motion

$$E = u_\mu \xi^\mu_{(t)}\,, \quad L_z = -u_\mu \xi^\mu_{(\varphi)}\,, \tag{5.4}$$

connected to the Killing vectors $\xi_{(t)}$ and $\xi_{(\varphi)}$, which can be interpreted as the specific energy and the angular momentum in direction of the symmetry axes. Here $u^\mu = dx^\mu/ds$ denotes the four-velocity. In the case of spherical symmetry, i.e. for $a = n = 0$, then these two constants of motion together with the restriction of the motion to the equatorial plane, which is possible without loss of generality, and the normalization

$g_{\mu\nu}dx^\mu dx^\nu = \varepsilon$ (where $\varepsilon = 0$ for light and $\varepsilon = 1$ for massive particles) allows to separate the geodesic equation. This yields then the differential equation

$$\left(\frac{dr}{ds}\right)^2 = E^2 - \frac{\Delta_r}{r^2}\left(\varepsilon + \frac{L^2}{r^2}\right), \tag{5.5}$$

and the energy and angular momentum take the form

$$E = g_{tt}\frac{dt}{ds}, \quad L = L_z = r^2\frac{d\varphi}{ds}. \tag{5.6}$$

If the spacetime is axially symmetric, we cannot in general restrict the motion of test particles to the equatorial plane. Therefore, we need an additional constant of motion. In 1968 Carter [4] found such a constant of motion, which can be derived as a separation constant. This constant is not connected to an obvious symmetry of the spacetime. With this constant the geodesic equation can again be separated and we get equations of motions in the form

$$\rho^4\left(\frac{dr}{ds}\right)^2 = ((r^2 + a^2 + n^2)E - aL_z)^2 - \Delta_r(\varepsilon r^2 + K), \tag{5.7}$$

$$\rho^4\left(\frac{d\theta}{ds}\right)^2 = \Delta_\theta(K - \varepsilon(n - a\cos\theta)^2) - \frac{(E(a\sin^2\theta + 2n\cos\theta) - L_z)^2}{\sin^2\theta}, \tag{5.8}$$

where K is the Carter constant. If $n = 0$ then $K = (aE - L_z)^2$ corresponds to motion restricted to the equatorial plane. From the expression of energy and angular momentum (5.4) we get the additional equations

$$\rho^2\frac{d\varphi}{ds} = \frac{a}{\Delta_r}((r^2 + a^2 + n^2)E - aL_z) - \frac{E(2n\cos\theta + a\sin^2\theta) - L_z}{\Delta_\theta\sin^2\theta}, \tag{5.9}$$

$$\rho^2\frac{dt}{ds} = \frac{r^2 + a^2 + n^2}{\Delta_r}((r^2 + a^2 + n^2)E - aL_z) - \frac{a\sin^2\theta + 2n\cos\theta}{\Delta_\theta\sin^2\theta}(E(2n\cos\theta + a\sin^2\theta) - L_z). \tag{5.10}$$

Note that the (5.7) and (5.8) are still coupled by the factor ρ^2. This issue was solved by Mino in 2003 [27] by introducing a new affine parameter λ defined by $\frac{ds}{d\lambda} = \rho^2$.

In the case of spherically symmetric spacetimes the differential equations for $r(\varphi)$ reads

$$\left(\frac{dr}{d\varphi}\right)^2 = \frac{r^4}{L^2}\left(E^2 - \frac{\Delta_r}{r^2}\left(\varepsilon + \frac{L^2}{r^2}\right)\right). \tag{5.11}$$

If the the cosmological constant vanishes then on the right hand side we have a polynomial of degree three or four. For a nonvanishing cosmological constant the degree of the polynomial is five or six. The same holds for the differential equations (5.7) and (5.8). If the polynomial is of degree three or four this kind of differential equations can be solved in terms of elliptic functions. In the other case the solution is given in terms of hyperelliptic functions.

5.3.1 Solutions in Terms of Elliptic Functions

For differential equations of the general type

$$\left(\frac{dx}{dy}\right)^2 = P_{3,4}(x)\,, \quad x(y_0) = x_0 \tag{5.12}$$

where $P_{3,4}$ is a polynomial of degree three or four, there are basically two (equivalent) solution methods based on the Jacobian elliptic function sn and on the Weierstrass elliptic function $\wp$. The first can be defined as the inverse of an elliptic integral,

$$z = \int_0^w \frac{dt}{\sqrt{(1-t^2)(1-k^2t^2)}} \quad \Rightarrow \quad \mathrm{sn}\,(z;k) = w \tag{5.13}$$

where $0 \le k \le 1$ is the modulus, $w \in [0, 1]$, and $z \in \mathbb{R}$. The Weierstrass elliptic function $\wp$ is given as a series

$$\wp\,(z; 2\omega_1, 2\omega_2) = \frac{1}{z^2} + \sum_{\omega_{nm}\in\Omega}\left(\frac{1}{(z-\omega_{nm})^2} - \frac{1}{\omega_{nm}^2}\right), \tag{5.14}$$

where $2\omega_1$, $2\omega_2$ are the periods of $\wp$ ($\frac{\omega_1}{\omega_2} \notin \mathbb{R}$) and $\Omega = \{\omega_{n,m} \in \mathbb{C} | \omega_{n,m} = 2n\omega_1 + 2m\omega_{2,n,m} \in \mathbb{Z}, (n,m) \neq (0,0)\}$. It solves the initial value problem (see e.g. [21, 26])

$$\left(\frac{dx}{dy}\right)^2 = 4x^3 - g_2 x - g_3\,, \quad x(0) = \infty \tag{5.15}$$

where $g_2 = 60 \sum_{\omega_{nm}\in\Omega} \omega_{nm}^{-4}$, $g_3 = 140 \sum_{\omega_{nm}\in\Omega} \omega_{nm}^{-6}$. Note that both sn and $\wp$ can be written in terms of the Riemann theta function

$$\theta[\tau v + w](z;\tau) = \sum_{m\in\mathbb{Z}^g} \exp(\pi i (m+v)^t(\tau(m+v) + 2z + 2w)), \tag{5.16}$$

where $z \in \mathbb{C}^g$, τ is a $g \times g$ symmetric matrix with positive definite imaginary part, $\tau v + w \in \mathbb{C}^g$ is the characteristic, and g is the genus, here $g = 1$.

The general differential equation (5.12) can be solved in terms of Jacobian elliptic functions by applying a substitution which converts the problem to the form

$$\left(\frac{d\tilde{x}}{dy}\right)^2 = (1-\tilde{x}^2)(1-k^2\tilde{x}^2). \tag{5.17}$$

This substitution depends on the degree of $P_{3,4}$ and its number of complex zeros as well as on the type of orbit you want to obtain. For a list see e.g. [1].

In order to find a solution of (5.12) in terms of the Weierstrass elliptic function one has to transform the problem to the standard form (5.15), which can be obtained by first converting a polynomial of degree four to degree three by $x = \xi^{-1} + x_P$ if $P_{3,4} = \sum_i a_i x^i$ and $P_{3,4}(x_P) = 0$ and subsequently, or if $P_{3,4}$ was of degree three in the first place, substituting $\xi = \frac{1}{b_3}(4z + \frac{b_2}{3})$ if $P = \sum_i b_i x^i$ is the polynomial of degree three. Note that one always have to take care of the initial condition, too.

5.3.2 *Solutions in Terms of Hyperelliptic Functions*

To generalize the solution methods outlined in the previous section we first need to consider the Jacobi inversion problem

$$y_i = \sum_{j=1}^{g} \int_{\infty}^{x_j} \frac{t^{i-1}dt}{\sqrt{P(t)}}, \quad i = 1, \ldots, g \tag{5.18}$$

where $P(t) = 4t^{2g+1} + \sum_{n=0}^{2g} a_n t^n$ is a polynomial of degree $2g+1$ and g is the genus. Note that for $g = 1$ and $a_2 = 0$ we recover (5.15). The g solutions $x_j(y_1, \ldots, y_g)$ of (5.18) can be given in terms of generalized Weierstrass functions. These are defined by the theta function via the Kleinian sigma function

$$\sigma(z; \omega_1, \omega_2) = C e^{iz^t \kappa z} \theta[K_\infty](z; \omega_1^{-1}\omega_2), \tag{5.19}$$

$$\wp_{ij}(z; \omega_1, \omega_2) = -\frac{\partial}{\partial z_i}\frac{\partial}{\partial z_j} \log \sigma(z; \omega_1, \omega_2), \tag{5.20}$$

where $z \in \mathbb{C}^g$, ω_i are $g \times g$ matrices such that $\tau = \omega_1^{-1}\omega_2$ is symmetric with positive definite imaginary part, $\kappa = \eta(2\omega_1)^{-1}$ with the periods of the second kind 2η, K_∞ is the vector of Riemann constants, and C is a constant which does not matter here (for further details see e.g. [2]; note that $K_\infty = \tau(\frac{1}{2}, \frac{1}{2})^t + (0, \frac{1}{2})^t$ if $g = 2$). The solutions of (5.18) are then given by the solutions of

$$x^g + \sum_{i=1}^{g} \wp_{gi}(y_1, \ldots, y_g) x^{i-1} = 0. \tag{5.21}$$

Let us consider now the case $g = 2$ and a general differential equation of the form

$$\left(\frac{dx}{dy}\right)^2 = P_{5,6}(x), \quad x(y_0) = x_0 \tag{5.22}$$

where $P_{5,6}$ is a polynomial of degree five or six. If it is of degree six it can be reformulated as $\tilde{x}\frac{d\tilde{x}}{dy} = \sqrt{P_5(\tilde{x})}$ by a substitution $x = \tilde{x}^{-1} + x_P$, where x_P is a zero of $P_{5,6}$ and P_5 is a polynomial of degree five. Therefore, equation (5.22) can be written in the form

$$t^{i-1}\frac{dt}{dy} = \sqrt{P(t)}, \quad i = 1, 2 \tag{5.23}$$

by an appropriate normalization. We may then find the solution to this equation as the limiting case $x_2 \to \infty$ of the Jacobi inversion problem (5.18) in the following way: first observe that

$$\begin{aligned} t := x_1 &= \lim_{x_2 \to \infty} \frac{x_1 x_2}{x_1 + x_2} = \lim_{x_2 \to \infty} \frac{\wp_{12}(y_1, y_2)}{\wp_{22}(y_1, y_2)} \\ &= \lim_{x_2 \to \infty} \frac{\sigma\sigma_{12} - \sigma_1\sigma_2}{\sigma_2^2 - \sigma\sigma_{22}}(y_1, y_2), \end{aligned} \tag{5.24}$$

where $\sigma_i(z)$ is the derivative of σ with respect to z_i. From (5.18) with $g = 2$ and $x_2 \to \infty$ we may identify either y_1 (if $i = 1$ in (5.23)) or y_2 (if $i = 2$) with our physical coordinate y in the differential equation (5.23). Fortunately, we get automatically rid of the other y_j, $j \neq i$, by the same limiting process $x_2 \to \infty$. This is because the set of zeros of the theta function $z \mapsto \theta[K_\infty]((2\omega)^{-1}z)$, which is a one dimensional submanifold of $\mathbb{C}^2$, is given by all vectors $z = (z_1, z_2)$ which can be written as $z_i = \int_\infty^x \frac{t^{i-1}dt}{\sqrt{P(t)}}$ with the same x (see e.g. [28]). This is exactly true for the vector (y_1, y_2), which means that we may write $y_j = f(y_i)$ for some function f. As the zeros of the theta function are also zeros of σ we can simplify (5.24) yielding

$$t = -\frac{\sigma_1}{\sigma_2}(y_1, f(y_1)) \quad \text{or} \quad t = -\frac{\sigma_1}{\sigma_2}(f(y_2), y_2). \tag{5.25}$$

Note that according to the given initial condition y_i is the physical coordinate minus a constant.

5.4 Analytical Solutions in the Literature

In this section we will collect applications of the methods outlined in Sect. 5.3 to geodesic motion in the Kerr-Newman-Taub-NUT-de Sitter spacetime as given in Sect. 5.2. For older literature we refer to [29] who collected most of the papers on geodesic motion in Kerr-Newman spacetime and subclasses, which were available at that time. Partly this is still quite complete but we also try to update his collection (with respect to analytical solutions). Note that we only consider analytical solutions to general timelike and lightlike geodesics (with an electric or magnetic charge, as applicable). In particular, we do not list the vast literature on equatorial motion in Kerr spacetime. Of course, we do not claim that our list is complete.

Schwarzschild: Regarding analytical solution methods the list of Sharp already contained the complete set of solutions. Most notably, this includes the works by Hagihara [20], who derived the analytical solutions in terms of Weierstrass elliptic functions, and Darwin [5, 6], who used Jacobian elliptic functions and integrals.

Reissner-Nordström: Surprisingly, the analytical solutions to the geodesic equation in Reissner-Nordström spacetime seem to be considered first only in 1983 by Gackstatter [8] although it can be handled completely analogously to the Schwarzschild case. He studied bound timelike geodesics and light in terms of Jacobian elliptic integrals and functions. Recently, Slezáková [30] gave a comprehensive analysis of arbitrary timelike, lightlike, and even spacelike geodesics. Grunau and Kagramanova [11] solved the equations of motion of electrically and magnetically charged particles in Reissner-Nordström spacetime in terms of Weierstrass elliptic functions.

Taub-NUT: Timelike geodesics were studied by Kagramanova et al. [22] in terms of Weierstrass elliptic functions.

Kerr: Most of the older literature on Kerr spacetime is concerned with the much simpler particular case of equatorial geodesics. We refer to Sharp [29] here for these early works. Note that in terms of the proper time (or the corresponding affine parameter for light) the equations of motion are still coupled. Therefore, before one introduced the Mino time [27] most of the analytical solutions implicitly included integrals over the latitude or the radius, see e.g. Kraniotis [23] or Slezáková [30] for a review. As notable exception, Čadež et al. [3] introduced already in 1998 a similar parameter (called P, see their (34)) as they considered the motion of light.

After the introduction of the Mino time, in 2009 Fujita and Hikida [7] used this new affine parameter to derive the analytical solution for bound timelike geodesics in terms of Jacobian elliptic functions. General timelike geodesics and lightlike motion were treated shortly after that by Hackmann [12] in 2010.

Note that Kraniotis [24] also derived analytical solutions for lightlike geodesics in terms of hypergeometric functions.

Kerr-Newman: Charged particle motion was considered by Hackmann and Xu [19] in terms of Weierstrass elliptic functions.

Schwarzschild-de Sitter also called Kottler space-time: Note that on the level of the differential equation, lightlike geodesics in Schwarzschild-de Sitter are identical with the lightlike equations of motion for Schwarzschild, as the cosmological constant can be absorbed in the definition of just a single parameter. Analytical solution are given e.g. in Gibbons et al. [9]. General timelike geodesics in Kottler spacetime can be treated in terms of hyperelliptic functions as elaborated by Hackmann and Lämmerzahl [15, 16].

Reissner-Nordström-de Sitter: The equations of motion for general timelike geodesics were solved in [13]. The motion of photons was very recently analytically calculated by Villanueva et al. [31] for a negative cosmological constants using Weierstrass elliptic functions.

Taub-NUT-de Sitter: The timelike motion was analyzed in [14].

Kerr-de Sitter: The equations of motions for timelike geodesics were analytically solved by Hackmann et al. [14, 17] in terms of hyperelliptic functions. Note that Kraniotis [24] also derived analytical solutions for lightlike geodesics in terms of hypergeometric functions.

Kerr-Newman-Taub-NUT-de Sitter: The general solution for timelike geodesic was shortly outlined in [14].

For all these solutions there are analytical expressions for the observables (i) perihelion shift, (ii) Lense-Thirring effect, and (iii) conicity of the orbit [18].

Acknowledgments We thank the German Research Foundation DFG for financial support within the Research Training Group 1620 "Models of Gravity".

References

1. M. Abramowitz, I.A. Stegun (eds.), *Handbook of Mathematical Functions* (Dover Publications Inc., New York, 1965)
2. V.M. Buchstaber, V.Z. Enolskii, D.V. Leykin, *Hyperelliptic Kleinian Functions and Applications*. Reviews in Mathematics and Mathematical Physics 10 (Gordon and Breach, 1997)
3. A. Čadež, C. Fanton, M. Calvani, New Astronomy **3**, 647 (1998)
4. B. Carter, Phys. Rev. **174**, 1559 (1968)
5. C. Darwin, Proc. R. Soc. London A **249**, 180 (1959)
6. C. Darwin, Proc. R. Soc. London A **263**, 39 (1961)
7. R. Fujita, W. Hikida, Class. Quantum Grav. **26**, 135002 (2009)
8. F. Gackstatter, Ann. Phys. **495**, 352 (1983)
9. G.W. Gibbons, C.M. Warnick, M.C. Werner, Class. Quantum Grav. **25**, 245009 (2008)
10. J.B. Griffiths, J. Podolsky, Int. J. Mod. Phys. **15**, 335 (2006)
11. S. Grunau, V. Kagramanova, Phys. Rev. D **83**, 044009 (2011)
12. E. Hackmann, Geodesic equations in black hole space-times with cosmological constant. Ph.D. thesis, Universität Bremen, 2010
13. E. Hackmann, V. Kagramanova, J. Kunz, C. Lämmerzahl, Phys. Rev. D **78**, 124018 (2008)
14. E. Hackmann, V. Kagramanova, J. Kunz, C. Lämmerzahl, Europhys. Lett. **88**, 30008 (2009)
15. E. Hackmann, C. Lämmerzahl, Phys. Rev. Lett. **100**, 171101 (2008)
16. E. Hackmann, C. Lämmerzahl, Phys. Rev. D **78**, 024035 (2008)
17. E. Hackmann, C. Lämmerzahl, V. Kagramanova, J. Kunz. Phys. Rev. D **81**, 044020 (2010)
18. E. Hackmann, C. Lämmerzahl, Phys. Rev. D **85**, 044049 (2012)
19. E. Hackmann, H. Xu, Phys. Rev. D **87**, 124030 (2013)
20. Y. Hagihara, Japanese J. Astr. Geophys. **8**, 67 (1931)
21. A. Hurwitz, *Vorlesungen über Allgemeine Funktionentheorie und elliptische Funktionen* (Springer, Berlin, 1964)
22. V. Kagramanova, J. Kunz, E. Hackmann, C. Lämmerzahl, Phys. Rev. D **81**, 124044 (2010)
23. G.V. Kraniotis, Class. Quantum Grav. **22**, 4391 (2005)
24. G.V. Kraniotis, Class. Quantum Grav. **28**, 085021 (2011)
25. V.S. Manko, E. Ruiz, Class. Quantum Grav. **22**, 3555 (2005)
26. A.I. Markushevich, *Theory of Functions of a Complex Variable, Vol. I, II, III* (Chelsea Publishing Co., New York, 1977). Translated and ed. by R.A. Silverman
27. Y. Mino, Phys. Rev. D **67**, 084027 (2003)

28. D. Mumford, *Tata Lectures on Theta, Vol. I and II* (Birkhäuser, Boston, 1983/1984)
29. N.A. Sharp, Gen. Rel. Grav. **10**, 659 (1979)
30. G. Slezáková, Geodesic Geometry of Black Holes, Ph.D. thesis, University of Waikato, 2006
31. J.R. Villanueva, J. Saavedra, M. Olivares, N. Cruz, Astrophys. Space Sci. **344**, 437 (2013)

Chapter 6
A Physical Derivation of the Kerr–Newman Black Hole Solution

Reinhard Meinel

Abstract According to the no-hair theorem, the Kerr–Newman black hole solution represents the most general asymptotically flat, stationary (electro-) vacuum black hole solution in general relativity. The procedure described here shows how this solution can indeed be constructed as the unique solution to the corresponding boundary value problem of the axially symmetric Einstein–Maxwell equations in a straightforward manner.

6.1 Introduction: From Schwarzschild to Kerr–Newman

The Schwarzschild solution, depending on a single parameter (mass M), represents the *general* spherically symmetric vacuum solution to the Einstein equations. Similarly, the Reissner–Nordström solution, depending on two parameters (M and electric charge Q), is the *general* spherically symmetric (electro-) vacuum solution to the Einstein–Maxwell equations. In contrast, the Kerr–Newman solution, depending on three parameters (M, Q and angular momentum J), is only a *particular* stationary and axially symmetric electro-vacuum solution to the Einstein–Maxwell equations. However, one can show under quite general conditions that the Kerr–Newman solution represents the most general asymptotically flat, stationary electro-vacuum *black hole* solution ("no-hair theorem"). Important contributions to the subject of black hole uniqueness were made by Israel, Carter, Hawking, Robinson and Mazur (1967–1982), for details see the recent review [3].

Assuming stationarity and axial symmetry, it is indeed possible to derive the Kerr–Newman black hole solution in straightforward manner, by solving the corresponding boundary value problem of the Einstein–Maxwell equations [7]. In the following sections, an outline of this work will be given. The method is a generalization of the technique developed for solving a boundary value problem of the vacuum Einstein equations leading to the global solution describing a uniformly rotating disc of dust in

R. Meinel (✉)
Theoretisch-Physikalisches Institut, Friedrich-Schiller-Universität Jena, Max-Wien-Platz 1, 07743 Jena, Germany
e-mail: meinel@tpi.uni-jena.de

P. Nicolini et al. (eds.), *1st Karl Schwarzschild Meeting on Gravitational Physics*, Springer Proceedings in Physics 170, DOI 10.1007/978-3-319-20046-0_6

terms of ultraelliptic functions [12, 13], see also [9]. It is based on the "integrability" of the stationary and axisymmetric vacuum Einstein and electro-vacuum Einstein–Maxwell equations via the "inverse scattering method", see [1]. In the pure vacuum case, the method was also used to derive the Kerr black hole solution [9, 10, 13].

6.2 Einstein–Maxwell Equations and Related Linear Problem

The stationary and axisymmetric, electro-vacuum Einstein–Maxwell equations are equivalent to the Ernst equations [4]

$$f\,\Delta\mathscr{E} = (\nabla\mathscr{E} + 2\bar{\Phi}\nabla\Phi)\cdot\nabla\mathscr{E}\,, \quad f\,\Delta\Phi = (\nabla\mathscr{E} + 2\bar{\Phi}\nabla\Phi)\cdot\nabla\Phi \tag{6.1}$$

$$\text{with}\quad f \equiv \Re\,\mathscr{E} + |\Phi|^2\,, \quad \Delta = \frac{\partial^2}{\partial\rho^2} + \frac{1}{\rho}\frac{\partial}{\partial\rho} + \frac{\partial^2}{\partial\zeta^2}\,, \quad \nabla = (\frac{\partial}{\partial\rho}, \frac{\partial}{\partial\zeta}). \tag{6.2}$$

The line element reads

$$\mathrm{d}s^2 = f^{-1}[\,h(\mathrm{d}\rho^2 + \mathrm{d}\zeta^2) + \rho^2\mathrm{d}\phi^2] - f(\mathrm{d}t + A\,\mathrm{d}\phi)^2, \tag{6.3}$$

where the coordinates t and ϕ are adapted to the Killing vectors corresponding to stationarity and axial symmetry:

$$\boldsymbol{\xi} = \frac{\partial}{\partial t}, \quad \boldsymbol{\eta} = \frac{\partial}{\partial\phi}. \tag{6.4}$$

We assume an asymptotic behaviour as $r \to \infty$ $(\rho = r\sin\theta,\ \zeta = r\cos\theta)$ given by

$$\Re\,\mathscr{E} = 1 - \frac{2M}{r} + \mathscr{O}(r^{-2})\,, \quad \Im\,\mathscr{E} = -\frac{2J\cos\theta}{r^2} + \mathscr{O}(r^{-3})\,, \quad \Phi = \frac{Q}{r} + \mathscr{O}(r^{-2}) \tag{6.5}$$

corresponding to asymptotic flatness and the absence of a magnetic monopole term (Q real). The metric functions h and A can be calculated from the complex Ernst potentials $\mathscr{E}(\rho,\zeta)$ and $\Phi(\rho,\zeta)$ according to

$$(\ln h)_{,z} = \frac{\rho}{f^2}(\mathscr{E}_{,z} + 2\bar{\Phi}\Phi_{,z})(\bar{\mathscr{E}}_{,z} + 2\Phi\bar{\Phi}_{,z}) - \frac{4\rho}{f}\Phi_{,z}\bar{\Phi}_{,z}\,, \tag{6.6}$$

$$A_{,z} = \frac{\mathrm{i}\rho}{f^2}[(\Im\,\mathscr{E})_{,z} - \mathrm{i}\bar{\Phi}\Phi_{,z} + \mathrm{i}\Phi\bar{\Phi}_{,z}] \quad (r \to \infty : h \to 1,\ A \to 0). \tag{6.7}$$

Here complex variables

$$z = \rho + i\zeta, \quad \bar{z} = \rho - i\zeta \tag{6.8}$$

have been used instead of ρ and ζ. Note that f has already been given in (6.2). The electromagnetic field tensor

$$F_{ik} = A_{k,i} - A_{i,k}\,, \quad A_i\,\mathrm{d}x^i = A_\phi \mathrm{d}\phi + A_t \mathrm{d}t \tag{6.9}$$

can also be obtained from the Ernst potentials:

$$A_t = -\Re\,\Phi, \qquad A_{\phi,z} = A\,A_{t,z} - \frac{i\rho}{f}(\Im\,\Phi)_{,z} \qquad (r \to \infty : A_\phi \to 0). \tag{6.10}$$

The Ernst equations (6.1) can be formulated as the integrability condition of a related Linear Problem (LP). We use the LP of [11] in a slightly modified form, which is advantageous in the presence of ergospheres:

$$\mathbf{Y}_{,z} = \left[\begin{pmatrix} b_1 & 0 & c_1 \\ 0 & a_1 & 0 \\ d_1 & 0 & 0 \end{pmatrix} + \lambda \begin{pmatrix} 0 & b_1 & 0 \\ a_1 & 0 & -c_1 \\ 0 & d_1 & 0 \end{pmatrix}\right]\mathbf{Y}, \tag{6.11}$$

$$\mathbf{Y}_{,\bar{z}} = \left[\begin{pmatrix} b_2 & 0 & c_2 \\ 0 & a_2 & 0 \\ d_2 & 0 & 0 \end{pmatrix} + \frac{1}{\lambda} \begin{pmatrix} 0 & b_2 & 0 \\ a_2 & 0 & -c_2 \\ 0 & d_2 & 0 \end{pmatrix}\right]\mathbf{Y} \tag{6.12}$$

with

$$\lambda = \sqrt{\frac{K - i\bar{z}}{K + iz}}, \tag{6.13}$$

$$a_1 = \bar{b}_2 = \frac{\mathscr{E}_{,z} + 2\bar{\Phi}\Phi_{,z}}{2f}, \quad a_2 = \bar{b}_1 = \frac{\mathscr{E}_{,\bar{z}} + 2\bar{\Phi}\Phi_{,\bar{z}}}{2f}, \tag{6.14}$$

$$c_1 = f\bar{d}_2 = \Phi_{,z}\,, \quad c_2 = f\bar{d}_1 = \Phi_{,\bar{z}}\,. \tag{6.15}$$

The integrability condition

$$\mathbf{Y}_{,z\bar{z}} = \mathbf{Y}_{,\bar{z}z} \tag{6.16}$$

is equivalent to the Ernst equations. The following points are relevant for the application of soliton theoretic solution methods:

- The 3×3 matrix $\mathbf{Y}$ depends not only on the coordinates ρ and ζ (or z and $\bar{z}$), but also on the additional complex "spectral parameter" K.
- Since $\bar{K}$ does not appear, we can assume without loss of generality that the elements of $\mathbf{Y}$ are holomorphic functions of K defined on the two-sheeted Riemann surface associated with (6.13), except from the locations of possible singularities.

- Each column of $\mathbf{Y}$ is itself a solution to the LP. We assume that these three solutions are linearly independent.
- For a given solution $\mathscr{E}$, Φ to the Einstein–Maxwell equations, the solution to the LP can be fixed (normalized) by prescribing $\mathbf{Y}$ at some point ρ_0, ζ_0 of the ρ-ζ plane as a (matrix) function of K in one of the two sheets of the Riemann surface.
- $\mathbf{Y}$ can be discussed in general as a unique function of ρ, ζ and λ.

Three interesting relations result directly from the structure of the LP (6.11), (6.12):

$$[f(\rho,\zeta)]^{-1}\det\mathbf{Y}(\rho,\zeta,\lambda) = C_0(K), \tag{6.17}$$

$$\mathbf{Y}(\rho,\zeta,-\lambda) = \begin{pmatrix} 1 & 0 & 0 \\ 0 & -1 & 0 \\ 0 & 0 & 1 \end{pmatrix} \mathbf{Y}(\rho,\zeta,\lambda)\mathbf{C}_1(K), \tag{6.18}$$

$$\left[\mathbf{Y}(\rho,\zeta,1/\bar{\lambda})\right]^{\dagger} \begin{pmatrix} [f(\rho,\zeta)]^{-1} & 0 & 0 \\ 0 & -[f(\rho,\zeta)]^{-1} & 0 \\ 0 & 0 & -1 \end{pmatrix} \mathbf{Y}(\rho,\zeta,\lambda) = \mathbf{C}_2(K), \tag{6.19}$$

where $C_0(K)$ as well as the matrices $\mathbf{C}_1(K)$ and $\mathbf{C}_2(K)$ do not depend on ρ and ζ.

6.3 Solving the Black Hole Boundary Value Problem

After formulating the black hole boundary value problem, we will use the LP to find its solution. The most important part comprises deriving the Ernst potentials on the axis of symmetry [7]. It is well known that these "axis data" uniquely determine the solution everywhere, see [5, 14]. A straightforward method for obtaining the full solution from the axis data is based on the analytical properties of $\mathbf{Y}$ as a function of λ [8].

6.3.1 Boundary Conditions

The event horizon $\mathscr{H}$ of a stationary and axisymmetric black hole is characterized by the conditions

$$\mathscr{H}: \quad \chi^i\chi_i = 0, \quad \chi^i\eta_i = 0, \tag{6.20}$$

where $\chi^i \equiv \xi^i + \Omega\eta^i$ and the constant Ω is the "angular velocity of the horizon" [2, 6]. Because of

$$\rho^2 = (\xi^i\eta_i)^2 - \xi^i\xi_i\eta^k\eta_k = (\chi^i\eta_i)^2 - \chi^i\chi_i\eta^k\eta_k \tag{6.21}$$

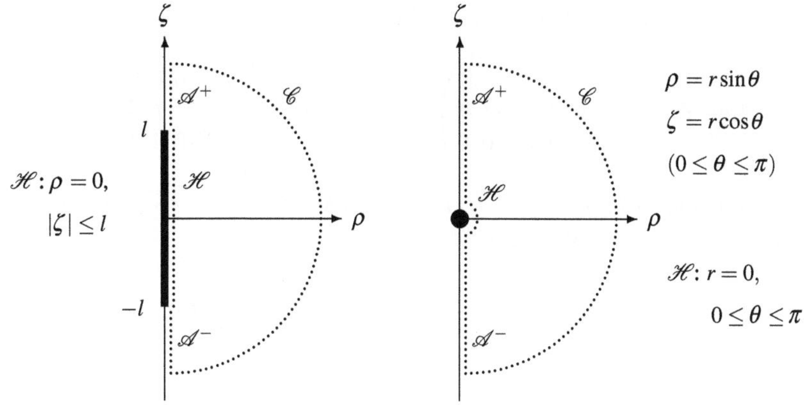

Fig. 6.1 In Weyl coordinates, the horizon is either a finite interval or a point on the ζ-axis (adapted from [7])

the horizon must be located on the ζ-axis of our Weyl coordinate system:

$$\mathscr{H}: \quad \rho = 0. \tag{6.22}$$

This results in two possibilities for a connected horizon[1]: (i) a finite interval on the ζ-axis and (ii) a point on the ζ-axis, see Fig. 6.1. Note that the two parts of the symmetry axis, $\mathscr{A}^+$ and $\mathscr{A}^-$, where the Killing vector η vanishes, are also characterized by $\rho = 0$. The black hole boundary value problem consists of finding a solution that is regular everywhere outside the horizon and satisfies (6.20) and (6.5).

6.3.2 Axis Data

At $\rho = 0$, the branch points $K = \mathrm{i}\bar{z}$ and $K = -\mathrm{i}z$ of (6.13) merge to $K = \zeta$ and for $K \neq \zeta$ holds $\lambda = \pm 1$. Consequently, the solution to the LP, for $\lambda = +1$, is of the form

$$\mathscr{A}^{\pm}: \quad \mathbf{Y}_{\pm} = \begin{pmatrix} \bar{\mathscr{E}} + 2|\Phi|^2 & 1 & \Phi \\ \mathscr{E} & -1 & -\Phi \\ 2\bar{\Phi} & 0 & 1 \end{pmatrix} \mathbf{C}_{\pm}, \tag{6.23}$$

$$\mathscr{H}: \quad \mathbf{Y}_{\mathrm{h}} = \begin{pmatrix} \bar{\mathscr{E}} + 2|\Phi|^2 & 1 & \Phi \\ \mathscr{E} & -1 & -\Phi \\ 2\bar{\Phi} & 0 & 1 \end{pmatrix} \mathbf{C}_{\mathrm{h}}. \tag{6.24}$$

[1] A connected horizon means a single black hole. We are not interested here in the problem of multi-black-hole equilibrium states.

We fix $\mathbf{C}_+(K)$ by the normalization condition

$$\lim_{K\to\zeta} \mathbf{Y}_+(\zeta, K) = \begin{pmatrix} 1 & 1 & 0 \\ 1 & -1 & 0 \\ 0 & 0 & 1 \end{pmatrix} \quad \Rightarrow \quad \mathbf{C}_+ = \begin{pmatrix} F & 0 & 0 \\ G & 1 & L \\ H & 0 & 1 \end{pmatrix} \tag{6.25}$$

and the functions $F(K)$, $G(K)$, $H(K)$ and $L(K)$, for $K = \zeta$, are given by the potentials $\mathscr{E} = \mathscr{E}_+$, $\Phi = \Phi_+$ on $\mathscr{A}^+$:

$$F(\zeta) = [f_+(\zeta)]^{-1}, \quad G(\zeta) = \left[|\Phi_+(\zeta)|^2 + \mathrm{i}\Im\,\mathscr{E}_+(\zeta)\right][f_+(\zeta)]^{-1}, \tag{6.26}$$

$$H(\zeta) = -2\bar{\Phi}_+(\zeta)[f_+(\zeta)]^{-1}, \quad L(\zeta) = -\Phi_+(\zeta) \tag{6.27}$$

and, vice versa,

$$\mathscr{E}_+(\zeta) = \frac{1 - \bar{G}(\zeta)}{F(\zeta)}, \quad \Phi_+(\zeta) = -\frac{\bar{H}(\zeta)}{2F(\zeta)}. \tag{6.28}$$

We can calculate $C_0(K)$, $\mathbf{C}_1(K)$ and $\mathbf{C}_2(K)$ of relations (6.17)–(6.19) for our normalization:

$$C_0 = -2F, \quad \mathbf{C}_1 = \begin{pmatrix} 0 & 1 & 0 \\ 1 & 0 & 0 \\ 0 & 0 & 1 \end{pmatrix}, \quad \mathbf{C}_2 = \begin{pmatrix} 0 & 2F & 0 \\ 2F & 0 & 0 \\ 0 & 0 & -1 \end{pmatrix}. \tag{6.29}$$

On $\mathscr{A}^+$, (6.19) reads

$$\left[\mathbf{C}_+(\bar{K})\right]^\dagger \begin{pmatrix} 0 & 2 & 0 \\ 2 & 0 & 0 \\ 0 & 0 & -1 \end{pmatrix} \mathbf{C}_+(K) = \begin{pmatrix} 0 & 2F & 0 \\ 2F & 0 & 0 \\ 0 & 0 & -1 \end{pmatrix}. \tag{6.30}$$

From continuity conditions at the "poles" of the horizon ($\rho = 0$, $\zeta = \pm l$ or $r = 0, \theta = 0, \pi$; see Fig. 6.1) and using the boundary conditions, one can calculate $\mathbf{C}_\mathrm{h}(K)$ and $\mathbf{C}_-(K)$ in terms of $\mathbf{C}_+(K)$, for details I refer to [7]. Closing the path of integration via infinity (curve $\mathscr{C}$: $\rho = R\sin\theta$, $\zeta = R\cos\theta$ with $0 \le \theta \le \pi$, $R \to \infty$), where $\mathbf{Y}$ is constant because of the LP and (6.5), but λ changes from ± 1 at $\theta = 0$ to ∓ 1 at $\theta = \pi$, we obtain with (6.18) and (6.29) an explicit expression for $\mathbf{C}_+(K)$ in terms of the parameters Ω, l (with $l = 0$ for a horizon at $r = 0$) and the values of the Ernst potentials at the poles. Using (6.28), we can calculate $\mathscr{E}_+$ and Φ_+. The number of free real parameters is reduced to four as a consequence of the constraint (6.30) and to three if no magnetic monopole is allowed. The final result is

$$F(K) = \frac{(K - L_1)(K - L_2)}{(K - K_1)(K - K_2)}, \quad G(K) = \frac{Q^2 - 2\mathrm{i}J}{(K - K_1)(K - K_2)}, \tag{6.31}$$

$$H(K) = -\frac{2Q(K - L_1)}{(K - K_1)(K - K_2)}, \quad L(K) = -\frac{Q}{K - L_1} \tag{6.32}$$

$$\text{with} \quad L_{1/2} = -M \pm \mathrm{i}\frac{J}{M}, \quad K_{1/2} = \pm\sqrt{M^2 - Q^2 - \frac{J^2}{M^2}} \tag{6.33}$$

and, correspondingly,

$$\mathcal{E}_+(\zeta) = 1 - \frac{2M}{\zeta + M - \mathrm{i}J/M}, \quad \Phi_+(\zeta) = \frac{Q}{\zeta + M - \mathrm{i}J/M} \tag{6.34}$$

together with the parameter relations

$$\frac{l^2}{M^2} + \frac{Q^2}{M^2} + \frac{J^2}{M^4} = 1 \quad \text{and} \quad \Omega M = \frac{J/M^2}{(1 + l/M)^2 + J^2/M^4}. \tag{6.35}$$

6.3.3 Solution Everywhere Outside the Horizon

Relation (6.18) together with the expression for $\mathbf{C}_1(K)$ in (6.29) is equivalent to the following structure of $\mathbf{Y}$:

$$\mathbf{Y}(\rho, \zeta, \lambda) = \begin{pmatrix} \psi(\rho, \zeta, \lambda) & \psi(\rho, \zeta, -\lambda) & \alpha(\rho, \zeta, \lambda) \\ \chi(\rho, \zeta, \lambda) & -\chi(\rho, \zeta, -\lambda) & \beta(\rho, \zeta, \lambda) \\ \varphi(\rho, \zeta, \lambda) & \varphi(\rho, \zeta, -\lambda) & \gamma(\rho, \zeta, \lambda) \end{pmatrix}, \tag{6.36}$$

where $\alpha(\rho, \zeta, \lambda) = \alpha(\rho, \zeta, -\lambda)$, $\beta(\rho, \zeta, \lambda) = -\beta(\rho, \zeta, -\lambda)$ and $\gamma(\rho, \zeta, \lambda) = \gamma(\rho, \zeta, -\lambda)$. The general solution of the LP for $K \to \infty$ and $\lambda = +1$ reads

$$\mathbf{Y}(\rho, \zeta, 1) = \begin{pmatrix} \bar{\mathcal{E}} + 2|\Phi|^2 & 1 & \Phi \\ \mathcal{E} & -1 & -\Phi \\ 2\bar{\Phi} & 0 & 1 \end{pmatrix} \mathbf{C}, \tag{6.37}$$

where $\mathbf{C}$ is a constant matrix. Equations (6.23), (6.25), (6.31), (6.32) imply $\mathbf{C} = \mathbf{1}$ and lead to the ansatz

$$\psi = 1 + k_1\left(\frac{1}{\kappa_1 - \lambda} - \frac{1}{\kappa_1 + 1}\right) + k_2\left(\frac{1}{\kappa_2 - \lambda} - \frac{1}{\kappa_2 + 1}\right), \tag{6.38}$$

$$\chi = 1 + l_1\left(\frac{1}{\kappa_1 - \lambda} - \frac{1}{\kappa_1 + 1}\right) + l_2\left(\frac{1}{\kappa_2 - \lambda} - \frac{1}{\kappa_2 + 1}\right), \tag{6.39}$$

$$\varphi = m_1\left(\frac{1}{\kappa_1 - \lambda} - \frac{1}{\kappa_1 + 1}\right) + m_2\left(\frac{1}{\kappa_2 - \lambda} - \frac{1}{\kappa_2 + 1}\right), \tag{6.40}$$

$$\alpha = \Phi + \frac{\alpha_0}{K - L_1}, \quad \beta = -\Phi\,\frac{\lambda(K + \mathrm{i}z)}{K - L_1}, \quad \gamma = 1 + \frac{\gamma_0}{K - L_1}, \tag{6.41}$$

where

$$\kappa_\mu = \sqrt{\frac{K_\mu - \mathrm{i}\bar{z}}{K_\mu + \mathrm{i}z}} \quad (\mathscr{A}^+ : \quad \kappa_\mu = +1)\,. \tag{6.42}$$

According to the LP, $\mathbf{Y}_{,z}\mathbf{Y}^{-1}$ and $\mathbf{Y}_{,\bar{z}}\mathbf{Y}^{-1}$ must be holomorphic functions of λ for all $\lambda \neq 0, \infty$. The regularity at $\lambda = \pm\kappa_\mu$ ($\mu = 1, 2$), the poles of the first two columns of $\mathbf{Y}$, is automatically guarantied, whereas regularity at $\lambda = \pm\lambda_\mu$ with $\lambda_\mu = \sqrt{(L_\mu - \mathrm{i}\bar{z})/(L_\mu + \mathrm{i}z)}$ ($\mathscr{A}^+$: $\lambda_\mu = +1$), where poles of the third column ($\mu = 1$) and zeros of $\det \mathbf{Y}$ ($\mu = 1, 2$) occur, see (6.17), (6.29), (6.31), is equivalent to a set of linear algebraic equations, which together with (6.23), (6.25), (6.31), (6.32) uniquely determine the unknowns $k_\mu(\rho, \zeta)$, $l_\mu(\rho, \zeta)$, $m_\mu(\rho, \zeta)$, $\alpha_0(\rho, \zeta)$, $\gamma_0(\rho, \zeta)$ and $\Phi(\rho, \zeta)$. With $\mathscr{E}(\rho, \zeta) = \chi(\rho, \zeta, 1)$, see (6.37), this leads to the result

$$\mathscr{E} = 1 - \frac{2M}{\tilde{r} - \mathrm{i}(J/M)\cos\tilde{\theta}}, \quad \Phi = \frac{Q}{\tilde{r} - \mathrm{i}(J/M)\cos\tilde{\theta}} \tag{6.43}$$

$$\text{with} \quad \rho = \sqrt{\tilde{r}^2 - 2M\tilde{r} + J^2/M^2 + Q^2}\,\sin\tilde{\theta}\,, \quad \zeta = (\tilde{r} - M)\cos\tilde{\theta}. \tag{6.44}$$

The "domain of outer communication" (the region outside the event horizon $\mathscr{H}$) is given by $\tilde{r} > \tilde{r}_\mathrm{h} = M + \sqrt{M^2 - J^2/M^2 - Q^2}$. The horizon itself is characterized by $\tilde{r} = \tilde{r}_\mathrm{h}$, and the axis of symmetry is located at $\tilde{\theta} = 0$ ($\mathscr{A}^+$) and $\tilde{\theta} = \pi$ ($\mathscr{A}^-$). Note that (6.35) implies $Q^2 + J^2/M^2 \le M^2$. The equality sign, corresponding to $l = 0$, is valid for the extremal Kerr–Newman black hole.

6.3.4 Full Metric and Electromagnetic Field

Using (6.2), (6.6), (6.7), (6.10) we can calculate the full metric and the electromagnetic four-potential:

$$\mathrm{d}s^2 = \frac{\Sigma}{\Delta}\,\mathrm{d}\tilde{r}^2 + \Sigma\,\mathrm{d}\tilde{\theta}^2 + \left(\tilde{r}^2 + a^2 + \frac{(2M\tilde{r} - Q^2)a^2\sin^2\tilde{\theta}}{\Sigma}\right)\sin^2\tilde{\theta}\,\mathrm{d}\phi^2 \tag{6.45}$$

$$- \frac{(2M\tilde{r} - Q^2)2a\sin^2\tilde{\theta}}{\Sigma}\,\mathrm{d}\phi\,\mathrm{d}t - \left(1 - \frac{2M\tilde{r} - Q^2}{\Sigma}\right)\mathrm{d}t^2 \tag{6.46}$$

$$\text{with} \quad \Delta = \tilde{r}^2 - 2M\tilde{r} + a^2 + Q^2, \quad \Sigma = \tilde{r}^2 + a^2\cos^2\tilde{\theta}, \quad a \equiv J/M \tag{6.47}$$

and

$$A_i\,\mathrm{d}x^i = \frac{Q\tilde{r}}{\Sigma}(a\sin^2\tilde{\theta}\,\mathrm{d}\phi - \mathrm{d}t). \tag{6.48}$$

This is the well-known Kerr–Newman solution in Boyer–Lindquist coordinates $\tilde{r}$ and $\tilde{\theta}$. For $Q = 0$ it reduces to the Kerr solution, $J = 0$ gives the Reissner–Nordström solution and $Q = J = 0$ leads back to the Schwarzschild solution.

References

1. V. Belinski, E. Verdaguer, *Gravitational Solitons* (Cambridge University Press, Cambridge, 2001)
2. B. Carter, Black hole equilibrium states, in *Black Holes*, ed. by C. DeWitt, B.S. DeWitt (Gordon and Breach Science Publishers, New York, 1973), pp. 57–214
3. P.T. Chruściel, J.L. Costa, M. Heusler, Stationary black holes: uniqueness and beyond. Living Rev. Relativity **15**, 7 (2012)
4. F.J. Ernst, New formulation of the axially symmetric gravitational field problem. II. Phys. Rev. **168**, 1415–1417 (1968)
5. I. Hauser, F.J. Ernst, A homogeneous Hilbert problem for the Kinnersley–Chitre transformations of electrovac space-times. J. Math. Phys. **21**, 1418–1422 (1980)
6. S.W. Hawking, G.F.R. Ellis, *The Large Scale Structure of Space-Time* (Cambridge University Press, Cambridge, 1973)
7. R. Meinel, Constructive proof of the Kerr–Newman black hole uniqueness including the extreme case. Class. Quant. Grav. **29**, 035004 (2012)
8. R. Meinel, R. Richter, Constructive proof of the Kerr–Newman black hole uniqueness: derivation of the full solution from scratch. arXiv:1208.0294 [gr-qc] (2012)
9. R. Meinel, M. Ansorg, A. Kleinwächter, G. Neugebauer, D. Petroff, *Relativistic Figures of Equilibrium* (Cambridge University Press, Cambridge, 2008)
10. G. Neugebauer, Rotating bodies as boundary value problems. Ann. Phys. (Leipzig) **9**, 342–354 (2000)
11. G. Neugebauer, D. Kramer, Einstein–Maxwell solitons. J. Phys. A **16**, 1927–1936 (1983)
12. G. Neugebauer, R. Meinel, General relativistic gravitational field of a rigidly rotating disk of dust: solution in terms of ultraelliptic functions. Phys. Rev. Lett. **75**, 3046–3047 (1995)
13. G. Neugebauer, R. Meinel, Progress in relativistic gravitational theory using the inverse scattering method. J. Math. Phys. **44**, 3407–3429 (2003)
14. N.R. Sibgatullin, *Oscillations and Waves in Strong Gravitational and Electromagnetic Fields* (Springer, Berlin, 1991)

Chapter 7
On the Black Holes in Alternative Theories of Gravity: The Case of Non-linear Massive Gravity

Ivan Arraut

Abstract It is already known that a positive Cosmological Constant Λ sets the scale $r_0 = \left(\frac{3}{2}r_s r_\Lambda^2\right)^{1/3}$, which depending on the mass of the source, can be of astrophysical order of magnitude. This scale was interpreted before as the maximum distance in order to get bound orbits. The same scale corresponds to the static observer position if we want to define the Black Hole temperature in an asymptotically de-Sitter space. r_0 also appears inside the non-linear theory of massive gravity (dRGT) as the Vainshtein radius for the Λ_3 version of the theory. I compare the role that this scale plays inside these different scenarios.

7.1 Introduction

The Schwarzschild de-Sitter (S-dS) space in static coordinates has been widely studied in the past. Its analytic extension for S-dS space has been performed by Bażański and Ferrari [1]. They interpreted the scale $r_0 = \left(\frac{3}{2}r_s r_\Lambda\right)^{1/3}$ as the distance where the 0-0 component of the S-dS metric takes a minimum value. As a consequence of this, it was found in [2] that r_0 represents a transition distance after which a photon suffers a gravitational blue shift when it moves away from a source. The same scale is used by Bousso and Hawking in order to find the appropriate expression for the temperature of a black hole immersed inside a de-Sitter space [3]. In such a case, the distance r_0 is interpreted as the position of the static observer in order to find the appropriate normalization for the time-like Killing vector. Then there exist a minimum temperature for the black hole given by $T = \frac{1}{2\pi r_\Lambda}$ [3, 4]. This analysis differs in some details from the one done in [5] where the Black Holes thermodynamics inside the S-dS space was analyzed in detail. The role of r_0 as a static radius was also analyzed in [6] inside the Kerr-de Sitter space. In Balaguera et al. [7], found that r_0 represents the maximum distance within which we can find bound orbits solutions for a test particle moving around a source. In the same manuscript, the velocity

I. Arraut (✉)
Theory Center, Institute of Particle and Nuclear Studies, The High Energy Accelerator Research Organization, Tsukuba, Ibaraki 305-0801, Japan
e-mail: ivanarr@post.kek.jp

P. Nicolini et al. (eds.), *1st Karl Schwarzschild Meeting on Gravitational Physics*, Springer Proceedings in Physics 170, DOI 10.1007/978-3-319-20046-0_7

bounds for a test particle inside the S-dS space were obtained, this work was then extended by Arraut et al. in [8] in order to incorporate other metric solutions. In [7], the authors also found that there exist a maximum angular momentum L_{max} for the test particle to be inside a bound orbit. If $L = L_{max}$, then there exist a saddle point for the effective potential at the distance r_x, this analysis was extended recently by Arraut [9]. In [10, 11], the scale r_0 was derived by using a different method and some conditions for the circular orbits and its stabilities were obtained. The scale r_0 plays a central role inside the Λ_3 version of the non-linear theory of massive gravity where it represents the distance below which non-linearities become important and General Relativity is restored [12, 13]. The paper is organized as follows: In Sect. 7.2, I introduce the basic aspects of the S-dS space in static coordinates and then derive the scale r_0 including its correction due to the angular momentum of a massive test particle moving around the source. In Sect. 7.2.1, I analyze the Black Hole temperature as it is defined by Bousso and Hawking. I explain the role of the scale r_0 in such a case. In Sect. 7.3, I introduce the S-dS solution derived from the non-linear theory of massive gravity and then I explain the role of r_0 in this theory.

7.2 The Schwarzschild de-Sitter Space

Schwarzschild-de Sitter metric in static coordinates, is given by:

$$ds^2 = -e^{\nu(r)}dt^2 + e^{-\nu(r)}dr^2 + r^2 d\theta^2 + r^2 \sin^2\theta d\phi^2 \tag{7.1}$$

where:

$$e^{\nu(r)} = 1 - \frac{r_s}{r} - \frac{r^2}{3r_\Lambda^2} \tag{7.2}$$

With $r_s = 2GM$ being the gravitational radius and $r_\Lambda = \frac{1}{\sqrt{\Lambda}}$ defines the Cosmological Constant scale. In this coordinate system, with the standard definition of the effective potential ($U_{eff}(r)$), the equation of motion of a massive test particle becomes:

$$\frac{1}{2}\left(\frac{dr}{d\tau}\right)^2 + U_{eff}(r) = \frac{1}{2}\left(E^2 + \frac{L^2}{3r_\Lambda^2} - 1\right) = C \tag{7.3}$$

where C is a constant depending on the initial conditions of motion. U_{eff} has 3 circular orbits. They correspond to the condition $\frac{dU_{eff}(r)}{dr} = 0$. For our present purposes, the most relevant scale corresponding to this condition is given by [9]:

$$r_0(\beta) = \left(\frac{3}{2}r_s r_\Lambda^2\right)^{1/3} - \frac{1}{4\beta^2}(3r_s r_\Lambda^2)^{1/3} \tag{7.4}$$

where we make explicit the angular momentum dependence through the parameter $\beta = L/L_{max}$, with $L_{max} = \frac{3^{2/3}}{4}(r_s^2 r_\Lambda)^{1/3}$ being the maximum angular momentum if we want to get bound orbits for massive test particles. This scale is the limit where the attractive effects due to gravity and the repulsive ones due to the Cosmological Constant (Λ) just cancel. This is the key point in the Bousso-Hawking definition of temperature as will be explained in the next section.

7.2.1 Black Hole Thermodynamics in an Asymptotically De-Sitter Space

In agreement with Bousso and Hawking, the appropriate way to define the black Hole thermodynamics is by normalizing the time-like Killing vector such that the static observer is located at the distance given by (7.4) with $\beta = 0$. We assume that the observer does not have any angular momentum. Then the surface gravity is defined as [3]:

$$\kappa_{BH,CH} = \left(\frac{(K^\mu \nabla_\mu K_\gamma)(K^\alpha \nabla_\alpha K^\gamma)}{-K^2}\right)^{1/2}_{r=r_{BH},r_{CH}} \tag{7.5}$$

The subindices BH and CH, correspond to the Black Hole Horizon and the Cosmological one respectively. The event horizons are obtained from the condition:

$$g^{rr}(r_c) = 0 \tag{7.6}$$

The two horizons become equal when the mass of the Black Hole reach its maximum value given by:

$$M_{max} = \frac{1}{3}\frac{m_{pl}^2}{m_\Lambda} \tag{7.7}$$

where m_{pl} corresponds to the Planck mass and $m_\Lambda = \sqrt{\Lambda}$. If the mass of a Black Hole is bigger than the value given by (7.7), then there is no radiation at all and we have a naked singularity. As $M = M_{max}$, the two event horizons take the same value ($r_{BH} = r_{CH} = r_\Lambda = 1/\sqrt{\Lambda}$), they are degenerate and a thermodynamic equilibrium is established. As has been explained by Bousso and Hawking [3], as $M \to M_{max}$, $V(r) \to 0$ between the two horizons (BH and Cosmological) and for that reason the Schwarzschild-like coordinates simply become inappropriate. In such a case we need a new coordinate system. In agreement with Ginsparg and Perry [14], the metric has to be written in a new coordinate system and is given explicitly by:

$$ds^2 = -r_\Lambda^2\left(1+\frac{2}{3}\varepsilon cos\chi\right)sin^2\chi d\psi^2 + r_\Lambda^2\left(1-\frac{2}{3}\varepsilon cos\chi\right)d\chi^2 + r_\Lambda^2(1-2\varepsilon cos\chi)d\Omega_2^2 \tag{7.8}$$

in these coordinates, the Black Hole horizon corresponds to $\chi = 0$ and the Cosmological horizon to $\chi = \pi$ [3]. This metric has been expanded up to first order in ε. Equation (7.8) is of course the appropriate metric to be used as the mass of the Black Hole is near to its maximum value given by (7.7). It has been found by Bousso and Hawking that the time-like Killing vector inside the definition (7.12) has to be normalized in agreement with:

$$\gamma_t = \left(1-\left(\frac{3r_s}{2r_\Lambda}\right)^{2/3}\right)^{-1/2} \tag{7.9}$$

With γ_t being the normalization factor for the time-like Killing vector defined as $K = \gamma_t\frac{\partial}{\partial t}$. In an asymptotically flat space, $\gamma_t \to 1$ when $r \to \infty$. But in the case of (7.9), the Killing vector is just normalized with respect to an observer at the position r_0 with $\beta = 0$ as has been defined previously. When the mass of the Black Hole reach its maximum value defined by $\varepsilon \to 0$ [3], the Black Hole temperature reach its minimum value given by:

$$2\pi T_{min=\kappa_{min}^{BH}} = \frac{1}{r_\Lambda} \tag{7.10}$$

where κ is the surface gravity.

7.3 Black Holes in dRGT Non-linear Theory of Massive Gravity

In agreement with Koyama, Niz, Tanisato, Gabadadze and colleagues, it is possible to construct Black Hole solutions inside the non-linear theory of massive gravity. Is natural to suspect is that such solution should be Schwarzschild de-Sitter, although other solutions are in principle possible. Inside this formalism, the solution given by (7.1) can be obtained but surrounded by a Stückelberg background halo of helicity 0 and ± 1. One possible solution proposed in [15–18] is given by:

$$ds^2 = -dt^2 + (dr \pm \sqrt{f}dt)^2 + r^2 d\Omega^2 \tag{7.11}$$

which is free of horizon singularities, such that the invariant $g^{\mu\nu}\partial_\mu\phi^a\partial_\nu\phi^b\eta_{ab}$ (defined inside the dRGT theory) remains finite when all the other standard relativistic invariants are also finite. The metric has to be a solution of the Einstein's equations, which in massive gravity are defined as:

$$G^{\mu}{}_{\nu} = -m^2 X^{\mu}{}_{\nu} \tag{7.12}$$

The solution (7.11), after the appropriate coordinate transformations, becomes the same solution given by (7.1) but surrounded by a Stückelberg background defined by:

$$\Phi^0 = \frac{1}{\kappa}(t + f(r)) \quad \Phi^r = \left(1 + \frac{1}{\alpha}\right) r \quad \Phi^\theta = \theta \quad \Phi^\phi = \phi \tag{7.13}$$

The previous results correspond to a family of solutions satisfying a specific relation between the two free parameters of the theory as has been explained in [15–19]. The scale r_0 defined before, inside the Λ_3 version of the theory, appears as the Vainshtein radius if we tune the mass of the graviton with the Λ scale. For distances satisfying the condition $r \ll r_0(\beta = 0)$, non-linearities become relevant and General Relativity is recovered, avoiding in such a way the DVZ discontinuity [20]. The non-linear solution inside the dRGT theory, admits perturbative expansions in terms of the mass of the graviton for distances satisfying $r \ll r_0(\beta = 0)$. On the other hand, the same solutions admit perturbative expansions in terms of the Newtonian Constant for distances $r \gg r_0(\beta = 0)$. Then in some sense, $r_0(\beta = 0)$ (zero angular momentum of a massive test particle), is a scale which marks the transition between a solution dominated by the Newtonian constant and the one dominated by the graviton mass in direct analogy with what happens in the standard theory of General Relativity. The main difference is that $r_0(\beta = 0)$ in massive gravity is related to the existence of a strong coupling scale $\Lambda_3 = (M_{pl} m^2)^{1/3}$ which appears in the Lagrangian of the theory [19].

7.4 Conclusions

The scale $r_0(\beta)$ is relevant for the definition of bound orbits and Black Hole temperature in asymptotically de-Sitter spaces. This inside the framework of General Relativity. On the other hand, the same scale appears inside the Λ_3 theory of massive gravity, where it marks a limit. For distances shorter than $r_0(\beta = 0)$, non-linearities become relevant and the standard results of General Relativity are recovered. For distances larger than $r_0(\beta = 0)$, non-linearities are not relevant, then the new degrees of freedom become important.

Acknowledgments The author would like to thank Gia Dvali for a very useful discussion during the Karl Schwarzschild meeting 2013 organized in FIAS, Frankfurt/Germany. This work is supported by MEXT (The Ministry of Education, Culture, Sports, Science and Technology) in Japan and KEK Theory Center.

References

1. S.L. Bażański, V. Ferrari, Analytic Extension of the Schwarzschild-de Sitter Metric, Il Nuovo Cimento, Vol. 91 B, N. 1, 11 Gennaio (1986)
2. I. Arraut, D. Batic, M. Nowakowski, Comparing two approaches to Hawking radiation of Schwarzschild-de Sitter black holes. Class. Quant. Grav. **26**, 125006 (2009)
3. R. Bousso, S.W. Hawking, Pair creation of black holes during inflation. Phys. Rev. D **54**, 6312–6322 (1996)
4. I. Arraut, *The Planck scale as a duality of the Cosmological Constant: S-dS and S-AdS thermodynamics from a single expression.* arXiv:1205.6905v3 [gr-qc]
5. G.W. Gibbons, S.W. Hawking, Cosmological event horizons, thermodynamics and particle creation. Phy. Rev. D **15**, 2738 (1977)
6. Z. Stuchlí, P. Slaný Equatorial circular orbits in the Kerr-de Sitter spacetimes. Phys. Rev. D **69**, 064001 (2004)
7. A. Balaguera Antolinez, C.G. Böhmer, M. Nowakowski, Scales set by the cosmological constant. Class. quant. Grav. 23, 485
8. I. Arraut, D. Batic, M. Nowakowski, Velocity and velocity bounds in static spherically symmetric metrics. Cent. Eur. J. Phys. **9**, 926 (2011)
9. I. Arraut, *On the astrophysical scales set by the Cosmological Constant.* arXiv:1305.0475 [gr-qc]
10. Z. Stuchlík, The motion of test particles in black hole backgrounds with non-zero cosmological constant. Bull. Astron. Inst. Czech. 34, 129 (1983)
11. Z. Stuchlík, Some properties of the Schwarzschild de-Sitter and Schwarzschild Anti de-Sitter spacetimes. Phys. Rev. D 60, 044006 (1999)
12. E. Babichev, C. Deffayet, An introduction to the Vainshtein mechanism. Class. Quant. Grav. **30**, 184001 (2013)
13. G. Chkareuli, D. Pirtskhalava, Vainshtein mechanism in Λ_3 theories. Phys. Lett. B **713**, 99–103 (2012)
14. P. Ginsparg, M.J. Perry, Nucl. Phys. B **222**, 245 (1983)
15. K. Koyama, G. Niz, G. Tasinato, Strong interactions and exact solutions in non-linear massive gravity. Phys. Rev. D **84**, 064033 (2011)
16. L. Berezhiani, G. Chkareuli, C. de Rham, G. Gabadadze, A.J. Tolley, On black holes in massive gravity. Phys. Rev. D **85**, 044024 (2012)
17. K. Koyama, G. Niz, G. Tasinato, Analytic solutions in non-linear massive gravity. Phys. Rev. Lett. **107**, 131101 (2011)
18. F. Sbisa, G. Niz, K. Koyama, G. Tasinato, Characterising vainshtein solutions in massive gravity. Phys. Rev. D **86**, 024033 (2012)
19. T.M. Nieuwenhuizen, Exact Schwarzschild-de Sitter black holes in a family of massive gravity models. Phys. Rev. D **84**, 024038 (2011)
20. K. Hinterbichler, Theoretical aspects of massive gravity. Rev. Mod. Phys. **84**, 671 (2012)

Chapter 8
The Near-Horizon Limit

Jiří Daněk

Abstract We present a new analytic coordinate system covering the whole global extension of the ultra-extreme Reissner-Nordström-de Sitter spacetime, analyse radial motions of two particles of different charges and demonstrate our results in an exact Penrose diagram. Further, we use the near-horizon limit to analyse the energy of particles' collisions near the ultra-extreme horizon, which lead to unbound collision energy in the center of mass system, and relate our results to the previously established behavior on simple and extreme horizon of Reissner-Nordström black hole.

8.1 Introduction

In the past black hole horizons enjoyed great attention because of the misunderstanding of coordinate singularities located on them in the original coordinate systems. With the understanding of coordinates' properties and discovery of equivalent non-singular coordinate systems, the peculiarity of the horizons faded a bit, but black hole horizons are still drawing attention due to their role in the global causal and geometric properties of spacetimes. The near-horizon limit is one way of investigating the direct proximity of an arbitrary horizon and lets us forget about the distant regions we are not interested in. Behaviour of free test particles can also reveal many interesting features of spacetimes and is another tool for investigating the near-horizon regions.

Recently, collision processes leading to unbound collision energies in the center of mass system were found to occur in the direct proximity of the extreme horizon of Kerr black hole [1]. Later it was proven that the extreme horizon of the static charged Reissner-Nordström black hole (RN) can also exhibit unbound collision energy and serve as a particle accelerator as well [7]. Similar investigation was also performed for the single inner horizon [6] of RN or for the single cosmological horizon of Reissner-Nordström-de Sitter spacetime (RNdS) [8].

J. Daněk (✉)
Institute of Theoretical Physics Faculty of Mathematics and Physics,
Charles University in Prague, Prague, Czech Republic
e-mail: jiri.danek@matfyz.cz

P. Nicolini et al. (eds.), *1st Karl Schwarzschild Meeting on Gravitational Physics*,
Springer Proceedings in Physics 170, DOI 10.1007/978-3-319-20046-0_8

Article is to search for unbound collision energy near the so far neglected ultra-extreme (U-E) horizon of the U-E RNdS (i.e., $9M^2 = 8Q^2 = 2/\Lambda$ [2]), the metric of which can be written as

$$\mathrm{d}s^2 = -f(r)\,\mathrm{d}t^2 + \frac{1}{f(r)}\mathrm{d}r^2 + r^2\left(\mathrm{d}\theta^2 + \sin^2\theta\mathrm{d}\phi^2\right), \tag{8.1}$$

with the lapse function $f(r) = -\frac{\Lambda}{3r^2}(r-N)(r-r_t)^3$ where $N = -3r_t$, $r_t = (2\Lambda)^{-1/2}$.

This paper is organized as follows: In Sect. 8.2 we will find a new continuous coordinate patch covering the whole global extension of RNdS. The geodesics of radial charged test particles will be investigated in Sect. 8.3. In Sect. 8.4 we will use the near-horizon limit to investigate the collision energy near the U-E horizon. Finally, we will demonstrate our results and collision in an exact Penrose diagram (Fig. 8.1).

8.2 Analytic Coordinates of Ultra-Extreme RNdS

Since the case of the U-E horizon is not sufficiently explored in the literature we suggest a new coordinate system of compact coordinates (U, V) continuous across the triple horizon. We will adapt a method used for Schwarzschild in [4].

Firstly, we need to give the relation for the tortoise coordinate $r_u^\star$. Due to the definition $\frac{\mathrm{d}r_u^\star}{\mathrm{d}r} = \frac{1}{f(r)}$ one gets the following partial fraction decomposition

$$\frac{\mathrm{d}r_u^\star}{\mathrm{d}r} = \frac{A_u}{(r-r_t)^3} + \frac{B_u}{(r-r_t)^2} + \frac{C_u}{r-r_t} + \frac{D_u}{r-N}, \tag{8.2}$$

with constants

$$A_u = -\frac{3r_t^3}{2},\quad B_u = -\frac{21r_t^2}{8},\quad C_u = -\frac{27r_t}{32} \text{ and } D_u = +\frac{27r_t}{32}. \tag{8.3}$$

Now we can solve the differential equation (8.2) and find a solution

$$r_u^\star = -\frac{1}{2}\frac{A_u}{(r-r_t)^2} - \frac{B_u}{(r-r_t)} + C_u \ln\left|\frac{r}{r_t} - 1\right| + D_u \ln\left|\frac{r}{N} - 1\right|. \tag{8.4}$$

Secondly, we will implicitly define new variables $\left(\hat{U}, \hat{V}\right)$ as

$$\begin{aligned} r_u^\star &= +h(\hat{U}) + h(\hat{V}),\\ t &= -h(\hat{U}) + h(\hat{V}), \end{aligned} \tag{8.5}$$

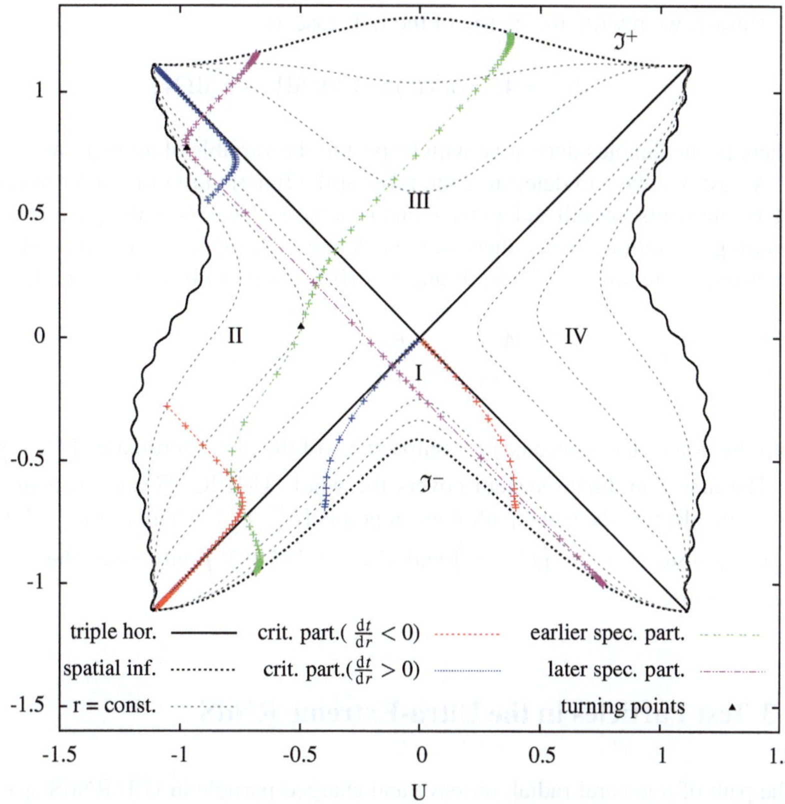

Fig. 8.1 We plotted the paths of four critical and two special particles in the global Penrose diagram of the U-E RNdS with crosses denoting equidistant values of proper time. We distinguish between critical particles with $\frac{dt}{dr} < 0$ and $\frac{dt}{dr} > 0$. As we can see, the points become denser as the critical particles approach the triple horizon, which demonstrates the inability of critical particles to reach the U-E horizon at a finite value of their proper time. On the other hand, special particles cross the horizon with a finite value of their proper time. All special particles have $X_{sp} > 0$ along their paths. In region I collision energy of special particle and critical particle with $\frac{dt}{dr} < 0$ diverges more strongly than in the case $\frac{dt}{dr} > 0$. The stronger divergence does not appear in region II since all future-oriented particles have $X_i > 0$ there and the PS effect is causally prohibited. Two special particles in the diagram differ only in their time of release and the differences in their paths originate from the nontrivial new coordinate path (U, V). Critical particles would reach infinities and the singularity, but we stopped their trajectories at suitable values of the radial coordinate r

where $h(x)$ is a suitable function, which will be discussed later. The new coordinates are directly connected to the classical Eddington-Finkelstein null coordinates $u = t - r_u^\star = -2h(\hat{U})$ and $v = t + r_u^\star = 2h(\hat{V})$.

Thirdly, we rewrite the metric of the U-E case as

$$\mathrm{d}s^2 = 4f(r)\, h'(\hat{U})h'(\hat{V})\mathrm{d}U\mathrm{d}V + r^2\mathrm{d}\Omega_2^2, \tag{8.6}$$

where prime denotes derivative with respect to the variable of a function.

At last, we have to determine the form of the function $h(x)$ in order to keep the metric elements smooth and nonzero on the horizon. There is no unique form of $h(x)$ ensuring our requirements, therefore, we define $h(x)$, which is injective on each of the distinct intervals $x \in \left(-\frac{\pi}{2}, 0\right)$ and $x \in \left(0, \frac{\pi}{2}\right)$ separately, to be of the form

$$h(x) = -\frac{1}{2}\frac{A_u}{\tan^2(x)} - \frac{B_u}{\tan(x)} + C_u \ln|\tan(x)| - \tan^2(x). \tag{8.7}$$

Introduction of tangents ensures compactness of the new coordinates $\left(\hat{U}, \hat{V}\right)$.

The new coordinate system covers the whole U-E RNdS and even its global extension despite the usual pathological points $\left(i\frac{\pi}{2}, j\frac{\pi}{2}\right)$, where $i, j \in \{-1, 0, 1\}$.

Definitions $U = \frac{\sqrt{2}}{2}\left(\hat{U} - \hat{V}\right)$ and $V = \frac{\sqrt{2}}{2}\left(\hat{U} + \hat{V}\right)$ only rotate the final Penrose diagram.

8.3 Test Particles in the Ultra-Extreme RNdS

The path of a general radial, massive and charged particle in U-E RNdS spacetime is given by the differential equation

$$\frac{\mathrm{d}t}{\mathrm{d}r} = \pm\left(E - \frac{Q\varepsilon}{r}\right) \times \left(f(r)\sqrt{\left(E - \frac{Q\varepsilon}{r}\right)^2 - f(r)}\right)^{-1}, \tag{8.8}$$

where the sign sets the direction of motion, E is the constant of motion connected to the time-like Killing vector and ε is the specific charge of the particle [5].

The equation simplifies enough to be solvable for two special particles listed in the following subsections.

The solutions involve integrals of the form $\mathrm{Int}_n(r, p) = \int \left((r - p)^n \sqrt{r^2 + \mathfrak{B}r + \mathfrak{A}}\right)^{-1} \mathrm{d}r$, the solutions of which are given as

$$\mathrm{Int}_1(r,p) = -\frac{1}{\sqrt{a}}\ln\left|\frac{2a+b(r-p)+2\sqrt{aR}}{r-p}\right| \quad \text{for } a>0, \tag{8.9}$$

$$\mathrm{Int}_2(r,p) = \frac{\sqrt{R}}{a(r-p)} - \frac{b}{2a}\mathrm{Int}_1(r,p),$$

$$\mathrm{Int}_3(r,p) = \left(\frac{-1}{2a(r-p)^2} + \frac{3b^2}{4a^2(r-p)}\right)\sqrt{R} + \left(\frac{3b^2}{4a^2} - \frac{c}{2a}\right)\mathrm{Int}_1(r,p),$$

with $R = r^2 + \mathfrak{B}r + \mathfrak{A}$, $b = 2p + \mathfrak{B}$ and $a = p^2 + \mathfrak{B}p + \mathfrak{A}$ according to [3].

8.3.1 Special Particle

The special particle is defined by its specific values $E = \frac{M}{Q\varepsilon}$ and $\varepsilon = \pm 1$ where sigh is chosen to satisfy condition $Q\varepsilon > 0$. The trajectory

$$\pm t_u^{sp}(r) = A_u^{sp}\mathrm{Int}_3(r,r_t) + B_u^{sp}\mathrm{Int}_2(r,r_t) + C_u^{sp}\mathrm{Int}_1(r,r_t) + D_u^{sp}\mathrm{Int}_1(r,N) + E_u^{sp}, \tag{8.10}$$

solves (8.8), where E_u^{sp} is one arbitrary constant,

$$A_u^{sp} = -\frac{\sqrt{3}r_t^4}{2} \quad B_u^{sp} = -\frac{19\sqrt{3}r_t^3}{8}, \quad C_u^{sp} = -\frac{45\sqrt{3}r_t^2}{32}, \text{ and } D_u^{sp} = +\frac{45\sqrt{3}r_t^2}{32}. \tag{8.11}$$

We also substituted $\mathfrak{A} = -\frac{2r_t^2}{3}$, $\mathfrak{B} = 0$ into the integrals (8.9).

8.3.2 Critical Particle

The critical particle is characterized by a fine-tuned value of its specific charge $\varepsilon = \frac{Er_t}{Q}$, for which (8.8) can be solved as

$$\pm t_u^{cr}(r) = A_u^{cr}\mathrm{Int}_3(r,r_t) + B_u^{cr}\mathrm{Int}_2(r,r_t) + C_u^{cr}\mathrm{Int}_1(r,r_t) + D_u^{cr}\mathrm{Int}_1(r,N) + E_u^{cr}, \tag{8.12}$$

with one arbitrary constant E_u^{cr}, constants $A_u^{cr} = \frac{E}{\sqrt{6r_t}}A_u$, $B_u^{cr} = \frac{E}{\sqrt{6r_t}}B_u$, etc. Expressions $\mathfrak{A} = 3r_t^2(2E^2-1)$, $\mathfrak{B} = 2r_t$ must be substituted into (8.9).

8.4 Collision with Unbound Collision Energy in C.M.S

A radial charged particle has a 4-velocity $u_i^\mu = \left(\frac{X_i}{f(r)}, \mp Z_i, 0, 0\right)$, where $X_i = \left(E_i - \frac{Q\varepsilon_i}{r}\right)$ and $Z_i = \sqrt{X_i^2 - f(r)}$. Let us examine the collision energy of two particles in their center of mass system. Let particle $i = 1$ be critical and particle $i = 2$ non-critical (e.g., special).

The total energy in the center of mass frame of the two colliding particles $E^2_{C.M.} = -\left(m_1 u_{1\mu} + m_2 u_{2\mu}\right)\left(m_1 u_1^\mu + m_2 u_2^\mu\right)$. If we assume both particles to have the same rest mass $m_1 = m_2 = m$, we can write the last formula as $\frac{E^2_{C.M.}}{2m^2} = 1 + \frac{X_1 X_2 - Z_1 Z_2}{f(r)}$.

In the near-horizon limit (i.e., $r \sim r_t$) we can expand these terms as

$$Z_1 \approx \left|1 - \frac{r_t}{r}\right| \left[|E_1| + \frac{\Lambda r^2}{6|E_1|}\left(1 - \frac{N}{r}\right)\left(1 - \frac{r_t}{r}\right)\right], \quad Z_2 \approx |X_{2(H)}| - \frac{f(r)}{2|X_{2(H)}|}, \tag{8.13}$$

with $X_{2(H)} = X_2(r = r_t) \neq 0$. Introducing a new near-horizon coordinate $\varepsilon_t := \frac{r - r_t}{r_t}$ and substituting the above expressions into $\frac{E^2_{C.M.}}{2m^2}$, we are led to the expression

$$\frac{E^2_{C.M.}}{2m^2} \approx 3\frac{\left|E_1 X_{2(H)}\right|(1 + \varepsilon_t)\delta_{s_1, -s_2}}{\operatorname{sign}(\varepsilon_t)\varepsilon_t^2} + \frac{1}{2}\left|\frac{X_{2(H)}}{E_1 \varepsilon_t}\right| + O(\varepsilon_t^0), \tag{8.14}$$

where $s_1 := \operatorname{sign} X_1$, $s_2 := \operatorname{sign} X_{2(H)}$ and $\delta_{i,j}$ is the Kronecker delta. There is a problem for $X_1 X_2 < 0$ on the static side as the location of the collision approaches the horizon (i.e., $\varepsilon_t \to 0^-$), since then $E^2_{C.M.} \to -\infty$. Fortunately, the so-called PS process with $X_1 X_2 < 0$ is prohibited here by a causality violation. Particle with $X_i > 0$ would move towards the future while particle with $X_j < 0$ would be past-oriented. On the nonstatic side, t is not a time-like coordinate and motions with $X_1 X_2 < 0$ can occur.

8.5 Conclusions

We have examined paths of radial charged particles and discovered that special particles are repulsed by the black hole charge and do not fall into the singularity. They can even reach three regions of globally extended U-E RNdS with a finite value of their proper time. Particle collisions involving the critical particle in the proximity of the U-E horizon exhibit resemblance to the collisions near an extreme horizon and result in infinite collision energies in C.M. The U-E horizon possesses properties of the inner and cosmological non-extreme horizons, since the order of collision energy divergence is not the same on both sides of the triple horizon.

Acknowledgments This work was supported by grant GAUK 398911 of the Charles University Grant Agency.

References

1. M. Bañados, J. Silk, S.M. West, Phys. Rev. Lett. **103** (2009)
2. D.R. Brill, S.A. Hayward, Class. Quantum Grav. **11** (1994)
3. I.S. Gradshteyn, I.M. Ryzhik, *Table of Integrals, Series, and Products*, 7th edn. (Elsevier Inc., Burlington, 2007)
4. J. Haláček, T. Ledvinka, in *WDS'12 Proceedings of Contributed Papers*, eds. by J. Šafránková, J. Pavlů. I.I.I. Part (Prague, Matfyzpress, 2012), pp. 105–110
5. M. Olivares, J. Saavedra, C. Leiva, J.R. Villanueva, Mod. Phys. Lett. A **26** (2011)
6. O.B. Zaslavskii, Phys. Rev. D **85** (2012)
7. O.B. Zaslavskii, JETP Lett. **92** (2010)
8. C. Zhong, S. Gao, JETP Lett. **94** (2011)

Chapter 9
Sourcing a Varying-Mass Black Hole in a Cosmological Background

Michele Fontanini and Daniel C. Guariento

Abstract Systems in which the local gravitational attraction is coupled to the expansion of the Universe have been studied since the early stages of General Relativity. The McVittie metric is an example of such systems, being an exact solution of the Einstein equations representing a black hole in a cosmological background. Here, by using imperfect fluids, we construct a generalization of the McVittie solution in which the mass function of the black hole increases with time, effectively describing an accreting compact object in an expanding Universe. A novel mechanism involving temperature gradients is the key ingredient that leads to this result while still avoiding phantom fields and fine-tuning.

9.1 A Brief Introduction to the McVittie Solution

In 1933 McVittie published [1] a solution to Einstein equations generalizing the Schwarzschild-de Sitter metric by allowing for a Friedmann-Lemaître-Robertson-Walker (FLRW in the following) cosmology. His metric is the unique spherically symmetric solution sourced by a perfect fluid with the following defining characteristics [2]:

1. shear-free,
2. asymptotically FLRW,
3. containing a black hole singularity at the center.

M. Fontanini (✉) · D.C. Guariento
Instituto de Física, Universidade de São Paulo, Caixa Postal 66.318,
São Paulo, SP 05315-970, Brazil
e-mail: fmichele@if.usp.br

D.C. Guariento
Perimeter Institute for Theoretical Physics, 31 Caroline St. N.,
Waterloo, ON N2L 2Y5, Canada
e-mail: carrasco@fma.if.usp.br

P. Nicolini et al. (eds.), *1st Karl Schwarzschild Meeting on Gravitational Physics*,
Springer Proceedings in Physics 170, DOI 10.1007/978-3-319-20046-0_9

In isotropic coordinates the McVittie metric takes the form

$$\mathrm{d}s^2 = -\frac{\left(1-\frac{m}{2ar}\right)^2}{\left(1+\frac{m}{2ar}\right)^2}\mathrm{d}t^2 + a^2\left(1+\frac{m}{2ar}\right)^4\left(\mathrm{d}r^2 + r^2\mathrm{d}\Omega^2\right), \tag{9.1}$$

where the m parameter represents the contribution to the Misner–Sharp mass coming from the central inhomogeneity [3], and $a \equiv a(t)$ corresponds to the scale factor. One can easily see that, taking the limit of small mass or large distances ($m \to 0$ or $r \to \infty$ respectively), the metric (9.1) above immediately takes the form of the FLRW line element. Similarly, taking $a(t) \to 1$ the Schwarzschild metric in isotropic coordinates is obtained. Despite its look, it is not obvious what (9.1) can represent, and in fact, since its first appearance an 80-year-long debate started with the aim of establishing whether or not this metric could really describe a black hole, rather than a naked singularity or yet something else. Many contributions to the subject have been made with contrasting points of view and results (see for instance [4, 5] and references therein), until the issue was resolved by two groups; in [6] first, and with a more detailed numerical analysis in [7] later.

The main interest behind looking at the McVittie solution (and others of its class [8]), is of course related to the fact that it represents a black hole in a quite generic cosmological background, with no approximations. In other words, the solution represents a good benchmark to study a cross-scale problem involving collapsed objects and global behaviors such as the cosmological expansion. Of course the McVittie solution is still very symmetrical and somehow simple, which is the reason why various generalizations of it were studied during the years. In particular, the most important successful attempt has been adding a time dependence for the mass parameter [5], thus formally obtaining an accreting or depleting black hole. The price to pay for such result though, has been the introduction of a phantom fluid and the need for a detailed balance between the latter and the regular fluid component.

In the following we want to show that it is possible to build a set of generalized McVittie solutions sourced by a realistic imperfect fluid, a key and novel element in the study of the problem. By introducing heat fluxes in fact, we can transfer energy into the central object and have its mass increase without disturbing the uniform density requirement of the metric.

9.2 Properties and Causal Structure of McVittie

In this section we introduce a more convenient set of coordinates to study the problem, and describe the causal structure of the McVittie solution in the cases of interest, namely when it can be shown to describe a black hole (or as we will see in short, a black-hole/white-hole couple).

9.2.1 Apparent Horizons in Canonical Coordinates

As already mentioned, the McVittie solution (9.1) reduces to familiar forms when the limits $a(t) \to 1$ and $m/r \to 0$ are taken, becoming the Schwarzschild (in isotropic coordinates), and FLRW metrics respectively. Despite the fact that these limits make perfect sense, the structure of the spacetime requires a deeper analysis and cannot be inferred simply by these analogies.

It is convenient to change coordinates since the isotropic ones for $a(t) = 1$ have the unfortunate property of covering the exterior of the black hole twice for $m/2 < r < \infty$ and $0 < r < m/2$. Canonical coordinates are then defined by the transformation (and its inverse)

$$r \to \hat{r} = \left(1 + \frac{m}{2a(t)r}\right)^2 a(t)r\,, \quad a(t)r = \frac{m}{2}\left[\frac{\hat{r}}{m} - 1 \pm \sqrt{\left(\frac{\hat{r}}{m} - 1\right)^2 - 1}\right]^{-1}, \tag{9.2}$$

which, being quadratic, defines two branches; the "–" branch is the physical one, since in it $\hat{r} \to \infty$ corresponds to $a(t)r \to \infty$, while $\hat{r} \to 2m$ corresponds to $a(t)r \to m/2$. The other branch instead, leading to a range $0 < a(t)r < m/2$, describes a patch terminating on spacelike curvature singularities on both past and future and can therefore be ignored.

For simplicity we drop the hat notation and keep $\hat{r} \to r$. In this new set of coordinates, the metric is no longer diagonal and takes the form

$$\mathrm{d}s^2 = -R^2\mathrm{d}t^2 + \left(\frac{\mathrm{d}r}{R} - Hr\mathrm{d}t\right) + r^2\mathrm{d}\Omega^2\,, \tag{9.3}$$

where for convenience we introduced the functions $R(r) \equiv \sqrt{1 - 2m/r}$, and $H(t) \equiv \dot{a}(t)/a(t)$.

In these coordinates it is easy to see what the surface $r = 2m$ (previously $m = 2a(t)r_{\mathrm{iso}}$ in isotropic coordinates) represents [6]; it is a spacelike 3-sphere of radius $2m$ ($\mathrm{d}r = 0$ and $g_{tt} = 4m^2H(t)^2 > 0$) that lies in the causal past of each point of the spacetime. Moreover, using the dot and prime notations for t and r derivatives, the scalar curvature on it $R_{\mathrm{Ricci}} = 12H + 6\dot{H}/R$ becomes singular; we then call $r = 2m$ the "McVittie big bang" (alternatively, for a contracting solution with $H < 0$ it would represent a big crunch).

Studying the apparent horizons of McVittie leads to the other salient features of the causal structure of the spacetime. Since they are defined as the surfaces where at least one congruence of null geodesics changes its focusing properties, they are therefore the locus of vanishing geodesic expansion, which in a spherical symmetric setup corresponds to the set of extrema of the area of a light front [6]. They can be found by searching for the zero slope points along radial light geodesics defined by the equation

$$\dot{r} = R(r)\left[rH(t) \pm R(r)\right]\,, \tag{9.4}$$

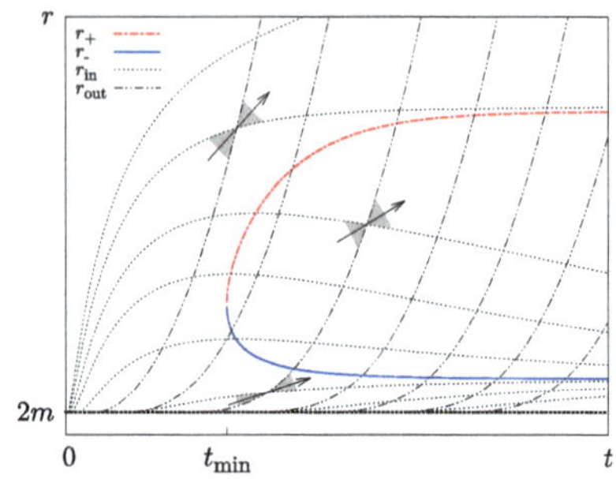

Fig. 9.1 Null geodesics (in- and outgoing), apparent horizons for the McVittie spacetime. Some light cones inside and outside the normal region enclosed by the two horizons are also shown

where the "−" sign corresponds to ingoing and the "+" to outgoing geodesics. For $H > 0$, outgoing geodesics do not define apparent horizons, and one is left to deal with a cubic equation (after rationalization) for the ingoing geodesics. Of the three possible solutions, one is always negative and can be discarded, while the other two become real after some time t_0 (branching time), and can be dubbed as $r_+(t)$ and $r_-(t)$, the "outer" and "inner" one respectively. Hence, two horizons branch off at the bifurcation 2-sphere $(t_0, r_\pm(t_0))$, and split the spacetime in two regions: a normal inner one defined at any time t as the radial values in between the horizons $r_-(t) < r(t) < r_+(t)$, and an outer antitrapping region for values of the radial coordinate outside these limits. It is worth noting that while the time coordinate t is everywhere timelike, the radial coordinate r remains spacelike only in the normal region as can be seen in Fig. 9.1.

Continuing with the peculiar features of the McVittie spacetime, one finds that (see for instance [6, 7, 9]) the outer horizon $r_+(t)$ asymptotes to a null surface defined by $\lim_{t\to\infty}(r_+(t), t)$ which for $H_0 \equiv \lim_{t\to\infty} H(t) > 0$ is a cosmological de Sitter-like horizon, while for $H_0 = 0$ becomes a null FLRW infinity at $r_+ = \infty$ and $t = \infty$. The inner horizon instead remains always finite, asymptotes to a null surface $(r_-(\infty), t = \infty)$ that for $H_0 > 0$ is either a regular black hole horizon [6], or a black-hole/white-hole bifurcation sphere [7, 10], located at a finite proper distance from all the points in the interior. When $H_0 = 0$ r_- becomes a null singularity.

With this information drawing a Penrose diagram of the spacetime reduces to integrating geodesics and is mostly a numerical effort. An example of McVittie spacetime is shown in Fig. 9.1 where some light cones are shown, and in Fig. 9.2, where the causal structure with the features discussed above is shown (in the case $H_0 > 0$, with and without the presence of a white hole).

9.2.2 The White Hole and $H(t)$

Besides the general non-convergent view about whether or not the metric (9.3) was describing a black hole, the authors of [7] also advocated the idea that the inner apparent horizon r_- would not asymptote to a black hole horizon, rather it would

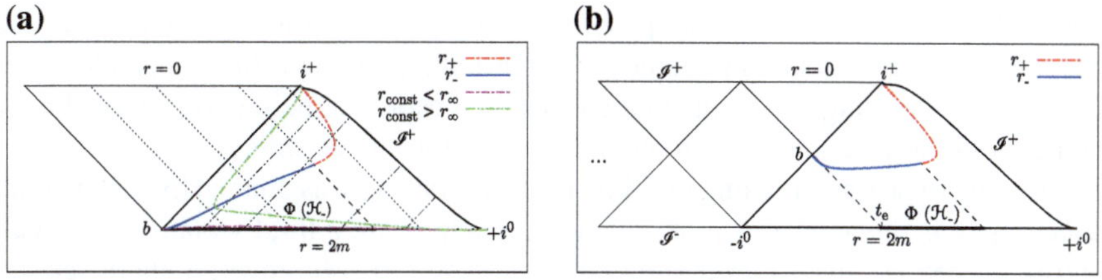

Fig. 9.2 **a** Causal structure for the black hole case. Constant radius curves are also shown. **b** Appearance of the white hole in the causal structure

separate the boundary of the spacetime into two sections, a black hole horizon in the future and a white hole one in the past. Their analysis was supported by extremely accurate numerics, and yet was in obvious disagreement with all previous works on the topic, presenting a difficult puzzle.

We recall that in order to establish the characteristics of the inner horizon in the limit $t \to \infty$ the properties of H are crucial, and that in particular for the existence of an event horizon it is crucial that $H_0 > 0$. What the authors in [6, 7] did not realize is that it is not just the limit of the expansion function that decides the asymptotes, but also how quickly $H(t)$ tends to H_0. In [9] in fact, it was shown that the asymptotic behavior of $r_-(t)$ can both produce a black hole horizon and a black-hole/white-hole separating surface. More in specific, the difference $\Delta H(t) \equiv H(t) - H_0$, appearing in an integral over all times that defines the causal past of the inner apparent horizon, determines the convergence or divergence of the interval of points on the McVittie big bang surface that following light geodesics reach r_-, and thus selects whether or not all ingoing geodesics enter the normal patch of the spacetime. In the case in which part of the geodesics starting from the big bang can stay outside the normal region, a white hole horizon appears to accommodate them. The illustration in Fig. 9.2 shows in an intuitive and graphical way how the white hole can appear.

It is interesting to note that even realistic and somehow simple models for the expansion can lead to both behaviors. For instance, in the ΛCDM model with no radiation the product mH_0, via the method of [9], completely determines the structure of the spacetime, except for one value of mH_0 for which numerics are needed. In [9] it is argued though that the method presented fails to give a definitive answer only for a physically irrelevant class of expansion functions.

9.3 Generalizing McVittie

The McVittie solution can be formally generalized to accommodate a time-varying mass function $m \to m(t)$ that corrects the contribution to the expansion function

$$\mathrm{d}s^2 = -R^2\mathrm{d}t^2 + \left[\frac{\mathrm{d}r}{R} - \left(H - M + \frac{M}{R}\right) r\mathrm{d}t\right]^2 + r^2\mathrm{d}\Omega^2\,, \qquad (9.5)$$

where $M(t) \equiv \dot{m}(t)/m(t)$. The problem then becomes finding an appropriate stress-energy tensor that can source it. In particular, adding a time dependence in m introduces off-diagonal terms in the Einstein tensor that, being proportional to $\dot{m}$ cannot be set to zero, and need to be matched on the source side.

One way to proceed is using multiple perfect fluids, as was done in [5, 11], but the price to pay is the appearance of a phantom component, a fluid violating the weak energy condition, given that it shows $w \equiv p/\rho < -1$. Moreover, a characteristic of the McVittie solution, also shared by generalized McVittie with time-dependent mass, is the so called "spatial Ricci isotropy", namely the equality of the radial and angular part of the Einstein tensor with mixed indices: $G^r_r = G^\theta_\theta$. Due to Ricci isotropy, the phantom and other fluids have to balance out in a fine-tuned fashion, leading to possible stability issues.

The alternative is to abandon the perfect fluid description and make use of the extra freedom encoded in an imperfect fluid, specifically the presence of heat fluxes.

9.3.1 Imperfect Fluid Source

An imperfect fluid is generically described by a stress-energy tensor that to first order in gradients takes the form

$$T^{\mu\nu} = (\rho + p)\, u^\mu u^\nu + p g_{\mu\nu} - \zeta h^{\mu\nu} u^\gamma_{;\gamma} - \chi \left(h^{\mu\gamma} u^\nu + h^{\nu\gamma} u^\mu \right) q_\gamma \,, \tag{9.6}$$

with

$$h_{\mu\nu} = g_{\mu\nu} + u_\mu u_\nu \,, \qquad q_\mu = \partial_\mu T + T u_{\mu;\gamma} u^\gamma \,, \tag{9.7}$$

and where u^μ is the four-velocity, $T(x^\mu)$ the temperature function, χ the heat conductivity, and ζ the bulk viscosity.

The starting point is considering the off-diagonal component of Einstein equations, now nontrivial due to a non-zero $\dot{m}$. This can be used to find the radial dependence of the temperature function

$$T(t,r) = \frac{T_\infty}{R} + \frac{1}{4\pi\chi} \frac{M}{R} \ln(R) \,, \tag{9.8}$$

where $T_\infty = T_\infty(t)$ is an arbitrary integration function at spatial infinity. In the limit of no time dependence of the mass we have $T \to T_\infty/\sqrt{-g_{tt}}$, namely the fluid is in thermal equilibrium (as expected). Using the remaining independent Einstein equations, it is possible to write an explicit form for the other functions entering the general stress-energy tensor (9.6) in terms of the functions appearing in the metric. The energy density reads

$$\rho(t,r) = \frac{3}{8\pi} \left[H - \frac{M}{R}(R-1) \right]^2 \,, \tag{9.9}$$

while the pressure can be written as

$$p(t,r) = \sqrt{6\pi\rho}\left(2\zeta - \frac{mM}{6\pi r R^3}\right) - \rho - \frac{mM^2}{4\pi r R^4} - \frac{1}{4\pi R}\left(\dot{H} + \dot{M}\frac{1-R}{R}\right) . \quad (9.10)$$

9.3.2 Apparent Horizons

Similarly to what happens with the causal structure of the original McVittie spacetime, the generalized one also contains a spacelike surface in the past of all timelike curves, the big bang singularity, that now lays at a non-fixed radial value (in canonical coordinates). As for the apparent horizons, now defined by

$$R\,(rH \pm R) + rM\,(1 - R) = 0\,, \quad (9.11)$$

it can be shown that for reasonable choices of the functions H and M (as discussed below), they still exist, branch off after an initial time t_0, and separate a normal region from an antitrapping one.

An important question has to be asked at this point: is the generalized McVittie metric still suitable to describe a black hole (now with variable mass) in an FLRW Universe? We do not have a direct answer to this very general question, but we can say for sure that there exist vast families of generalized solutions described by specific choices of H and M that do describe a black hole (and possibly a black/white-hole pair) like in the case of the original McVittie metric.

In fact, any choice of H that is allowed by the original metric, and any choice of M that vanishes fast enough with t, provide the same description for the asymptotics of the apparent horizons. The analysis of [6, 7] then applies unchanged.

9.3.3 An Explicit Model

We present now a toy model in which we generate families of viable solutions by patching together generalized McVittie (time dependent) and regular McVittie (constant) mass functions. A possible choice for the functions H and M can be

$$H(t) = \frac{2}{3t} + H_0\,, \quad m(t) = \begin{cases} 1, \quad t \le t_i\,; \quad \text{and} \qquad 2, \quad t \ge t_f\,; \\ \frac{1}{2}\left[3 + \sin\left(\omega t + \phi\right)\right] \quad t_i < t < t_f\,, \end{cases} \quad (9.12)$$

where, with H_0 positive, this approximates the cosmology of pressureless dust with the addition of a cosmological constant. As for the mass function, we build it variating only in a window of times, for some choice of t_i and t_f, with ω and ϕ chosen to smoothly connect to constant values outside.

When $\dot{m}$ ceases to be zero, ρ and T acquire gradients toward the singularity where they themselves go to infinity. The presence of this density gradient in the dynamical case avoids the rather artificial setup of the original McVittie, whose requirement of a homogeneous ρ supported by pressure gradients was physically difficult to justify. Conversely, the pressure, besides showing discontinuities which are a feature of the oversimplification introduced in this special case (the model above is smooth only up to first derivative of the mass, while the pressure contains second derivatives), behaves much like in the static-mass case, going to infinity at the singularity.

9.4 Remarks

We have seen how generalized McVittie can be built satisfying the requirements to ensure that an event horizon appears in the spacetime, while still allowing the mass function of the central object to vary. To achieve this, we require the physically reasonable assumption that the black hole stops accreting after some time.

It is interesting to notice the effect on the apparent horizons due to accretion: as can be seen in Fig. 9.3, they tend to get closer together. This feature requires more study, and it of course implies the presence of limits of validity for mass increasing models. In fact, taking an extreme point of view, one can increase the effect of M coupled to H and reach a critical point at which r_- and r_+ touch a second time after the initial branching t_0. Figure 9.4 shows this possibility featuring two disconnected normal regions separated by a region which is everywhere antitrapping.

We have seen then that McVittie can be generalized using a realistic fluid as source for the exact solution; that, in the class of models that have a mass function $m \to m_0$ constant at large times, the black-hole (or black/white-hole) interpretation of the

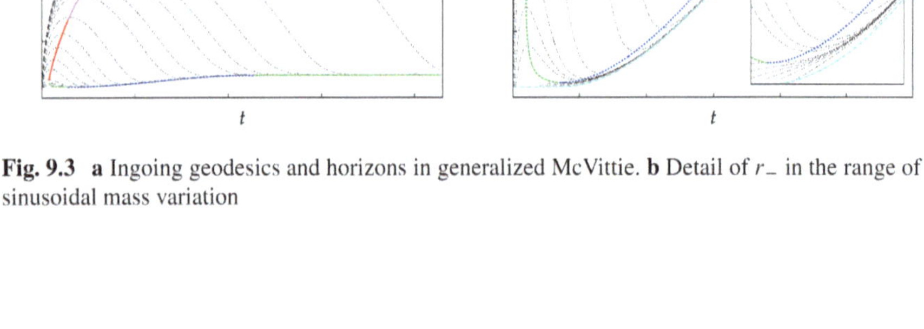

Fig. 9.3 **a** Ingoing geodesics and horizons in generalized McVittie. **b** Detail of r_- in the range of sinusoidal mass variation

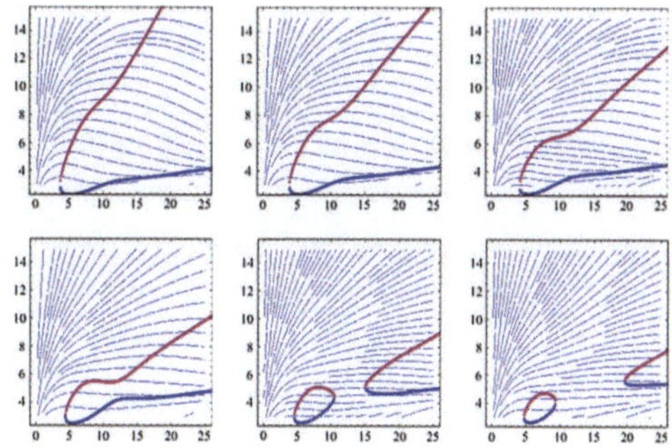

Fig. 9.4 Effect of a radical choice for the mass function with increasing cosmological constant. Once the apparent horizons merge two disconnected normal regions appear

metric still applies; and that some bounds have to be posed to the mass function even at finite times to retain a physical interpretation of the inner region defined by the apparent horizons.

Acknowledgments This work is supported by FAPESP.

References

1. G.C. McVittie, Mon. Not. Roy. Astr. Soc. **93**, 325–339 (1933)
2. A.K. Raychaudhuri, *Theoretical Cosmology* (Clarendon Press, Oxford, 1979)
3. M. Carrera, D. Giulini, Rev. Mod. Phys. **82**, 169–208 (2010)
4. B. Nolan, Class. Quant. Grav. **16**, 1227–1254 (1999)
5. V. Faraoni, A. Jacques, Phys. Rev. D **76**, 063510 (2007)
6. N. Kaloper, M. Kleban, D. Martin, Phys. Rev. D **81**, 104044 (2010)
7. K. Lake, M. Abdelqader, Phys. Rev. D **84**, 044045 (2011)
8. P. Kustaanheimo, B. Qvist, CPM **13**, 1–11 (1948) (reprinted in Gem. Rel. Gravit. **30**, 663–673 (1998))
9. A.M. da Silva, M. Fontanini, D.C. Guariento, Phys Rev. D **87**, 064030 (2013)
10. D.C. Guariento, M. Fontanini, A.M. da Silva, E. Abdalla, Phys. Rev. D **86**, 124020 (2012)
11. V. Faraoni, A.F.Z. Moreno, R. Nandra. Phys. Rev. D **85**, 083526 (2012)

Chapter 10
Tidally Distorted Black Holes

Norman Gürlebeck

Abstract According to the no-hair theorem, static black holes are described by a Schwarzschild spacetime provided there are no other sources of the gravitational field. This requirement, however, is in astrophysical realistic scenarios often violated, e.g., if the black hole is part of a binary system or if they are surrounded by an accretion disk. In these cases, the black hole is distorted due to tidal forces. We show that the subsequent formulation of the no-hair theorem holds nonetheless: The contribution of the distorted black hole to the multipole moments that describe the gravitational field close to infinity is that of a Schwarzschild black hole. This implies that there is no multipole moment induced in the black hole and that its second Love numbers, which measure the distortion, vanish as was already shown in approximations to general relativity. But here we proof this property of black holes in full general relativity.

10.1 Introduction

The no-hair theorem states that any isolated static black hole is necessarily a Schwarzschild black hole and that there is only one free parameter describing the spacetime—the mass M.[1] The metric of a Schwarzschild black hole is given in canonical Weyl coordinates $(\rho, \zeta, \varphi, t)$ by

$$\begin{aligned} ds^2 &= e^{2k_S-2U_S}\left(d\rho^2 + d\zeta^2\right) + e^{-2U_S}\rho^2 d\varphi^2 - e^{2U_S}dt^2, \\ U_S &= \frac{1}{2}\log\left[\frac{r_+ + r_- - 2M}{r_+ + r_- + 2M}\right], \quad k_S = \frac{1}{2}\log\left[\frac{(r_+ + r_-)^2 - 4M^2}{4r_+ r_-}\right], \\ r_\pm^2 &= \rho^2 + (\zeta \pm M)^2. \end{aligned} \tag{10.1}$$

[1] This means that the black hole has actually *one* hair.

N. Gürlebeck (✉)
Zentrum für angewandte Raumfahrttechnologie und Mikrogravitation,
Universität Bremen, Am Fallturm, 28359 Bremen, EU, Germany
e-mail: norman.guerlebeck@zarm.uni-bremen.de

P. Nicolini et al. (eds.), *1st Karl Schwarzschild Meeting on Gravitational Physics*,
Springer Proceedings in Physics 170, DOI 10.1007/978-3-319-20046-0_10

The horizon of this spacetime is located at the symmetry axis ($\rho = 0, \zeta \in [-M, M]$). In fact, in canonical Weyl coordinates the horizon can always be located at $\rho = 0$, see [4].

For the no-hair theorem to hold, the black hole has to be isolated, i.e., the exterior of the black hole has to be an asymptotically flat vacuum. However, this is not valid in many astrophysical situations, like for black holes in binary systems or if the black hole is surrounded by an accretion disk. In the exterior field of such additional sources, the black hole is distorted, i.e., the inner geometry of the horizon changes. This is measured by the Love numbers of the first kind or the multipole moments of isolated horizons, see [1, 7]. Nonetheless, we show in this paper that this does not imply that the black holes "grow more hair". More precisely: Although the total multipole moments of the spacetime measured at infinity change, this is solely due to the external sources and not to a different contribution of the black holes themselves. In fact, distorted black holes have only a mass monopole, cf. [14].

It is of general interest, particularly for inspirals treated in an adiabatic regime, to quantify distortions of black holes and neutron stars. The imprints of the distortions in the gravitational wave forms emitted by inspiraling binaries give information on the equation of state [10, 16] of neutron stars. Similarly, they can be used to experimentally reveal if a constituent of a binary system is a black hole. Quantifying the distortion is done with the help of the Love numbers of first and second kind, h_r and k_r, cf. [3, 8, 18]. Roughly speaking, the h_r measure the changes in the shape of the horizon and the k_r measure the change in the asymptotic multipole moments caused by the distortion due to an external source, see [3, 5, 7, 8, 17] for their use in general relativity. They were also applied to establish equation-of-state-independent relations between certain physical parameters describing neutron stars, see [6, 19, 21, 22] but also [9, 15]. The here considered black hole case is solved analytical in full general relativity and, thus, it serves as a test for the various approximation schemes employed for neutron stars.

10.2 Preliminaries

In this section, we will repeat the notions that are necessary in the present paper. We use geometric units, in which $G = c = 1$, where c is the velocity of light and G Newton's gravitational constant. The metric has the signature $(-1, 1, 1, 1)$. Greek indices run from 0 to 3 and Latin indices run from 1 to 3.

10.2.1 Distorted Black Holes

In case the exterior sources are static and axially symmetric or allow for a quasi-static description, the general metric near the horizon $\mathscr{H}$ of distorted black hole was found by Geroch and Hartle in [12]. In a neighborhood of $\mathscr{H}$, we assume vacuum, which is physical reasonable if the matter should satisfy the energy conditions, cf. [2]. Thus, there exists a surface $\mathcal{S}_{\mathcal{H}}$, which encloses $\mathcal{H}$ and no other sources. If $\mathcal{S}_{\mathcal{H}}$ is sufficiently close to $\mathcal{H}$, the metric in between reads

$$ds^2 = \mathrm{e}^{2k-2U_S-2U_D}\left(d\rho^2 + d\zeta^2\right) + \rho^2 \mathrm{e}^{-2U_S-2U_D} d\varphi^2 - \mathrm{e}^{2U_S+2U_D} dt^2. \quad (10.2)$$

The function U_D is determined by a Laplace equation

$$\left(\frac{\partial^2}{\partial\rho^2} + \frac{1}{\rho}\frac{\partial}{\partial\rho} + \frac{\partial^2}{\partial\zeta^2}\right) U_D = 0. \quad (10.3)$$

The horizon of this spacetime lies still on the ζ-axis. The canonical Weyl coordinates allow a shift in the $\zeta-$coordinate. We employ this freedom to place the horizon symmetrically with respect to that coordinate, i.e., that the "north/south pole" of the horizon are characterized by $\zeta_{N/S} = \pm\zeta_{\mathcal{H}}$. At these points, U_D has to take the same value to avoid struts, see [12], which we want to exclude for simplicity. The function k is obtained by a line integration once U_D is known:

$$k_{,\zeta} = 2\rho\,(U_D + U_S)_{,\rho}\,(U_D + U_S)_{,\zeta}\,, \quad k_{,\rho} = \rho\left(\left(U_{S,\rho} + U_{D,\rho}\right)^2 - \left(U_{S,\zeta} + U_{D,\zeta}\right)^2\right). \quad (10.4)$$

The metric (10.2) does neither describe directly the asymptotic behavior nor the metric in the interior of the external source. In fact, the exterior sources do not have to be specified for the subsequent conclusions and they could also include other black holes. We only require that the spacetime is asymptotically flat and that all external sources are enclosed by a surface $\mathcal{S}_{\text{ext}}$, which does not contain $\mathcal{H}$ and which does not extend to infinity.

10.2.2 The Source Integrals

To disentangle the contributions of the black hole and the external sources to the asymptotic multipole moments, the source integrals proved to be the essential tool. They were recently derived in [13]. With these it is possible, to define the asymptotics of the spacetime including the Geroch multipole moments by evaluating quasi-local surface or volume integrals. The respective surfaces and volumes need only to envelope or contain all regions with a non-vanishing stress-energy tensor. Here we need only the surface integrals and introduce the required quantities, subsequently.

The source integrals can be derived for arbitrary static and axially symmetric spacetimes. Then, the metric can be written in the Weyl form under standard technical assumptions:

$$ds^2 = \mathrm{e}^{2k-2U}\left(d\rho^2 + d\zeta^2\right) + W^2 \mathrm{e}^{-2U} d\varphi^2 - \mathrm{e}^{2U} dt^2, \quad (10.5)$$

where the functions U, k and W depend on ρ and ζ. Note that the metric functions U and W can be expressed by the timelike Killing vector ξ^α and the spacelike Killing vector η^α:

$$e^{2U} = -\xi_\alpha \xi^\alpha, \quad W^2 = -\eta_\alpha \eta^\alpha \xi_\beta \xi^\beta. \tag{10.6}$$

The Weyl multipole moments $U^{(r)}$ are defined as the expansion of U along the axis of symmetry close to infinity, i.e.,

$$U = \sum_{r=0}^{\infty} \frac{U^{(r)}}{|\zeta|^{r+1}}. \tag{10.7}$$

As was shown in [11], from these the Geroch multipole moments m_r can be determined by non-linear algebraic relations. To calculate m_r, the $U^{(k)}$ need to be known for $0 \leq k \leq r$. Thus, it is sufficient for us to consider here the $U^{(r)}$.

Furthermore, it is beneficial to introduce the functions

$$\begin{aligned} N_-^{(r)}(x,y) &= \sum_{k=0}^{\lfloor \frac{r}{2} \rfloor} \frac{2(-1)^{k+1} r! x^{2k+1} y^{r-2k}}{4^k (k!)^2 (r-2k)!}, \\ N_+^{(r)}(x,y) &= \sum_{k=0}^{\lfloor \frac{r-1}{2} \rfloor} \frac{2(-1)^{k+1} r! x^{2k+2} y^{r-2k-1}}{4^k (k!)^2 (r-2k-1)!(2k+2)}. \end{aligned} \tag{10.8}$$

As was shown in [13], these functions obey the equations

$$N_{+,x}^{(r)} - N_{-,y}^{(r)} = 0, \quad N_{+,y}^{(r)} + N_{-,x}^{(r)} - \frac{N_-^{(r)}}{x} = 0. \tag{10.9}$$

Additionally, let us introduce the 1-form

$$Z_\alpha = \varepsilon_{\alpha\beta\gamma\delta} W^{,\beta} W^{-1} \eta^\gamma \xi^\delta, \tag{10.10}$$

where $\varepsilon_{\alpha\beta\gamma\delta}$ is the volume form of the spacetime. In vacuum, Z_α is exact and it is hypersurface orthogonal in the entire spacetime. Since the surfaces of interest, $\mathscr{S}_\mathscr{H}$ and $\mathscr{S}_{\text{ext}}$, lie in the vacuum region or its boundaries, we can introduce a scalar Z via $Z_{,\alpha} = Z_\alpha$, for technical details and a more general treatment see [13].

With this notation at hand, we can express the Weyl multipole moments by

$$\begin{aligned} U^{(r)} &= \int_{\mathscr{S}_\mathscr{H}} \eta_a^{(r)} \hat{n}^a \mathrm{d}\mathscr{S}_\mathscr{H} + \int_{\mathscr{S}_{\text{ext}}} \eta_a^{(r)} \hat{n}^a \mathrm{d}\mathscr{S}_{\text{ext}}, \\ \eta_a^{(r)} &= \frac{1}{8\pi} \frac{e^U}{W} \left(N_-^{(r)} U_{,a} - N_{+,W}^{(r)} Z_{,a} U + N_{+,Z}^{(r)} W_{,a} U \right), \end{aligned} \tag{10.11}$$

where $\hat{n}^a$ denotes the outward pointing unit normal to the surfaces $\mathscr{S}_{\mathscr{H}}$ and $\mathscr{S}_{\text{ext}}$ and the functions $N_{\pm}^{(r)}$ depend on $(x, y) = (W, Z)$. $\mathrm{d}\mathscr{S}_{\mathscr{H}}$ and $\mathrm{d}\mathscr{S}_{\text{ext}}$ are the proper area elements of $\mathscr{S}_{\mathscr{H}}$ and $\mathscr{S}_{\text{ext}}$, respectively. In vacuum, we can always choose canonical Weyl coordinates such that $W = \rho$ and $Z = \zeta$.

10.3 The Induced Multipole Moments of Distorted Black Holes

With (10.11), we can covariantly identify the contribution of the different sources to the asymptotic Weyl multipole moments. The first term in (10.11), which we denote $U_{\mathscr{H}}^{(r)}$, gives the contribution of the distorted black hole and the second term, $U_{\text{ext}}^{(r)}$, the contribution of the external sources. The induced multipole moment of a distorted black hole is now simply defined as $U_{\text{ind}}^{(r)} = U_{\mathscr{H}}^{(r)} - U_S^{(r)}$, where the $U_S^{(r)}$ denote the Weyl multipole moments of a undistorted Schwarzschild black hole. They coincide with the Newtonian multipole moments of a line mass of uniform density, see [20]. Using a parametrization of $\mathscr{S}_{\mathscr{H}}$ for constant angles ($s \in [s_N, s_S] \mapsto (\rho(s), \zeta(s), \varphi = \text{const.})$, cf. [13], and (10.2), $U_{\mathscr{H}}^{(r)}$ reads

$$U_{\mathscr{H}}^{(r)} = \frac{1}{8\pi} \int_{s_N}^{s_S} \int_0^{2\pi} \left[N_-^{(r)} (U_S + U_D)_{,n} - \left(N_{-,W}^{(r)} Z_{,n} - N_{+,Z}^{(r)} W_{,n} \right) (U_S + U_D) \right] \mathrm{d}\varphi \mathrm{d}s, \tag{10.12}$$

where we denote by $f_{,n}$ the normal derivative $-f_{,\rho} \frac{\mathrm{d}}{\mathrm{d}s}\zeta(s) + f_{,\zeta} \frac{\mathrm{d}}{\mathrm{d}s}\rho(s)$ along the surface $\mathscr{S}_{\mathscr{H}}$. Thus, the induced multipole moment is given by

$$U_{\text{ind}}^{(r)} = \frac{1}{8\pi} \int_{s_N}^{s_S} \int_0^{2\pi} \left[N_-^{(r)} U_{D,n} - N_{+,W}^{(r)} Z_{,n} U_D + N_{+,Z}^{(r)} W_{,n} U_D \right] \mathrm{d}\varphi \mathrm{d}s. \tag{10.13}$$

Applying the divergence theorem, we can rewrite $U_{\text{ind}}^{(r)}$:

$$U_{\text{ind}}^{(r)} = \frac{1}{8\pi} \int_{V_{\mathscr{H}}} \frac{1}{\rho} \left[U_{D,\rho} \left(N_{-,\rho}^{(r)} + N_{+,\zeta}^{(r)} - \frac{N_-^{(r)}}{\rho} \right) + U_{D,\zeta} \left(N_{-,\zeta}^{(r)} - N_{+,\rho}^{(r)} \right) \right] \mathrm{d}V_{\mathscr{H}}, \tag{10.14}$$

which vanishes by virtue of (10.9). $V_{\mathscr{H}}$ is the coordinate volume enclosed by $\mathscr{S}_{\mathscr{H}}$ and $\mathscr{H}$ in canonical Weyl coordinates. Thus, the induced multipole moments vanish and the contribution of the distorted black hole to the asymptotic Weyl multipole moments is the same as for the Schwarzschild black hole. With the results in [11,

13], this can readily be translated to Geroch's multipole moments. This generalizes the no-hair theorem to black holes that are distorted by external matter. This result simplifies the evaluation of the source integrals in [13] considerably, since only the mass of the individual black holes has to be calculated to know all $U^{(r)}_{\mathscr{H}}$.

The vanishing of the induced multipole moments implies that the second Love numbers k_r vanish, too, because they are proportional to $U^{(r)}_{\text{ind}}$. This is in compliance with [3, 5, 8, 17]. But here no approximations and linearizations were used and the result holds in full general relativity. Note that $k_r = 0$ is specific to black holes and does not hold for neutron stars, cf. [3]. Nonetheless, the source integrals of the Weyl multipole moments are still tailored to calculate their k_r, since the contributions from the individual sources to the Weyl multipole moments are separated and a definition of an induced multipole moment becomes possible also in full GR.

Acknowledgments I thank Abhay Ashtekar for pointing out this interesting question. I also thank J. Steinhoff for fruitful discussions. Furthermore, I gratefully acknowledge support from the DFG within the Research Training Group 1620 "Models of Gravity".

References

1. A. Ashtekar, B. Krishnan, Isolated and dynamical horizons and their applications. Living Rev. Relat. **7**(10) (2004). http://www.livingreviews.org/lrr-2004-10
2. J.M. Bardeen, Rapidly rotating stars, disks, and black holes, in *Black Holes (Les Astres Occlus)*, eds. by C. Dewitt, B.S. Dewitt (1973)
3. T. Binnington, E. Poisson, Relativistic theory of tidal love numbers. Phys. Rev. D **80**, 084018 (2009)
4. B. Carter, Black hole equilibrium states, in *Black Holes (Les Astres Occlus)*, eds. by C. Dewitt, B.S. Dewitt (1973), pp. 57–214
5. S. Chakrabarti, T. Delsate, J. Steinhoff, New perspectives on neutron star and black hole spectroscopy and dynamic tides. arXiv:1304.2228 [gr-qc]
6. S. Chakrabarti, T. Delsate, N. Gürlebeck, J. Steinhoff, I-Q relation for rapidly rotating neutron stars. Phys. Rev. Lett. **112**, 201102 (2014)
7. T. Damour, O.M. Lecian, Gravitational polarizability of black holes. Phys. Rev. D **80**, 044017 (2009)
8. T. Damour, A. Nagar, Relativistic tidal properties of neutron stars. Phys. Rev. D **80**, 084035 (2009)
9. D.D. Doneva, S.S. Yazadjiev, N. Stergioulas, K.D. Kokkotas, Astrophys. J. Lett. **781**, L6 (2014)
10. E. Flanagan, T. Hinderer, Constraining neutron star tidal love numbers with gravitational wave detectors. Phys. Rev. D **77**, 021502(R) (2008)
11. G. Fodor, C. Hoenselaers, Z. Perjés, Multipole moments of axisymmetric systems in relativity. J. Math. Phys. **30**, 2252 (1989)
12. R. Geroch, J.B. Hartle, Distorted black holes. J. Math. Phys. **23**, 680 (1982)
13. N. Gürlebeck, Source integrals for multipole moments of static and axially spacetimes. Phys. Rev. D **90**, 024041 (2014)
14. N. Gürlebeck, Phys. Rev. Lett. **114**, 151102 (2015)
15. B. Haskell, R. Ciolfi, F. Pannarale, L. Rezzolla, On the universality of I-Love-Q relations in magnetized neutron stars. MNRAS Lett. **438**, L71-L75 (2014)
16. T. Hinderer, Tidal love numbers of neutron stars. Astrophys. J. **677**, 1216 (2008). Erratum. Astrophys. J. **697**, 964 (2009)

17. B. Kol, M. Smolkin, Black hole stereotyping: induced gravito-static polarization. JHEP **2**, 10 (2012)
18. A. Love, *Some Problems of Geodynamics* (Cornell University Library, Ithaca, 1911)
19. A. Maselli, V. Cardoso, V. Ferrari, L. Gualtieri, P. Pani, Equation-of-state-independent relations in neutron stars. Phys. Rev. D **88**, 023007 (2013)
20. H. Stephani, D. Kramer, M. MacCallum, C. Hoenselaers, E. Herlt, *Exact Solutions of Einstein's Field Equations* (Cambridge University Press, Cambridge, 2003)
21. K. Yagi, N. Yunes, I-Love-Q. Sci. **341**, 365 (2013)
22. K. Yagi, N. Yunes, I-Love-Q relations in neutron stars and their applications to astrophysics, gravitational waves, and fundamental physics. Phys. Rev. D **88**, 023009 (2013)

Chapter 11
Self–completeness in Alternative Theories of Gravity

Maximiliano Isi, Jonas Mureika and Piero Nicolini

Abstract The possible existence of a minimum black hole horizon radius suggests that the trans-Planckian regime of gravity may be semiclassical. We explore the extension of this "self-completeness" of gravity to the beyond-Einstein formalisms of Randall-Sundrum extra dimensions and the generalized uncertainty principle.

11.1 Introduction

Gravity is problematic from the perspective of quantization. For instance, graviton path integrals in (3 + 1)-D are readily divergent. As part of the effort to solve this, it has been shown [6–8, 12, 13] that gravity may be considered "self–complete," in that there exists a minimum horizon scale hiding the singularity. Specifically, this distance is defined by the confluence of the classical Schwarzschild radius and the Compton wavelength:

$$r_H = \lambda_C \quad \Longrightarrow \quad \frac{2GM_{\mathrm{BH}}}{c^2} = \frac{h}{cM_{\mathrm{BH}}}. \tag{11.1}$$

M. Isi (✉) · J. Mureika
Department of Physics, Loyola Marymount University,
Los Angeles, CA 90045-2659, USA
e-mail: misi@lion.lmu.edu

J. Mureika
e-mail: jmureika@lmu.edu

P. Nicolini
Frankfurt Institute for Advanced Studies, Ruth-Moufangstraße 1,
60438 Frankfurt am Main, Germany
e-mail: nicolini@fias.uni-frankfurt.de

P. Nicolini
Institut für Theoretische Physik, Johann Wolfgang Goethe-Universität
Frankfurt am Main, Max-von-Lauestraße 1, 60438 Frankfurt am Main, Germany

P. Nicolini et al. (eds.), *1st Karl Schwarzschild Meeting on Gravitational Physics*,
Springer Proceedings in Physics 170, DOI 10.1007/978-3-319-20046-0_11

In terms of the Planck mass, $M_{\rm Pl} = \sqrt{\hbar c/G}$, this gives a minimum mass:

$$M_{\rm BH} \geq \sqrt{\frac{hc}{2G}} = \sqrt{\pi} M_{\rm Pl}, \tag{11.2}$$

below which the effective length scale increases as $M_{\rm BH}^{-1}$. That is, Planck-mass holes are the smallest resolvable black objects.

11.2 Randall-Sundrum

The Randall-Sundrum model posits our Universe is a n-dimensional brane in a bulk with an infinite extra dimension at a distance ℓ, the AdS curvature radius [11]. Einstein's equations in the bulk are

$$\tilde{G}_{AB} = \tilde{\kappa}^2 \left[-\tilde{\Lambda} \tilde{g}_{AB} + \delta(\chi) (-\lambda g_{AB} + T_{AB}) \right], \tag{11.3}$$

where the coupling $\tilde{\kappa} = 8\pi/\tilde{M}_{\rm Pl}^{\ 3}$ is a function of the reduced $(n+1)$-dimensional Planck mass $\tilde{M}_{\rm Pl}$. The hierarchy problem is resolved by assuming gravity originates in the extra brane, causing our effective gravitational constant to be $G_4 = G_5/\ell$, where G_5 is the "true" coupling strength [3].

In the case of an electrically neutral black hole, the induced Einstein equations on the brane yield a Reissner-Nordström-like solution of the form [3][1]

$$ds_4^2 = -\left(1 - \frac{2G_4 m}{c^2 r} + \frac{Q}{r^2}\right) c^2 dt^2 + \frac{dr^2}{1 - \frac{2G_4 m}{c^2 r} + \frac{Q}{r^2}} + r^2 d\Omega^2, \tag{11.4}$$

with $d\Omega^2 = d\theta^2 + sin^2\theta d\phi^2$ and the term Q is the *tidal charge*, resulting from leakage into the bulk. Knowing that $G_4 = \hbar c/M_{\rm Pl}^2$, we can write (11.4) in terms of the Planck mass to obtain the (outer) black hole horizon:

$$r_H = \frac{\hbar}{c} \frac{M_{\rm BH}}{M_{\rm Pl}^2} \left(1 + \sqrt{1 - \frac{c^2}{\hbar^2} \frac{M_{\rm Pl}^4}{M_{\rm BH}^2} Q}\right). \tag{11.5}$$

For the external horizon to be greater than the Schwarzschild radius, we require a negative Q. Otherwise, both radii would be smaller than the regular Schwarzschild horizon and we would get the usual self–completeness condition from (11.2). Furthermore, $Q < 0$ is arguably a more "physical" choice [4]. Regardless of Q, $r_H(M_{\rm BH})$ becomes linear for large enough mass (Fig. 11.1).

[1] Note that in [3] β represents the tidal charge and Q for the electric charge, which we take to be null.

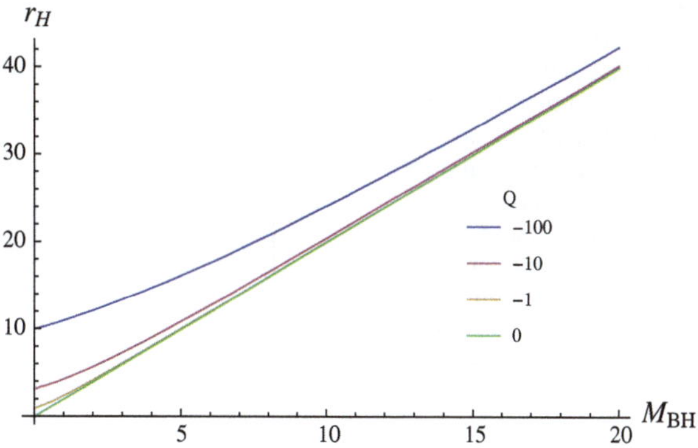

Fig. 11.1 Black hole horizon radius as a function of mass in the Randall-Sundrum model (Planck units). The plot displays (11.5) for different values (negative) of the tidal charge Q

Equating the Compton wavelength to the black hole horizon radius, we obtain an expression for the minimum black hole mass as a function of Q:

$$M_{\rm BH} \geq \frac{\pi M_{\rm Pl}}{\sqrt{\pi - c^2 M_{\rm Pl}^2 Q / 4\hbar^2}} \tag{11.6}$$

However, Q is not an independent variable. For small length scales, compared to the AdS radius ($r \ll \ell$), the tidal charge becomes a linear function of the brane separation distance:

$$Q \underset{r \ll \ell}{\approx} -\frac{M_{BH}}{M_{\rm Pl}^2} \frac{\hbar}{c} \ell. \tag{11.7}$$

Consequently, the minimum mass is also a function of ℓ. After some basic algebra, we find:

$$M_{\rm BH} \geq A \left(B + B^{-1} - 1 \right), \tag{11.8}$$

$$A \equiv \frac{4\pi}{3} \frac{\hbar}{c\ell}, \quad B \equiv \left[\frac{3\pi}{2} \left(\frac{M_{\rm Pl}}{A} \right)^2 - 1 + \frac{M_{\rm Pl}}{A} \sqrt{\left(\frac{3\pi}{2} \frac{M_{\rm Pl}}{A} \right)^2 - 3\pi} \right]^{1/3}. \tag{11.9}$$

The meaning of (11.8) can be illuminated by means of an expansion in powers of $M_{\rm Pl}$

$$M_{\rm min} = \sqrt{\pi}\, M_{\rm Pl} - \frac{c\ell}{8\hbar} M_{\rm Pl}^2 + \frac{5}{128} \left(\frac{c\ell}{\hbar} \right)^2 \pi^{-1/2} M_{\rm Pl}^3 + \mathcal{O}(M_{\rm Pl}^4). \tag{11.10}$$

Again, we recover (11.2) for vanishing ℓ, as expected (Fig. 11.2). On the other hand, note that $M_{\rm min} \to 0$ as $\ell \to \infty$. Furthermore, because (11.8) is continuous for all positive values of ℓ (which we require in order to have $Q < 0$), Randall-Sundrum gravity can always be considered self–complete.

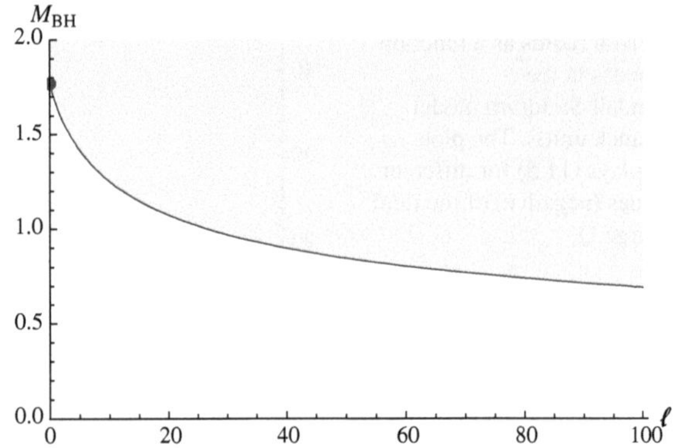

Fig. 11.2 Minimum black hole mass (11.8) in Planck-mass units as a function of brane separation (*solid line*). The *shaded region* indicates the allowed values of the mass. As the correction is removed (viz. $\ell \to 0^+$), the minimum mass is again $\sqrt{\pi} M_{\rm Pl}$ (indicated by a *dot*), agreeing with (11.2)

11.3 Generalized Uncertainty Principle

If additional momentum dependent terms exist in the usual commutation relation, this will result in a modified uncertainty relation of the form $\Delta x \Delta p \geq \frac{\hbar}{2}\left(1+\beta(\Delta p)^2\right)$. Such modification is known by the name of *generalized uncertainty principle* (GUP). In turn, this translates into a non-zero commutator between the coordinate operators:

$$[\mathbf{x}_i, \mathbf{x}_j] = 2i\hbar\beta(\mathbf{p}_i\mathbf{x}_j - \mathbf{p}_j\mathbf{x}_i). \tag{11.11}$$

Because the commutator does not vanish unless $\beta = 0$, the GUP introduces a non-zero minimal uncertainty in position, which translates into the existence of a minimal length. Furthermore, this results in a momentum integration measure

$$\int \frac{\mathrm{d}^n p}{1+\beta \mathbf{p}^2} |p\rangle\langle p| = 1, \tag{11.12}$$

which presents a UV cutoff of $\sqrt{\beta}$ [10]. This has important consequences for black hole evaporation and results in remnant formation.

The GUP replaces the Dirac delta in the description of point particles of regular quantum mechanics with a wider Gaussian distribution, $e^{-|\mathbf{x}|\sqrt{\beta}}$. As shown in [9], we can reproduce these non-local effects by means of the GUP-inspired metric

$$\mathrm{d}s^2 = -\left(1 - 2\frac{GM}{c^2 r}\gamma(2; r/\sqrt{\beta})\right)\mathrm{d}t^2 - \left(1 - 2\frac{GM}{c^2 r}\gamma(2; r/\sqrt{\beta})\right)^{-1}\mathrm{d}r^2 + r^2\mathrm{d}\Omega^2 \tag{11.13}$$

where $\gamma(s; x) = \int_0^x t^{s-1}e^{-t}\mathrm{d}t$ is the lower incomplete gamma function. The metric coefficient $1/g_{rr}$ is shown in Fig. 11.3. Note that the extremal case happens at $M_{\rm BH} \approx 1.66\sqrt{\beta}/G$ and $r_H \approx 1.73\sqrt{\beta}$.

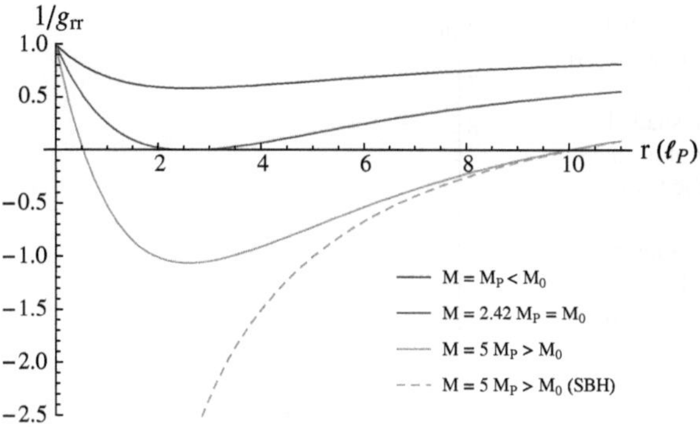

Fig. 11.3 Metric coefficient for GUP-inspired metric (11.13). Notice naked singularity, extremal and regular black hole cases. The Schwarzschild (SBH) case for $M = 5M_{Pl}$ is shown for comparison

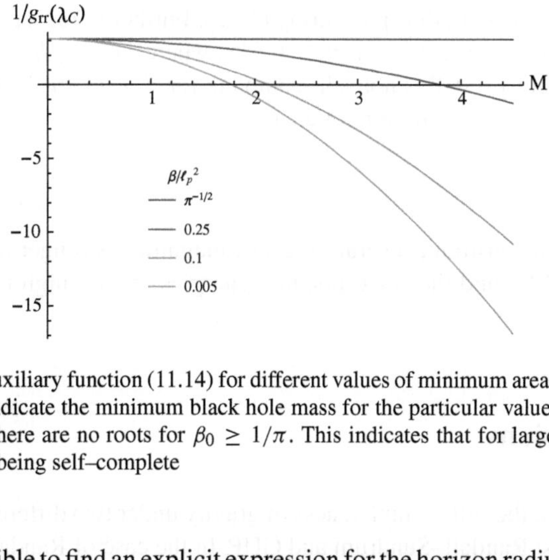

Fig. 11.4 GUP auxiliary function (11.14) for different values of minimum area β, shown in Planck units. The roots indicate the minimum black hole mass for the particular value of β_0, where $\beta_0 = \beta/\ell_P^2$. Note that there are no roots for $\beta_0 \geq 1/\pi$. This indicates that for large enough minimum areas, GUP stops being self–complete

It is not possible to find an explicit expression for the horizon radius corresponding to (11.13). However, we can naively attempt to study the self–completeness of this metric by numerically solving $1/g_{rr} = 0$ under the constraint $r = \lambda_C(M)$, viz.

$$1 - 2\frac{GM}{c^2\lambda_C}\gamma(2; \lambda_C/\sqrt{\beta}) = 0 \tag{11.14}$$

Rather than taking the usual expression for λ_C, we follow [1, 2] by correcting the Compton wavelength to account for GUP effects: $\lambda_{GUP} = \frac{\hbar}{Mc}(1+\beta M^2)$. Note that this step is not strictly necessary (see [9] for a more rigorous approach). The RHS of (11.14) is plotted in Fig. 11.4 The roots of this function can be interpreted as the values of M_{BH} at which the horizon radius coincides with the modified Compton wavelength for a given β, i.e. a minimum black hole mass.

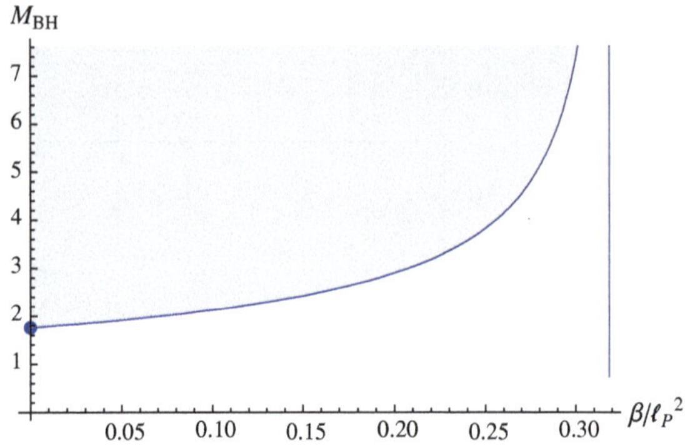

Fig. 11.5 Black hole mass in Planck units for varying β. The allowed values correspond to the shaded region. The minimum-mass curve (*solid*), which was obtained numerically, presents an asymptote at $\beta = \ell_P^2/\pi \approx 0.318$. For $\beta = 0$, we recover the GR constraint (*dot*)

The allowed black hole masses are shown in Fig. 11.5. The GUP corrections can be undone by letting $\beta \to 0$, thus recovering (11.2). Furthermore, we find that (11.14) has no positive roots for $\beta/\ell_P^2 \geq 1/\pi \approx 0.318$, where $\ell_P = \sqrt{\hbar G/c^3}$ is the Planck length. Consequently, GUP is not self–complete for $\beta \geq \ell_P^2/\pi$. This can be turned into an upper bound on the minimum area:

$$\beta < \ell_P^2/\pi. \tag{11.15}$$

Note that this is a constraint several orders of magnitude stronger than those found in [5] of $\beta_0 < 10^{21}$ and the corresponding energies are too high to be tested with current experiments.

11.4 Conclusions

We have explored the self–completeness of gravity under two different and independent frameworks: Randall–Sundrum and GUP. In the case of Randall-Sundrum, we have shown that gravity should be self–complete regardless of the AdS curvature radius and found an closed-form solution for the minimum mass, (11.8). This is not the case for GUP: under this formalism, gravity is only self–complete as long as the minimum area β satisfies (11.15). Such condition could be understood as a constraint on GUP. This, however, is a heuristic analysis and should be complemented by a more formal treatment (c.f. [9]).

Acknowledgments MI and JM would like to thank the generous hospitality of the Frankfurt Institute for Advanced Studies, at which this work was initiated. MI would like to thank Loyola Marymount University Honors Program for continued support. This work has been supported by the project "Evaporation of microscopic black holes" (PN) of the German Research Foundation (DFG), by the Helmholtz International Center for FAIR within the framework of the LOEWE program (Landesoffensive zur Entwicklung Wissenschaftlich-Ökonomischer Exzellenz) launched by the State of Hesse (PN), partially by the European Cooperation in Science and Technology (COST) action MP0905 "Black Holes in a Violent Universe" (PN).

References

1. R.J. Adler, Am. J. Phys. **78**, 925 (2010)
2. B. Carrr et al., Generalized uncertainty principle and self-dual black holes. arXiv:1107.0708
3. A. Chamblin et al., Phys. Rev. D **63**, 064015 (2001). arXiv:hep-th/0008177
4. N. Dadhich, V. Rezania et al., Phys. Lett. B **487**, 1–6 (2000). arXiv:hep-th/0003061
5. S. Das, E.C. Vagenas, Phys. Rev. Lett. **101** (2008). arXiv:0810.5333
6. G. Dvali, C. Gomez, Self-completeness of Einstein gravity. arXiv:1005.3497
7. G. Dvali, S. Folkerts, C. Germani, Phys. Rev. D **84**, 024039 (2011). arXiv:1006.0984
8. G. Dvali, C. Gomez, Ultra-high energy probes of classicalization. arXiv:1205.2540
9. M. Isi, J. Mureika, P. Nicolini, JHEP **11**, 139 (2013). arXiv:1310.8153
10. A. Kempf et al., Phys. Rev. D **52**, 1108 (1995). arXiv:hep-th/9412167
11. L. Randall, R. Sundrum, Phys. Rev. Lett. **83**, 3370–3373 (1999). arXiv:hep-ph/9905221
12. E. Spallucci, S. Ansoldi, Phys. Lett. B **701**, 471 (2011). arXiv:1101.2760
13. E. Spallucci, A. Smailagic, Phys. Lett. B **709**, 266 (2012). arXiv:1202.1686

Chapter 12
Gravitational Collapse to Black Holes and More

Daniele Malafarina

Abstract Are black holes the only possible end state of complete gravitational collapse? Or is there any alternative outcome, either within general relativity or coming from some quantum theory of gravity? If so, could the existence of such exotic objects be in principle inferred from observations?

12.1 Introduction

There is strong observational evidence of the existence of many black hole candidates in the universe. These range from compact objects of the order of a few solar masses to supermassive ones that can be as massive as a few billion suns. Nevertheless we don't know yet if these candidates must all necessarily be black holes nor we know if black holes are the only possible final outcome of realistic gravitational collapse as predicted by General Relativity (GR) for very massive objects. Furthermore GR is expected to require modifications due to quantum effects in the strong field regime that is reached towards the final stages of collapse. Therefore before concluding that all collapse processes in the universe above a certain threshold lead to the formation of a black hole we need to understand more carefully the limits of classical models and how they might relate to realistic collapse.

One possible approach to a description of collapse in the strong field regime is provided by a semiclassical treatment of quantum corrections, where we solve Einstein's equations for an effective energy momentum tensor that takes into account the modifications due to quantum gravitational effects (see for example [1, 2]). We see that following this method the classical singularity at the end of collapse is

D. Malafarina (✉)
Center for Field Theory and Particle Physics & Department of Physics,
Fudan University, 220 Handan Road, Shanghai 200433, China
e-mail: daniele@fudan.edu.cn

D. Malafarina
Department of Physics, SST Nazarbayev University,
53 Kabanbay Batyr avenue, Astana 010000, Kazakhstan
e-mail: daniele.malafarina@nu.edu.kz

P. Nicolini et al. (eds.), *1st Karl Schwarzschild Meeting on Gravitational Physics*,
Springer Proceedings in Physics 170, DOI 10.1007/978-3-319-20046-0_12

replaced by a bounce and the final Schwarzschild black hole doesn't form. Instead we have the formation of two 'evaporating' trapped regions, one before and one after the bounce, that may leave the portion of the spacetime where the bounce occurs visible to far away observers.

It is clear that such a scenario might have important observational consequences for astrophysical black hole candidates. Even more so if the dispersal of the cloud after the bounce leaves behind some kind of exotic compact object. These exotic remnants might have observational properties that distinguish them from black holes of the same mass thus allowing for the possibility of detection in future observations (see for example [3–5]). If true, this would open an important observational window on the strong field regime of gravity and possibly new physics.

12.2 Classical Collapse and Quantum Bounce

Relativistic gravitational collapse has been studied for many decades (see [6] and references therein) and it is well known now that under some basic assumptions (i.e. energy conditions) it ends necessarily with the formation of a spacetime singularity [7, 8]. The standard paradigm for gravitational collapse leading to the formation of a black hole is given by the Oppenheimer-Snyder model, which describes collapse of an homogeneous dust (i.e. pressureless) cloud [9]. The model can be easily generalized to the case of collapse of homogeneous perfect fluids with a linear equation of state and it leads to the same final outcome, namely a simultaneous singularity covered by a horizon at all times. The energy momentum tensor is given by $T^{\mu}_{\nu} = \mathrm{diag}(\rho(t), p(t), p(t), p(t))$ and the pressure p can be related to the energy density ρ via and equation of state of the form $p = \lambda\rho$ (with $\lambda \in [-1, 1]$ to satisfy energy conditions). The metric in comoving coordinates is given by

$$ds^2 = -dt^2 + \frac{a^2}{1+kr^2}dr^2 + r^2a^2d\Omega^2 \,, \tag{12.1}$$

where the dimensionless scale factor $a(t)$ determines the evolution of the cloud and it is taken to be 1 at the initial time $t_i = 0$. Also k is a constant related to the initial velocity of the infalling particles that we will assume to be zero (a case known as marginally bound collapse) and $d\Omega^2$ represents the line element on the unit two-sphere. The energy density and pressure are given by $\rho = 3M/a^3$, $p = -\dot{M}/a^2\dot{a}$, with $M(t) = M_0/a^{3\lambda}$ being a mass profile related to the Misner-Sharp of the system $F(r, t) = r^3M(t)$, which represents the amount of matter enclosed within the shell r at the time t [10]. From the Misner-Sharp mass we can derive the equation of motion for a as

$$M(t) = a\dot{a}^2 \,. \tag{12.2}$$

Dust collapse is recovered for $\lambda = 0$ which implies $p = 0$ and in this case the cloud matches to a vacuum Schwarzschild exterior with mass parameter M_{sc} across the collapsing boundary surface given by $R_b(t) = r_b a(t)$. In the perfect fluid case the matching can be done with an exterior radiating Vaidya spacetime. The total mass of the collapsing system at any given time is then given by the matching conditions as $F(r_b, t) = r_b^3 M(t)$ that in the case of dust reduces to $r_b^3 M_0 = 2M_{sc}$ [11–14]. Solving equation (12.2) to obtain the scale factor a solves the system of Einstein's equations completely. The solution can be written as

$$a(t) = \left(1 - \frac{3(\lambda+1)}{2}\sqrt{M_0}t\right)^{2/3(\lambda+1)} . \tag{12.3}$$

From this solution we see that all matter falls into the central singularity at the time $t_s = 2/3(\lambda+1)\sqrt{M_0}$. The evolution of the apparent horizon inside the cloud can be described by a curve $t_{ah}(r)$ which describes the time at which the shell r becomes trapped, and is governed by the equation $a(t_{ah}(r)) = r^2 M$. We then see that the cloud becomes trapped at the time

$$t_{ah}(r_b) = t_s - \frac{2}{3(\lambda+1)}\left(r_b^{3(\lambda+1)} M_0\right)^{1/(3\lambda+1)} < t_s . \tag{12.4}$$

The above picture is derived within GR by solving Einstein equations $G_{\mu\nu} = 8\pi T_{\mu\nu}$. We know that GR is a valid theory of gravity in the weak field regime and therefore we assume Einstein's equations to hold at the initial time of collapse. Nevertheless as we approach the formation of the singularity the energy density grows arbitrarily high and we can expect that correction due to quantum-gravitational effects become important in the strong field limit. We thus expect Einstein's equations to be modified in some way. We can in general suppose that we can write $G_{\mu\nu} + G_{\mu\nu}^{\text{corr}} = 8\pi T_{\mu\nu}$, where the specific form of $G_{\mu\nu}^{\text{corr}}$ will depend on the theory of quantum-gravity employed. In order to investigate how such quantum-gravity corrections occurring in the strong field regime affect the collapse scenario we shall use a semiclassical approach. This consists in treating the term $G_{\mu\nu}^{\text{corr}}$ as an 'unphysical' matter field to be added on the right hand side of Einstein's equations. In this way we solve the usual equations for relativistic collapse for an effective matter source that takes into account the original $T_{\mu\nu}$ plus the quantum corrections in the strong field. The simplest correction that can be implemented and that can be derived by some semiclassical framework in Loop Quantum Cosmology is given by $\rho^{\text{corr}} = -\rho^2/\rho_{\text{cr}}$ (see for example [15–17]). This choice yields and effective density of the form

$$\rho_{\text{eff}} = \rho + \rho^{\text{corr}} = \rho\left(1 - \frac{\rho}{\rho_{\text{cr}}}\right) , \tag{12.5}$$

where ρ_{cr} is a critical density that depends on the specific form of the quantum-gravity theory and it is of the order of the Planck density. Einstein's equations then provide the form for the effective pressure and effective mass as

$$M_{\rm eff}=M\left(1-\frac{\rho}{\rho_{\rm cr}}\right)\,,\quad p_{\rm eff}=p\left(1-\frac{2\rho}{\rho_{\rm cr}}\right)-\frac{\rho^2}{\rho_{\rm cr}}\,. \tag{12.6}$$

Solving the equation of motion (12.2) with the effective mass in place of the classical mass gives the scale factor for the quantum corrected scenario as

$$a(t)=\left[K+\left(\sqrt{1-K}-\frac{3(\lambda+1)}{2}\sqrt{M_0}t\right)^2\right]^{1/3(\lambda+1)}\,, \tag{12.7}$$

where we have defined the critical scale $K=a_{\rm cr}^{3(\lambda+1)}=3M_0/\rho_{\rm cr}$. Note that as K goes to zero (i.e. $\rho_{\rm cr}$ goes to infinity) we recover the classical solution. The scale factor reaches the minimal value $a_{\rm cr}$ at the time of the bounce $t_{\rm cr}$ given by

$$t_{\rm cr}=\frac{2\sqrt{1-K}}{3(\lambda+1)\sqrt{M_0}}\,, \tag{12.8}$$

when $\dot{a}=0$ and $\rho=\rho_{\rm cr}$ (note that at the time of the bounce we have $\rho_{\rm eff}=0$).

The structure of trapped surfaces inside the cloud for the effective model of collapse can be obtained from the condition

$$1-\frac{r^2M_{\rm eff}}{a}=0\,. \tag{12.9}$$

It is immediately clear that at the time of the bounce for which $a=a_{\rm cr}\neq 0$ and $M_{\rm eff}=0$ the radius of the apparent horizon to satisfy (12.9) must be at infinity, and therefore outside the boundary. So we conclude that in the limit of asymptotic freedom at the instant of the bounce the cloud is not hidden behind any horizon. Nevertheless, since the early stages of collapse follow the classical scenario, a trapped region must form as collapse passes the Schwarzschild boundary. The trapped region must then 'evaporate' before the bounce occurs. After the bounce the cloud re-expands and the dynamics follows the time reversal of the collapse picture. Therefore another trapped region forms in the expanding phase only to disappear (as a white hole) when the cloud's boundary becomes greater than the Schwarzschild radius (see Fig. 12.1). This picture is in agreement with similar scenarios studied in the framework of asymptotic safe gravity (see for example [18–22]), where the gravitational interaction is supposed to vanish in the limit of small distances and high densities (as towards the end of collapse). Seen from this context the time of the bounce, where $M_{\rm eff}$ vanishes and the metric is given by the flat Minkowski spacetime, represents the instant where asymptotic safety is achieved.

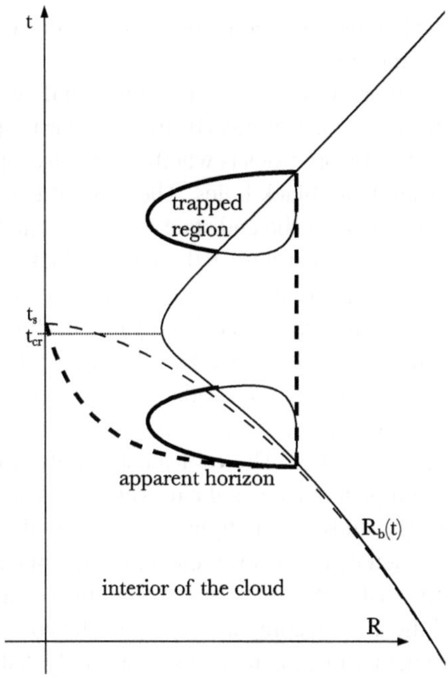

Fig. 12.1 Comparison between classical and 'quantum corrected' homogeneous collapse. The *dashed lines* represent the classical picture, the *continuous lines* represent the modified scenario. Initially collapse proceeds following the classical evolution as determined by Einstein's equations. The trapped region forms initially at the collapsing boundary of the cloud $R_b(t)$. The event horizon in the exterior Schwarzschild geometry (*vertical dashed line*) forms in the classical scenario at the instant of formation of the trapped region. At the same instant the apparent horizon forms in the interior (*curved thick dashed line*). The apparent horizon meets the simultaneous singularity at the center of the cloud at the time of formation of the singularity itself t_s. The singularity is entirely hidden from far away observers. In the quantum corrected scenario the singularity is replaced by a bounce at the time t_{cr}, the trapped region (the closed *continuous curve*, *thin* in the exterior and *thick* in the interior) 'evaporates' as the cloud shrinks. At the instant of the bounce we have a flat metric and no trapped surfaces. The evolution after the bounce follows the time reversal of the collapse scenario

12.3 Concluding Remarks

We have seen that once quantum-gravitational effects occurring in the strong field regime are considered the standard picture of relativistic gravitational collapse is changed. The matter cloud bounces without forming a singularity. The event horizon that is the signature of the Schwarzschild black hole doesn't form and is replaced by two trapped regions that 'live' for a finite time. Although this time is very short for the co-moving observers it can be long for observers at spatial infinity thus allowing for the existence of astrophysical objects that 'mimic' a standard black hole. Nevertheless the lifetime of these objects is limited. An important question that arises naturally at this point is whether the expanding cloud, after it has dispersed away, leaves behind a compact remnant of some sort.

Exotic compact objects as leftovers from gravitational collapse have been considered for many years in various settings (from gravastar to quark stars and so on) and the crucial question for astrophysics is whether these hypothetical objects can be detected and distinguished from black holes. Therefore it is worth asking if similar objects can form in the universe as the end state of collapse and if they would posses some observational features that distinguish them from black holes of the same mass. In classical GR there are many solutions describing static configurations with non vanishing energy momentum that can be used as toy models for compact objects. Such interior solutions, that match to the Schwarzschild metric at a finite boundary, can be obtained from relativistic gravitational collapse under certain simple assumptions and may be considered as a first step towards the theoretical description of compact objects composed of exotic matter. The tool most widely used in astrophysics to study the properties of compact objects in the universe is the accretion disk that most of these objects posses. By studying the light curves emitted by matter in the accretion disk it is possible to determine several features of the black hole candidates that exist at the center of the disk. Therefore by simulating the same light curves from accretion disks around theoretical compact objects as the ones mentioned above it is possible to determine whether these can be observationally distinguished from black holes.

For example in [23, 24] the properties of accretion disks around a class of perfect fluid interiors that was obtained by Tolman in [25] was investigated. It was found that the luminosity spectrum as received by distant observers exhibits a tail at high frequencies that is absent in the case of black holes. This is supported by simulations of the kα iron line of absorption that is expected to be present in the disk, which shows very different behaviour from the black hole case [3–5]. This analysis supports the claim that certain compact objects for which we have observational data are likely to posses a horizon, at least a temporary one as described previously. Nevertheless we can not exclude a priori, given the extremely small sample of available candidates, that more exotic objects without horizon will be found in the future.

References

1. C. Bambi, D. Malafarina, L. Modesto, Phys. Rev. D **88**, 044009 (2013)
2. C. Barcelo, S. Liberati, S. Sonego, M. Visser, Phys. Rev. D **77**, 044032 (2008)
3. C. Bambi, D. Malafarina, Phys. Rev. D **88**, 064022 (2013)
4. Z. Li, C. Bambi, JCAP **1303**, 031 (2013)
5. C. Bambi, Phys. Rev. D **87**, 023007 (2013)
6. P.S. Joshi, D. Malafarina, Int. J. Mod. Phys. D **20**(14), 2641 (2011)
7. S.W. Hawking, R. Penrose, Proc. Roy. Soc. Lond. A **314**, 529 (1970)
8. S.W. Hawking, G.F.R. Ellis, *The Large Scale Structure of Space-Time* (Cambridge University Press, Cambridge, 1973)
9. J.R. Oppenheimer, H. Snyder, Phys. Rev. **56**, 455 (1939)
10. C. Misner, D. Sharp, Phys. Rev. **136**, B571 (1964)

11. W. Israel, Nuovo Cimento B **44**, 1 (1966)
12. W. Israel, Nuovo Cimento B **48**, 463 (1966)
13. F. Fayos, X. Jaen, E. Llanta, J.M.M. Senovilla, Phys. Rev. D **45**, 2732 (1992)
14. F. Fayos, J.M.M. Senovilla, R. Torres, Phys. Rev. D **54**, 4862 (1996)
15. Y. Tavakoli, J. Marto, A. Dapor, arXiv:1303.6157 [gr-qc]
16. R. Goswami, P.S. Joshi, P. Singh, Phys. Rev. Lett. **96**, 031302 (2006)
17. M. Bojowald, R. Goswami, R. Maartens, P. Singh, Phys. Rev. Lett. **95**, 091302 (2005)
18. C. Bambi, D. Malafarina, L. Modesto, Eur. Phys. J. C **74**, 2767 (2014)
19. R. Torres, Phys. Lett. B **733**, 21 (2014)
20. R. Casadio, S.D.H. Hsu, B. Mirza, Phys. Lett. B **695**, 317 (2011)
21. P. Hajicek, Nucl. Phys. B **603**, 555 (2001)
22. V.P. Frolov, arXiv:1402.5446 [gr-qc]
23. P.S. Joshi, D. Malafarina, R. Narayan, Class. Quantum Grav. **28**, 235018 (2011)
24. P.S. Joshi, D. Malafarina, R. Narayan, Class. Quantum Grav. **31**, 015002 (2014)
25. R.C. Tolman, Phys. Rev. **55**, 364 (1939)

Chapter 13
Experimental Tests of Pseudo-Complex General Relativity

Thomas Schönenbach, Gunther Caspar, Peter O. Hess, Thomas Boller, Andreas Müller, Mirko Schäfer and Walter Greiner

Abstract Based on a recently proposed extension to General Relativity (GR), called pseudo-complex General Relativity (pc-GR), we present a selection of several tests of GR near compact massive objects. The investigated phenomena are the redshift, the orbital frequency of a test particle and the innermost stable circular orbit (ISCO) around a massive object. We observe that the redshift and orbital frequency are in general lower in pc-GR compared to Einstein's GR. Also the orbital frequency for prograde motion now exhibits a maximum, which is not present in GR. In addition the concept of an innermost stable circular orbit does not hold in pc-GR as it arises in GR. All modifications due to pc-GR correction terms appear only at small distances (mostly below three Schwarzschild radii) and thus can only be observed by measurements in regions of strong gravity. Those and more results have been published already in [1].

13.1 Introduction

Up to now no experimental evidence suggests that the theory of General Relativity does not hold. However there exist singularities, e.g. in the centre of black holes. Also the occurrence of event horizons brings the problem that there exists a region of spacetime which is not accessible to even nearby external observers. There is a wide

T. Schönenbach (✉) · G. Caspar · M. Schäfer · W. Greiner
Frankfurt Institute for Advanced Studies, Ruth-Moufangstraße 1,
60438 Frankfurt am Main, Germany
e-mail: schoenenbach@fias.uni-frankfurt.de

P.O. Hess
Instituto de Ciencias Nucleares, Universidad Nacional Autónoma de México,
Circuito Exterior, C.U., A.P. 70-543, 04510 Mexico, DF, Mexico

T. Boller
Max-Planck-Institut für Extraterrestrische Physik, Giessenbachstraße,
85748 Garching, Germany

A. Müller
Exzellenzcluster Origin and Structure of the Universe,
Technische Universität München, Boltzmannstraße 2, 85748 Garching, Germany

P. Nicolini et al. (eds.), *1st Karl Schwarzschild Meeting on Gravitational Physics*,
Springer Proceedings in Physics 170, DOI 10.1007/978-3-319-20046-0_13

spectrum of proposed extensions or modifications to GR (just to name some [2–7]), one of which is the pseudo-complex General Relativity [8, 9]. It includes in addition to the standard GR a Stress-Energy-Tensor modelling a dark energy. Due to this modification, gravity is effectively weakened in regions close to compact massive objects. Basis for the present work is a Kerr like solution of Einstein's equations in pc-GR

$$\begin{aligned} g_{00} &= \frac{r^2 - 2mr + a^2\cos^2\vartheta + \frac{B}{2r}}{r^2 + a^2\cos^2\vartheta} \\ g_{11} &= -\frac{r^2 + a^2\cos^2\vartheta}{r^2 - 2mr + a^2 + \frac{B}{2r}} \\ g_{22} &= -r^2 - a^2\cos^2\vartheta \\ g_{33} &= -(r^2 + a^2)\sin^2\vartheta - \frac{a^2\sin^4\vartheta\left(2mr - \frac{B}{2r}\right)}{r^2 + a^2\cos^2\vartheta} \\ g_{03} &= \frac{-a\sin^2\vartheta\, 2mr + a\frac{B}{2r}\sin^2\vartheta}{r^2 + a^2\cos^2\vartheta}, \end{aligned} \tag{13.1}$$

where $m = GM$ is the gravitational radius of the compact object with mass M, G is the gravitational constant and all modifications are modelled by the parameter B. For $B > \frac{64}{27}\,\mathrm{m}^3$ the solution does not contain any event horizon.

We use the metric signature $(+, -, -, -)$, the convention $a = \frac{-\kappa J}{m}$ for the specific angular momentum or spin parameter of the central object and units with the speed of light c set to one.

13.2 Results

As already mentioned, General Relativity is one of the best tested theories in physics, although regions of strong gravity are hard to access experimentally. However, there are some measurements which can be taken as testbeds for GR in strong regimes. Here we will focus on orbits around massive objects and the redshift close to these objects.

13.2.1 Orbital Frequencies

Orbiting gas clouds around compact massive objects can serve as a good testbed for GR. Flares around the black hole candidate Sgr A* in our galaxy already were observed [10, 11]. At the moment another cloud is starting to pass around Sgr A* [12], part of might start to orbit the massive object.

We model such clouds as massless point particles moving on geodesics. Their orbital frequency in an equatorial plane ($\mathrm{d}\vartheta = 0,\ \vartheta = \pi/2$) is given by [1]

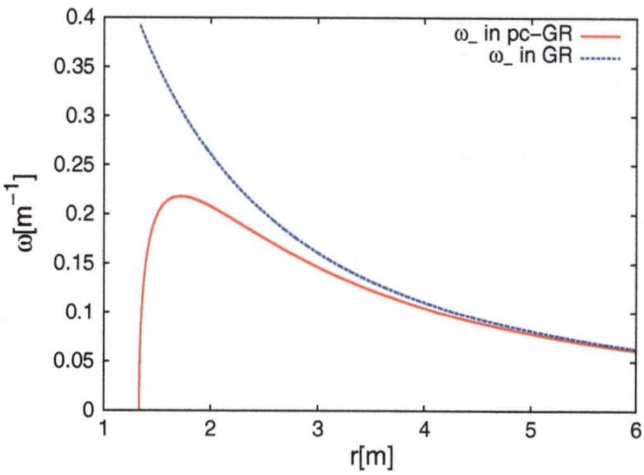

Fig. 13.1 Orbital frequency as a function of r, for stable geodesic prograde (rotating in the same way as the central massive object) circular motion. This plot is made with parameter values of $a = -0.995$ m and $B = \frac{64}{27}$ m^3. ©MNRAS http://mnras.oxfordjournals.org/content/430/4/2999/ [1]

$$\omega_{\pm} = \frac{1}{-a \mp \sqrt{\frac{2r}{h(r)}}}, \tag{13.2}$$

where we introduced the auxiliary function

$$h(r) = \frac{2m}{r^2} - \frac{3B}{2r^4}. \tag{13.3}$$

In (13.2), ω_- describes prograde motion of a particle around a central object, whereas ω_+ describes retrograde motion, respectively. Figure 13.1 displays the orbital frequency of a particle in prograde stable motion around a massive object. For retrograde motion differences between GR and pc-GR stay very small, thus we omit this case here.

For large distances (greater than three Schwarzschildradii) almost no difference between GR and pc-GR is observable. However for smaller distances we can see a clear divergence between the plots. In pc-GR the orbital frequency exhibits a maximum, whose position is independent of the spin parameter a. Thus one is able to determine a value for the pc-parameter B, if a maximum in the orbital frequency is measured, without the knowledge of the spin parameter a. This is important, as the spin parameter is not easily measurable [13].

In Fig. 13.2 we see the limiting curves for orbital motion around a massive object. They are given by

$$\bar{\omega}_{\pm} = \frac{af(r) \pm \sqrt{D(r)}}{-\left(r^2 + a^2\right) - a^2 f(r)}, \tag{13.4}$$

with the two auxiliary functions

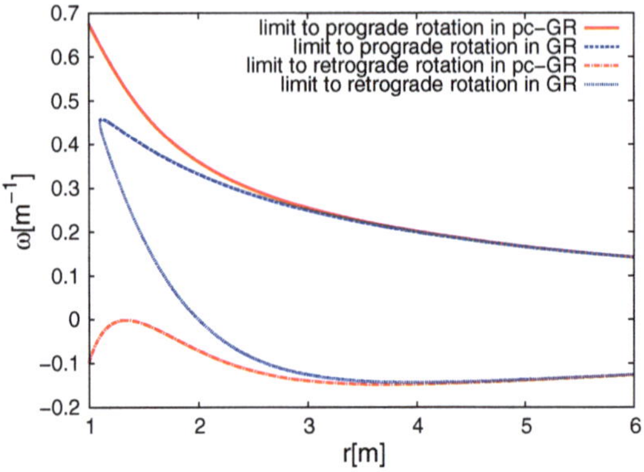

Fig. 13.2 Limits on the orbital frequencies of circular motion in the equatorial plane. The frequencies of particles moving on circular orbits must lie between the limiting curves. The parameters are $a = -0.995$ m and $B = \frac{64}{27}$ m^3. ©MNRAS http://mnras.oxfordjournals.org/content/430/4/2999/ [1]

$$f(r) = \frac{2m}{r} - \frac{B}{2r^3}$$
$$D(r) = r^2 + a^2 - 2mr + \frac{B}{2r} \,. \tag{13.5}$$

Again for large distances the differences between GR and pc-GR become negligible. Observable is the phenomenon of *frame-dragging* in the GR scenario.[1] Curves for pro- and retrograde motion approach and finally converge. In the pc-GR case this behaviour is strongly suppressed.

13.2.2 Redshift

The gravitational redshift of an object is a rather easily measurable feature. Nevertheless it carries valuable information about the observed object. It is defined by

$$z := \frac{1}{\sqrt{g_{00}} - 1} \,. \tag{13.6}$$

We investigated two cases here, the view entirely in the equatorial plane (which is equivalent to the Schwarzschild case, $\vartheta = \pi/2$) in Fig. 13.3 and the view directly on the pole of a rotating object ($\vartheta = 0$) in Fig. 13.4. Only in the inner region around a compact massive object, differences are visible.

[1] Do not confuse this with the effect of frame-dragging for weak gravitational fields which, according to the predictions of [14], has been confirmed experimentally, see [15, 16].

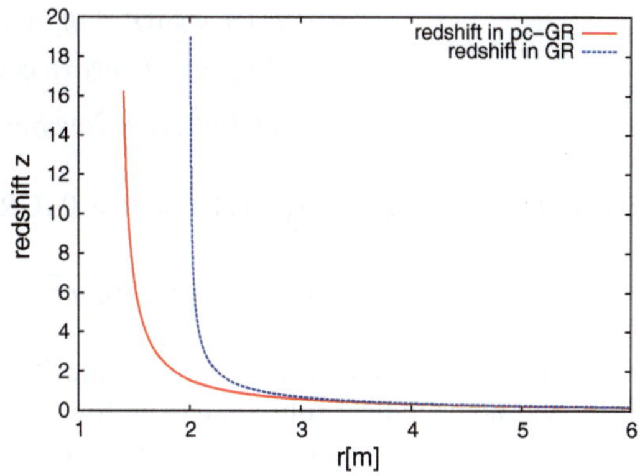

Fig. 13.3 Redshift of an emitter at the position r in the outside field of a spherically symmetric, uncharged and static mass (Schwarzschild metric) and also for the field at the equator of a rotating mass ($\vartheta = \frac{\pi}{2}$). B is set to $\frac{64}{27}$ m^3 and $a = -0.995$ m. ©MNRAS http://mnras.oxfordjournals.org/content/430/4/2999/ [1]

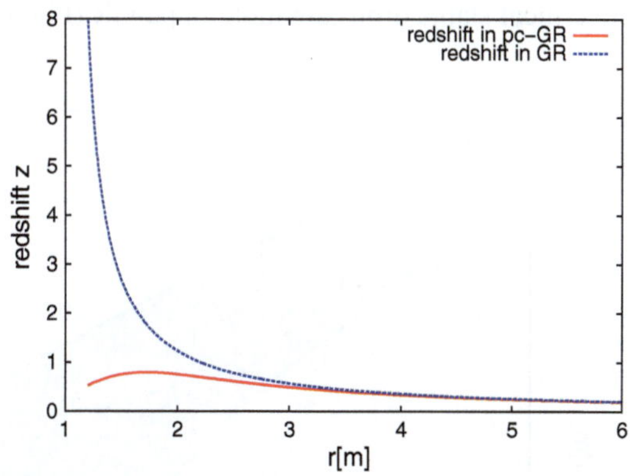

Fig. 13.4 Redshift for an emitter at the position r in the outside field of an axially symmetric, uncharged and rotating mass (Kerr metric) at the poles (e.g. $\vartheta = 0$ or $\vartheta = \pi$). B is again chosen to be $\frac{64}{27}$ m^3 and $a = -0.995$ m. ©MNRAS http://mnras.oxfordjournals.org/content/430/4/2999/ [1]

13.2.3 Innermost Stable Circular Orbit

A feature emerging in GR, which was not present in the Newtonian Theory, is the existence of critical stable orbits. Beyond the so called innermost stable circular orbit (ISCO) in GR there is no stable circular motion possible around a compact massive object. Those orbits can be found by setting $\partial^2 V/\partial r^2 = 0$, where V is the effective potential, given as [1]

$$V(r, \tilde{E}, \tilde{L}) = \frac{\tilde{L}^2}{2r^2} - \left(\frac{m}{r} - \frac{B}{4r^3}\right)\left(1 + \frac{\left(\tilde{L} + a\tilde{E}\right)^2}{r^2}\right) + \frac{(1-\tilde{E}^2)a^2}{2r^2} + \frac{1}{2}, \tag{13.7}$$

with $\tilde{E}$ and $\tilde{L}$ being the normalised energy at infinity and angular momentum of the test particle. This equation can be rewritten into

$$\begin{aligned} & g''_{33}(g_{00}+\omega g_{03})^2 + g''_{00}(g_{03}+\omega g_{33})^2 \\ & \quad - 2g''_{03}(g_{00}+\omega g_{03})(g_{03}+\omega g_{33}) \\ & \quad + D''(\omega^2 g_{33} + 2\omega g_{03} + g_{00}) = 0\,, \end{aligned} \tag{13.8}$$

where $D = \left(-g_{00}g_{33} + g_{03}^2\right)$ [1]. For $B = 0$ (GR-limit) it reduces to

$$r^2 - 6mr \pm 8a\sqrt{mr} - 3a^2 = 0\,, \tag{13.9}$$

which is the same as in [17]. Without setting $B = 0$, (13.8) has a rather complicated form and no analytical solution can be found anymore. However, it is possible to solve it numerically, leading to two solutions for a co-rotating particle, but only up to values of $a \approx -0.42$ m. For higher absolute values of the spin parameter there are no critical stable orbits anymore in pc-GR (Fig. 13.5).

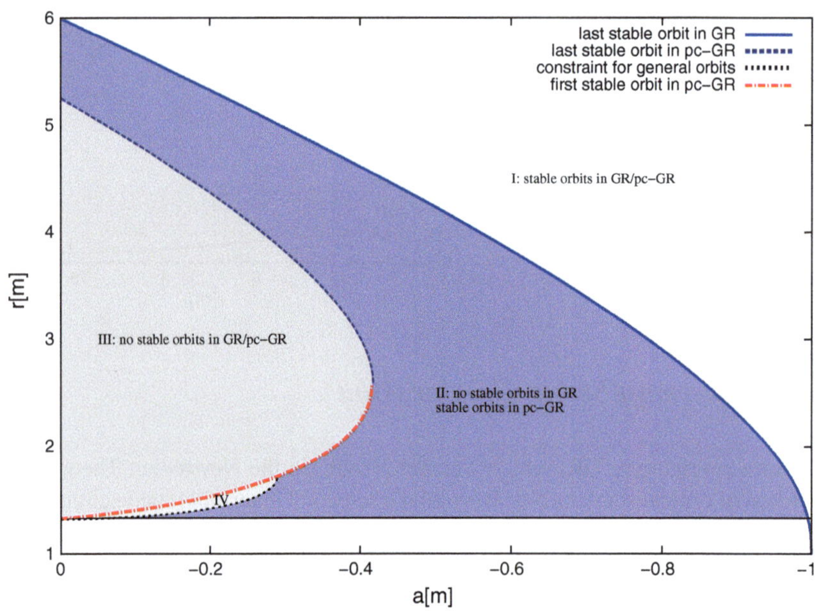

Fig. 13.5 Critical stable orbits for prograde test masses (The curves describe from *top* to *bottom*: 1. The last stable orbit in standard GR (*thick solid*), 2. the 'last' stable orbit in pc-GR (*dashed*), 3. the 'first' stable orbit in pc-GR (*dash-dotted*), 4. the limit to general orbits given by (13.4) (*dotted*) and 5. the point where the pc-equations become imaginary ($r = \frac{4}{3}$, *thin solid*)). We now have to distinguish between four different areas. In the *unshaded* area (*I*) orbits are stable both in GR and pc-GR, whereas in the *dark shaded* area (*II*) orbits are only stable in pc-GR. Both *lighter shaded* areas (*III* and *IV*) do not contain stable orbits at all. The plot is done for a value of $B = \frac{64}{27}\,\mathrm{m}^3$. ©MNRAS http://mnras.oxfordjournals.org/content/430/4/2999/ [1]

13.3 Conclusion and Outlook

We investigated a set of different phenomena occurring in GR and their alterations in pc-GR. The key observations are a reduced orbital frequency with a spin independent maximum of a test particle around a compact massive object (13.2), the weakened effect of frame-dragging, the reduced redshift (13.6) and the modification of the concept of a last stable orbit (13.8). These phenomena can serve as testbeds to discriminate between GR and pc-GR. In future we will extend this list of observables by calculating iron Kα emission line profiles and images of accretion disks to be able to discriminate between both theories.

Acknowledgments T.S. acknowledges support from Stiftung Polytechnische Gesellschaft Frankfurt am Main. P.O.H acknowledges financial support from DGAPA PAPIIT (IN 103212) and CONACyT.

References

1. T. Schönenbach, G. Caspar, P.O. Hess, T. Boller, A. Müller, M. Schäfer, W. Greiner, Astron. Soc. **430**, 2999 (2013)
2. A. Einstein, Ann. Math. **46**, 578 (1945)
3. A. Einstein, Rev. Mod. Phys. **20**, 35 (1948)
4. C.Mantz, T. Prokopec, arXiv:gr-qc/0804.0213v1 (2008)
5. D. Lovelook, Annalidi Matematica Pura **83**, 43 (1969)
6. M. Born, Proc. Roy. Soc. A **165**, 291 (1938)
7. M. Born, Rev. Mod. Phys. **21**, 463 (1969)
8. P.O. Hess, W. Greiner, Int. J. Mod. Phys. E **18**, 51 (2009)
9. G. Caspar, T. Schönenbach, P.O. Hess, M. Schäfer, W. Greiner, Int. J. Mod. Phys. E **21**, 1250015 (2012)
10. R. Genzel, R. Schödel, T. Ott, A. Eckart, T. Alexander, F. Lacombe, D. Rouan, B. Aschenbach, Nature 425, (2003)
11. B. Aschenbach, N. Grosso, D. Porquet, P. Predehl, A&A **417**(1), 71 (2004)
12. S. Gillessen, R. Genzel, T.K. Fritz, E. Quataert, C. Alig, A. Burkert, J. Cuadra, F. Eisenhauer, O. Pfuhl, K. Dodds-Eden, C.F. Gammie, T. Ott, Nature **481**, 51–54 (2012)
13. L.W. Brenneman, C.S. Reynolds, Astrophys. J. **652**(2), 1028 (2006)
14. J. Lense, H. Thirring, Phys. Z. **19**, 156 (1918)
15. I. Ciufolini, E.C. Pavlis, Nature **431**, 958 (2004)
16. C.W.F. Everitt, D.B. DeBra, B.W. Parkinson, J.P. Turneaure, J.W. Conklin, M.I. Heifetz, G.M. Keiser, A.S. Silbergleit, T. Holmes, J. Kolodziejczak, M. Al-Meshari, J.C. Mester, B. Muhlfelder, V.G. Solomonik, K. Stahl, P.W. Worden, W. Bencze, S. Buchman, B. Clarke, A. Al-Jadaan, H. Al-Jibreen, J. Li, J.A. Lipa, J.M. Lockhart, B. Al-Suwaidan, M. Taber, S. Wang, Phys. Rev. Lett. **106**, 221101 (2011)
17. J.M. Bardeen, W.H. Press, S.A. Teukolsky, Astrophys. J. **178**, 347 (1972)

Chapter 14
Magnetic Field Amplification in Hypermassive Neutron Stars via the Magnetorotational Instability

Daniel M. Siegel and Riccardo Ciolfi

Abstract Mergers of binary neutron stars likely lead to the formation of a hypermassive neutron star (HMNS), which is metastable and eventually collapses to a black hole. This merger scenario is thought to explain the phenomenology of short gamma-ray bursts (SGRBs). The very high energies observed in SGRBs have been suggested to stem from neutrino-antineutrino annihilation and/or from very strong magnetic fields created during or after the merger by mechanisms like the magnetorotational instability. Here, we report on results that show for the first time the development of the magnetorotational instability in HMNSs in three-dimensional, fully general-relativistic magnetohydrodynamic simulations. This instability amplifies magnetic fields exponentially and could be a vital ingredient in solving the SGRB puzzle.

14.1 Introduction

A significant fraction of neutron star–neutron star (NS–NS) binary mergers can lead to the formation of a hypermassive neutron star (HMNS) that eventually collapses to a stellar-mass black hole, surrounded by a hot and dense accretion torus (e.g. [1, 2]). Besides being among the most promising sources for the first direct detection of gravitational waves with advanced ground based interferometers such as Advanced LIGO and Virgo [3, 4], the inspiral and coalescence of NS–NS binaries is also thought to be the progenitor system for short gamma-ray bursts (SGRBs), the most luminous explosions observed in the universe (see, e.g., [5, 6] for a review). The association of SGRBs with NS–NS coalescence is supported on both observational [7, 8] and theoretical [9, 10] grounds. The observed SGRB fluxes, their cosmological distances and their duration of <2 s require highly relativistic motion with Lorentz factors of up to

D.M. Siegel (✉) · R. Ciolfi
Max-Planck-Institut für Gravitationsphysik (Albert-Einstein-Institut),
Am Mühlenberg 1, 14476 Potsdam, Germany
e-mail: daniel.siegel@aei.mpg.de

R. Ciolfi
e-mail: riccardo.ciolfi@aei.mpg.de

P. Nicolini et al. (eds.), *1st Karl Schwarzschild Meeting on Gravitational Physics*,
Springer Proceedings in Physics 170, DOI 10.1007/978-3-319-20046-0_14

several hundreds to resolve the so-called compactness problem [5], which states that in the absence of high Lorentz factors, SGRBs would show a thermal spectrum—in contradiction to the observed non-thermal spectra. Apart from neutrino-antineutrino annihilation as one possibility, the enormous amounts of energy needed to generate such extreme velocities have been suggested to stem from strong magnetic fields produced during/after the merger process by mechanisms such as the Kelvin-Helmholtz (KH) [11–13] and the magnetorotational instability (MRI) [14, 15], or by high field strengths generated in the torus after the central black hole has formed (as in [9]). One advantage of the former mechanisms is that they do not depend on the creation and properties of a torus and that they would already act prior to the collapse. While the amount of amplification through the KH instability, triggered when the two stars enter into contact, is controversial and maybe limited to only one order of magnitude [11, 13], the MRI triggered in the metastable differentially rotating HMNS appears to constitute a promising amplification mechanism, especially in the light of recent results indicating a rather stiff equation of state [16, 17]. The latter results suggest that HMNSs are indeed a likely outcome of NS–NS mergers [18, 19], and that they are probably longer lived than previously thought, providing more time for a potential MRI to act. However, simulating the MRI in three dimensions under the extreme physical conditions of HMNS interiors is a challenge and had not been accomplished until very recently [20].

Here, we elaborate on these recent results that have shown for the first time direct evidence for the MRI in the interior of a HMNS in global, three-dimensional and fully general-relativistic magnetohydrodynamic simulations.

14.2 Capturing the MRI in HMNSs

The magnetorotational instability [21, 22] can be triggered in differentially rotating magnetized fluids [23] and refers to modes that grow exponentially from initial seed perturbations. From a linear perturbation analysis of the Newtonian MHD equations, one can estimate the characteristic timescale $\tau_{\rm MRI}$ and wavelength $\lambda_{\rm MRI}$ for the fastest-growing mode with wavevector $\mathbf{k}$ by

$$\tau_{\rm MRI} \sim \Omega^{-1}, \qquad \lambda_{\rm MRI} \sim \left(\frac{2\pi}{\Omega}\right)\left(\frac{\mathbf{B}\cdot\mathbf{e_k}}{\sqrt{4\pi\rho}}\right), \tag{14.1}$$

where Ω denotes the angular velocity of the fluid, ρ the density, $\mathbf{B}$ the magnetic field, and $\mathbf{e_k}$ the unit vector in direction of $\mathbf{k}$ [23, 24]. We note that there is no general analytic description of the MRI within general-relativistic MHD to date. Nevertheless, we can compare our fully general-relativistic numerical results with the above analytical estimates, provided that the Newtonian predictions are translated into general relativity by employing equivalence principle arguments [20].

Resolving the MRI in an MHD simulation is a challenge, as $\lambda_{\rm MRI}$ is typically much smaller than the characteristic length scale of the astrophysical system under study. Local simulations of only a small part of the system (e.g. [25]) or simulations in axisymmetry [14, 15] are usually conducted in order to render the problem computationally affordable. In the case of HMNSs, capturing the MRI for realistic (i.e. relatively low) magnetic field strengths is particularly demanding due to the extremely high densities and high angular velocities involved (cf. (14.1)).

Here, we discuss global three-dimensional simulations, which start from a typical axisymmetric and differentially rotating HMNS model of mass $M = 2.23\,M_{\odot}$ and central angular velocity $\Omega_c = 2\pi \times 7\,\text{kHz}$ (representing the outcome of a NS–NS merger) and employ a four-level nested-boxes grid hierarchy, with the finest refinement level covering the HMNS at all times. In order to capture the MRI and despite the very high resolutions used here (with finest grid spacing of $h = 44\,\text{m}$), high initial magnetic field strengths of $B^{\text{in}} = (1{-}5) \times 10^{17}$ G have to be employed, assuming that these field strengths have previously been generated by compression during the merger, the KH instability, magnetic winding and previous MRI activity. However, it is important to point out that these field strengths are still very small in terms of the average magnetic-to-fluid pressure ratio, which is between $(0.045 - 1.2) \times 10^{-2}$. In order to reduce the computational costs, a reflection symmetry across the $z = 0$ plane and a $\pi/2$ rotation symmetry around the z-axis have been applied. By performing two additional simulations, removing either the reflection symmetry or replacing the rotation symmetry by a π symmetry, we have verified that these discrete symmetries do not significantly influence our results. For instance, the relative differences for the maximum of the toroidal field strength (as plotted in Fig. 14.3) are well below 10^{-3} up to $t \approx 0.4$ ms when the star starts collapsing to a black hole.

14.3 Discussion of Simulation Results

Figure 14.1 provides a representative overview of the HMNS evolution: the initial axisymmetric configuration, which shows a highly flattened HMNS due to rapid rota-

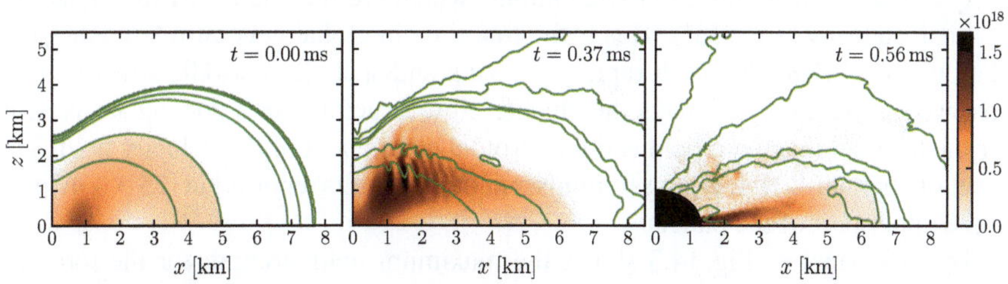

Fig. 14.1 Three characteristic stages of the HMNS evolution represented by a cut in the x–z plane, showing the colour-coded norm of the magnetic field (in G) and selected density contours: initial HMNS model, pronounced MRI development, and early post-collapse phase with a black hole (horizon is masked) surrounded by a magnetized torus

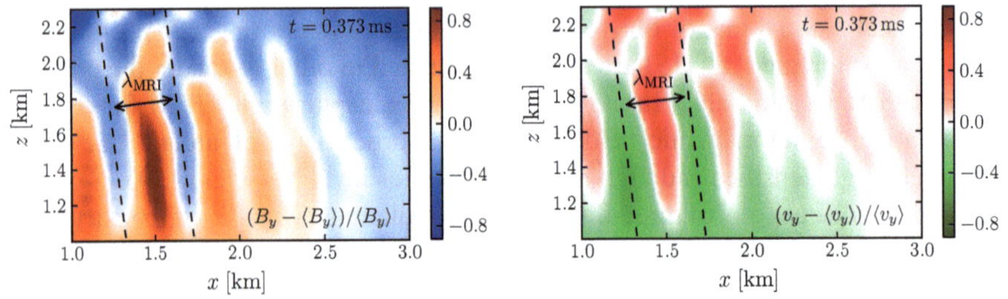

Fig. 14.2 Channel flow structures as seen in the toroidal magnetic field (*left*) and the toroidal component of the 3-velocity field (*right*). Shown are the relative deviations from the mean value in the region considered here. The wavelength of the fastest-growing MRI mode can be clearly identified

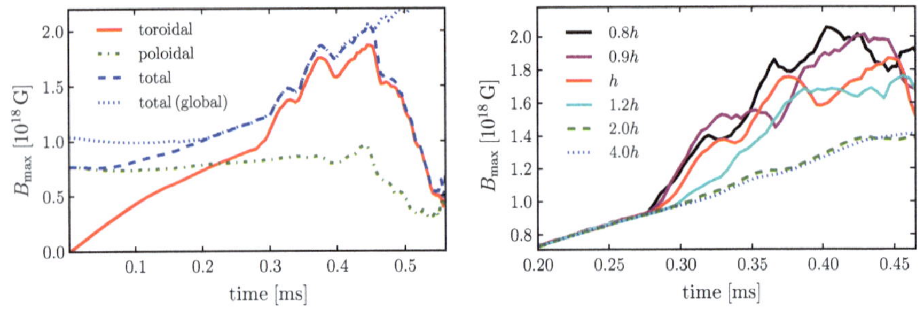

Fig. 14.3 *Left* Maximum magnetic field strength for the toroidal, poloidal and total magnetic field in the region $(x, z) = [1.0, 3.0] \times [1.0, 2.3]$ km, and the corresponding total evaluated in the entire x–z plane. *Right* Maximum toroidal magnetic field strengths in the same region for runs with different resolution (but otherwise identical), with finest grid spacings between (4–0.8) h, where $h = 44$ m refers to the fiducial value

tion; the stage of a developed MRI, indicated by the ripples in the magnetic field and density; the early post-collapse phase showing a black hole surrounded by a magnetized and geometrically thick torus. The "ripple patterns" seen in this simulation are similar to the coherent channel flow structures (WKB-like modes) observed in local axisymmetric Newtonian MRI simulations, which are the characteristic signatures of this instability (e.g. [25]). This is the first time in global general-relativistic simulations of HMNSs that such rapidly-growing and spatially-periodic structures are observed. Figure 14.2 displays the channel flow structures in a zoomed-in version for the toroidal magnetic field and the toroidal component of the velocity field. The onset of channel-flow merging (reminiscent of the results reported in [25]) is evident in the upper part of the panels.

The left panel of Fig. 14.3 shows the maximum field strength for the toroidal, poloidal and total magnetic field in the region $(x, z) \in [1.0, 3.0] \times [1.0, 2.3]$ km, where the instability develops most prominently, along with the analogous quantity for the total field evaluated over the entire x–z plane. The purely poloidal initial seed field geometry is lost very soon due to magnetic winding, which leads to a linear increase in the toroidal field strength. At $t \approx 0.3\,\text{ms}$, exponential magnetic field amplification due to the MRI sets in and lasts up to $t \approx 0.4\,\text{ms}$ when the star starts collapsing to a black hole. It is important to point out that even during this very short time frame, the MRI contributes significantly to a global magnetic field amplification of the system (compare local with global maximum of the total field).

The right panel of Fig. 14.3 demonstrates that the onset of the instability is well resolved. While any sign of the instability is absent in the case where there are less than five grid points per wavelength of the fastest-growing mode (as given by (14.1)), we gradually recover the growth rate of the fastest-growing mode with increasing resolution. The extracted values for this growth rate agree within error bars among the three highest-resolution runs, and the resulting value of $\tau_{\text{MRI}} = (8.2 \pm 0.4) \times 10^{-2}\,\text{ms}$ is in order-of-magnitude agreement with the Newtonian analytic prediction from (14.1) for the considered region once translated to our general-relativistic setting, $\tau_{\text{MRI}} = (4\text{–}5) \times 10^{-2}\,\text{ms}$. Also the wavelength of the fastest-growing mode as measured with a Fourier analysis of the magnetic field in the selected region ($\lambda_{\text{MRI}} \approx 0.4\,\text{km}$) is in order-of-magnitude agreement with the corrected Newtonian analytic prediction from (14.1) for this region, $\lambda_{\text{MRI}} \approx (0.5\text{–}1.5)\,\text{km}$. For further details and verification of additional properties of the MRI as expected from local Newtonian simulations in other astrophysical systems, we refer to [20].

14.4 Conclusion

The simulations discussed here have shown for the first time direct evidence for the MRI in HMNSs in three-dimensional, fully general-relativistic MHD simulations. This evidence manifests itself, e.g., in the presence of coherent channel flow structures, which have not been previously observed in three-dimensional general-relativistic MHD simulations. The growth of these structures coincides with exponential growth in the toroidal field strength. Furthermore, the two characteristic quantities of the instability, the growth time and wavelength of the fastest-growing mode, were measured directly, and the resulting values are in order-of-magnitude agreement with the simplified Newtonian analytical estimates, once the latter have been corrected for coordinate effects due to the general relativistic framework of the simulations.

It is interesting to note that these simulations also represent the first detailed observation of the MRI in the strong gravity regime, where the characteristic length scale of spacetime curvature becomes comparable to the wavelength of the fastest-growing MRI mode. Despite the fact that the existence of WKB-type modes becomes less obvious in this context, these modes are observed and the idealized Newtonian analytic description still seems to provide reasonable predictions.

The global magnetic field amplification due to the MRI leads to very strong magnetic fields that at least for some time stay in the vicinity of the newly formed black hole, and can thus potentially contribute to power a relativistic jet launched by the black hole torus system. Therefore, the presence of the MRI in HMNSs is of great astrophysical importance, as it could be a vital ingredient in solving the compactness problem of SGRBs.

One open question is to understand how much amplification can be achieved in a HMNS, before the dominant saturation mechanisms take place. In our model, the relatively short life of the HMNS limits the amplification to less than one order of magnitude. At the time of the collapse there is still no sign of saturation, which suggests the possibility of much higher magnetic field amplification in longer-lived models. This will be the focus of future studies.

Acknowledgments DMS greatly acknowledges the award of the first Karl Schwarzschild Prize sponsored by Springer for the best talk in the student section of the first Karl Schwarzschild Meeting, held in Frankfurt, Germany, July 2013. DMS also thanks the organizers of this meeting for travel support.

References

1. M. Shibata, K. Taniguchi, Phys. Rev. D **73**, 064027 (2006)
2. L. Rezzolla, L. Baiotti, B. Giacomazzo et al., Class. Quantum Grav. **27**(11), 114105 (2010)
3. G.M. Harry et al., Class. Quantum Grav. **27**, 084006 (2010)
4. T. Accadia, F. Acernese, F. Antonucci et al., Class. Quantum Grav. **28**(11), 114002 (2011)
5. T. Piran, Rev. Mod. Phys. **76**, 1143 (2004)
6. N. Gehrels, E. Ramirez-Ruiz, D.B. Fox, Ann. Rev. Astron. and Astrophys. **47**, 567 (2009)
7. S.D. Barthelmy, G. Chincarini, D.N. Burrows et al., Nature **438**, 994 (2005)
8. N. Gehrels, C.L. Sarazin, P.T. O'Brien et al., Nature **437**, 851 (2005)
9. L. Rezzolla, B. Giacomazzo, L. Baiotti et al., Astrophys. J. **732**(11), L6 (2011)
10. S. Rosswog, T. Piran, E. Nakar, Mon. Not. R. Astron. Soc. **430**, 2585 (2013)
11. R.H. Price, S. Rosswog, Science **312**, 719 (2006)
12. M. Anderson, E.W. Hirschmann, L. Lehner et al., Phys. Rev. Lett. **100**, 191101 (2008)
13. B. Giacomazzo, R. Perna, Astrophys. J. Lett. **771**, L26 (2013)
14. M.D. Duez, Y.T. Liu, S.L. Shapiro et al., Phys. Rev. D **73**, 104015 (2006)
15. M.D. Duez, Y.T. Liu, S.L. Shapiro et al., Phys. Rev. Lett. **96**(3), 031101 (2006)
16. P.B. Demorest, T. Pennucci, S.M. Ransom et al., Nature **467**, 1081 (2010)
17. J. Antoniadis, P.C.C. Freire, N. Wex et al., Science **340**, 448 (2013)
18. K. Hotokezaka, K. Kyutoku, H. Okawa et al., Phys. Rev. D **83**(12), 124008 (2011)
19. K. Hotokezaka, K. Kiuchi, K. Kyutoku et al., Phys. Rev. D **87**(2), 024001 (2013)
20. D.M. Siegel, R. Ciolfi, A.I. Harte et al., Phys. Rev. D **87**(12), 121302(R) (2013)
21. E.P. Velikhov, Sov. Phys. JETP **36**, 995 (1959)
22. S. Chandrasekhar, Proc. Natl. Acad. Sci. **46**, 253 (1960)
23. S.A. Balbus, J.F. Hawley, Astrophys. J. **376**, 214 (1991)
24. S.A. Balbus, Astrophys. J. **453**, 380 (1995)
25. M. Obergaulinger, P. Cerdá-Durán, E. Müller et al., Astron. Astrophys. **498**, 241 (2009)

Chapter 15
Extracting Information on the Equation of State from Binary Neutron Stars

Kentaro Takami, Luciano Rezzolla and Luca Baiotti

Abstract Recently Bauswein and Janka [6, 7] found that the typical frequency of a hypermassive neutron star, which is called f_2 in this paper, is a simple function of the average rest-mass density, essentially independently of the equation of state considered. While expected, this result is very important to decide the system mass from observed gravitational waves. However in their simulations, the Einstein equations were solved by assuming conformal flatness and employing a gravitational radiation-reaction scheme within a post-Newtonian framework. Besides this mathematical approximation, there is also a numerical one in the use of smooth-particle hydrodynamics code, which is well-know to be particularly dissipative and that rapidly suppresses the amplitude of the bar-mode deformation and rapidly yields to an almost axisymmetric system. Therefore we have reinvestigated the calculations in their work improving on the two approximations discussed above (*i.e.*, conformal flatness and smooth-particle hydrodynamics) to obtain an accurate description both during the inspiral and after the merger. Then we have found another typical frequency with a clear peak, which is called $f_{\rm LI}$ in this paper. Finally we show the relations between the initial masses and the $f_{\rm LI}$ and f_2 frequencies of the gravitational waves emission from a hypermassive neutron stars.

K. Takami (✉) · L. Rezzolla
Max-Planck-Institut für Gravitationsphysik (Albert-Einstein-Institut),
Am Mühlenberg 1, 14476 Potsdam, Germany
e-mail: kentaro.takami@aei.mpg.de

Luca Baiotti
Institute of Laser Engineering, Osaka University, Suita, Osaka 567-0086, Japan

P. Nicolini et al. (eds.), *1st Karl Schwarzschild Meeting on Gravitational Physics*,
Springer Proceedings in Physics 170, DOI 10.1007/978-3-319-20046-0_15

15.1 Introduction

Binary systems of compact objects, such as neutron-star binaries, neutron star-black hole binaries and binary black holes, inspiral and merge as a result of the emission of gravitational waves (GWs). They are the most promising sources of GWs in a series of advanced detectors such as LIGO [1], Virgo [2] and KAGRA [3], which will be operated within next five years. Especially binary neutron star mergers (BNSs) may be the most common source with realistic detection rate of $\sim$40 yr^{-1} [4].

It is known that GWs from hypermassive neutron stars (HMNSs) formed after the merger have a typical peak frequency (see the right panel of Fig. 15.2), which has been recently identified as the frequency of the fundamental quadrupolar fluid mode (f_2 mode) [5]. Recently, Bauswein and Janka [6, 7] found that the typical GW frequency of a HMNS is a simple function of the average mass density, essentially independent of the EOS considered, by performing a large number of simulations. This result is very important to deduce the system mass from the observed GWs.

However in their simulations, the Einstein equations were solved by assuming the conformal flatness condition (CFC) and employing a gravitational radiation-reaction scheme within a post-Newtonian framework. Besides this approximation in the equations, their results may be affected by the numerical technique they use to solve the equations of relativistic hydrodynamics, namely smoothed-particle hydrodynamics (SPH), whose well-known large dissipativity probably suppresses rapidly the amplitude of the bar-mode deformation and so rapidly leads to an almost axisymmetric system.

In this work, we reinvestigate their work by improving on the two approximations discussed above (*i.e.*, CFC and SPH) to obtain an accurate description both during the inspiral and after the merger. Unless explicitly stated, we use units in which $c = G = M_\odot = 1$.

15.2 Methodology

All of our calculations have been performed in full general relativity. The evolution of the spacetime is obtained by using a recently developed constraint-damping CCZ4 formalism [8] with a "1 + log" slicing condition and a "Gamma-driver" shift condition. The general-relativistic hydrodynamics equations are solved using the `Whisky` code [9], with the Marquina flux formula and a PPM reconstruction. The grid hierarchy, with a reflection symmetry condition across the $z = 0$ plane, a π-symmetry condition across the $x = 0$ plane and a moving-mesh refinement, is handled by the `Carpet` mesh refinement driver [10], where we have used six refinement levels, the finest having a resolution of $0.15\,M_\odot \simeq 0.221$ km. Thanks to this, we can extract accurate gravitational waveforms in the very far zone from the center of the system (typically at a radius $r_{\rm ex} = 500\,M_\odot \simeq 738$ km). In this work, we analyse only the $\ell = m = 2$ mode of GWs, which is dominated one.

In order to close the system of equations for the hydrodynamics, we employ a "hybrid" equation of state (EOS) based on a piecewise polytropic (PP) EOS augmented by a Γ-law EOS. That is, pressure and specific internal energy are

$$p = p_{\rm ad} + p_{\rm th}, \qquad \varepsilon = \varepsilon_{\rm ad} + \varepsilon_{\rm th}, \tag{15.1}$$

where

$$\begin{aligned} p_{\rm th} &= (\Gamma_{\rm th} - 1)\,\rho\,\varepsilon_{\rm th}, \\ p_{\rm ad} &= K_i\,\rho^{\Gamma_i} \quad (\rho_{i-1} \le \rho < \rho_i, \ 1 \le i \le n), \\ \varepsilon_{\rm ad} &= \varepsilon_i + \frac{K_i}{\Gamma_i - 1}\,\rho^{(\Gamma_i - 1)}, \quad K_{l+1} = K_l\,\rho_l^{(\Gamma_l - \Gamma_{l+1})} \quad (1 \le l \le n-1). \end{aligned}$$

The detail parameters used by this work are shown in Table 15.1.

We use quasi-equilibrium initial data of irrotational equal-mass BNSs with coordinate separation $D = 45$ km, which are computed by the multi-domain spectral-method code LORENE [12] under the assumption of a conformally at spacetime metric. This configurations undergoes ~3.5 orbits in the inspiral phase and then creates a HMNS.

15.3 Results

Figure 15.1 shows the spectrogram of the $\ell = m = 2$ plus polarization of GW, h_+. Clearly there are basically two dominating frequencies, one correspond to the orbital frequency in the inspiral phase ($t < 0$) and the other is the typical GW frequency (f_2 mode) of the HMNS after the merger ($t > 0$). As expected, the f_2 mode undergoes a large variation in time. This behavior is rather different from the one reported in [6, 7] (compare Fig. 15.1 with Fig. 15.2 of [6]). The waveform at 20 Mpc is shown in the left panel of Fig. 15.2. It consist of the part of the inspiral, merger, HMNS and black hole. In the right panel, we show the power spectral density (PSD) of the effective amplitude which is defined by

$$\tilde{h}(f) \equiv \sqrt{\frac{|\tilde{h}_+(f)|^2 + |\tilde{h}_\times(f)|^2}{2}}, \qquad \tilde{h}_{\rm A}(f) \equiv \begin{cases} 2\int h_{\rm A}(t)e^{-i2\pi f t}dt & (f \ge 0) \\ 0 & (f < 0) \end{cases}, \tag{15.2}$$

where "A" indicates two polarization modes, + or ×. Beside a f_2 peak, we can clearly see another peak labeled $f_{\rm LI}$ ("LI" means "Last Inspiral") in the figure. These typical peak frequencies will be observable by advanced LIGO.

Table 15.1 The values are based on parametrized PP EOSs proposed by [11]

EOS	n	K_1	Γ_1	Γ_2	Γ_3	Γ_4	ρ_1	ρ_2	ρ_3	M_{max}	R_{max}
polyt.	1	123.647	2.0	–	–	–	–	–	–	1.821	8.490
SLy	4	8.951×10^{-2}	1.357	3.005	2.988	2.851	2.367×10^{-4}	8.114×10^{-4}	1.619×10^{-3}	2.061	6.729
APR4	4	8.951×10^{-2}	1.357	2.830	3.445	3.348	2.448×10^{-4}	8.114×10^{-4}	1.619×10^{-3}	2.200	6.687
H4	4	8.951×10^{-2}	1.357	2.909	2.246	2.144	1.437×10^{-4}	8.114×10^{-4}	1.619×10^{-3}	2.028	7.858
GNH3	4	8.951×10^{-2}	1.357	2.664	2.194	2.304	1.078×10^{-4}	8.114×10^{-4}	1.619×10^{-3}	1.977	7.630
⋮	⋮	⋮	⋮				⋮			⋮	

In the thermal part, we adopt $\Gamma_{th} = \Gamma_1$ for all models. The gravitational mass M_{max} and the radius R_{max} of the maximum-mass TOV configuration are also shown

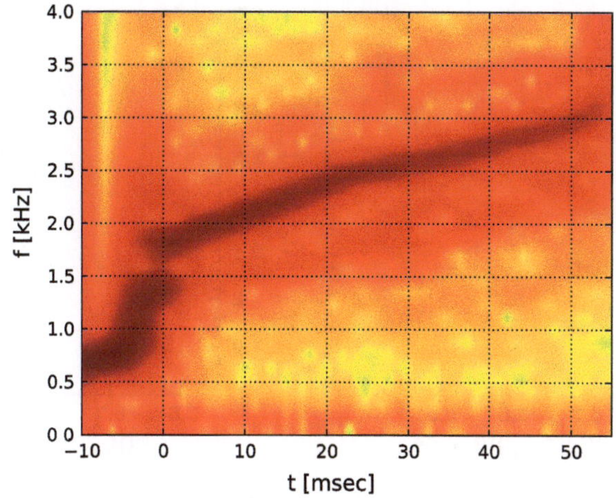

Fig. 15.1 The spectrogram of the $\ell = m = 2$ plus polarization of GW, h_+, for the model with $M_{\mathrm{ADM}} \approx 2.731 M_\odot$ and polyt. EOS, which is equivalent to two NSs with $M \approx 1.378 M_\odot$ at infinity. The waveform is aligned at the time of the merging [13]

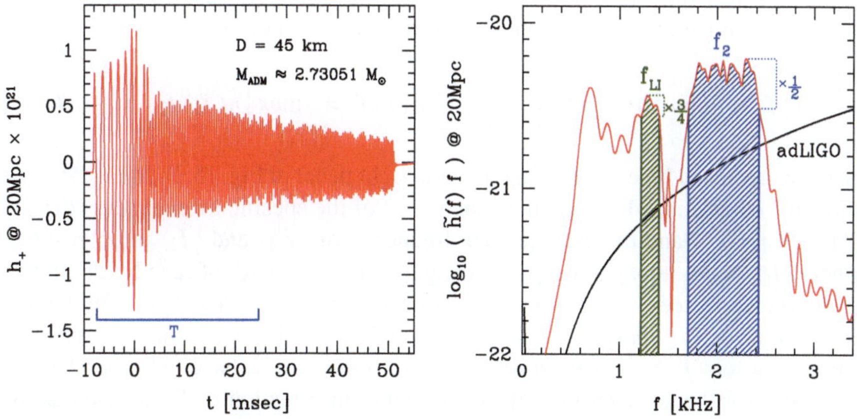

Fig. 15.2 Time evolution of the GW amplitude h_+ with the phase of inspiral, merger, HMNS and BH (*left*) and the PSD of the effective amplitude $\tilde{h}(f)f$ in the time interval $T = [-1500, 5000]\ M_\odot \approx [-7.39, 24.63]$ ms (*right*) for the same model of Fig. 15.1. In the *right panel*, the *black solid line* shows the sensitivity curves, $\sqrt{S_h(f)f}$, of advanced LIGO [14], the *green* and *blue shaded* regions show f_{L} and f_{R}, which are used for the definition of the effective peak frequencies

Because the peak frequencies change in time, we define its average as

$$\langle f \rangle = \frac{\int_{f_{\mathrm{L}}}^{f_{\mathrm{R}}} \tilde{h}(f) f^2\, df}{\int_{f_{\mathrm{L}}}^{f_{\mathrm{R}}} \tilde{h}(f) f\, df} \tag{15.3}$$

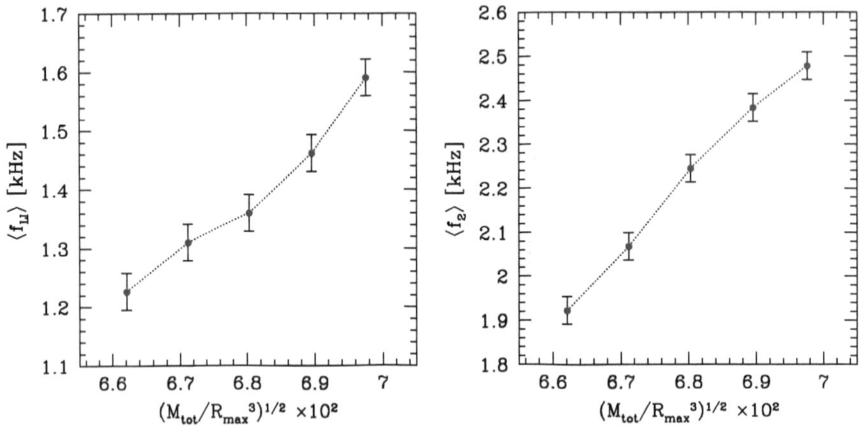

Fig. 15.3 Correlation between the square root of the specific average mass density $\sqrt{M_{\rm tot}/R_{\rm max}{}^3}$ and the average peak frequency of $\langle f_{\rm LI}\rangle$ (*left*), $\langle f_2\rangle$ (*right*)

within the time interval $T = [-1500, 5000]\ M_\odot \approx [-7.39, 24.63]$ ms, where $f_{\rm L}$ and $f_{\rm R}$ are defined by the frequencies that satisfy the condition

$$\tilde{h}(f)f \;=\; \max\left[\tilde{h}(f_{\rm LI})f_{\rm LI}\right]\times\frac{3}{4}\,, \quad \tilde{h}(f)f \;=\; \max\left[\tilde{h}(f_2)f_2\right]\times\frac{1}{2} \quad (15.4)$$

(see the green and blue shaded region in the right panel of Fig. 15.2). In Fig. 15.3 we show the correlations between the square root of the specific average mass density $(M_{\rm tot}/R_{\rm max}{}^3)^{1/2}$ and the average peak frequency of $\langle f_{\rm LI}\rangle$ and $\langle f_2\rangle$ in polyt. EOS, where $M_{\rm tot} = M_1 + M_2$ and $M_{1,2}$ are the gravitational masses of each NS at infinity. We can clearly see a simple linear relation for both $\langle f_{\rm LI}\rangle$ and $\langle f_2\rangle$. The similar correlation for $\langle f_2\rangle$ have been already pointed out by [6, 7], while we report the correlation for $\langle f_{\rm LI}\rangle$ for the fast time. Definitely both cases have tight correlations, and these can be a powerful tool to extract the information of BNSs such as the system mass from the observed GWs.

15.4 Final Remark

This work is still in progress. We will provide soon complete results for several binaries obeying a variety of hybrid EOSs, although we have reported the results for only one EOS, *i.e.*, polyt. EOS, in this paper.

Acknowledgments This work was supported in part by the DFG grant SFB/Transregio 7 and by "CompStar", a Research Networking Programme of the ESF. The simulations were performed on SuperMUC at LRZ-Munich and on Datura at AEI-Potsdam.

References

1. G.M. Harry et al., Class. Quantum Grav. **27**, 084006 (2010)
2. T. Accadia et al., Class. Quantum Grav. **28**, 114002 (2011)
3. Y. Aso, Y. Michimura, K. Somiya, M. Ando, O. Miyakawa, T. Sekiguchi, D. Tatsumi, H. Yamamoto, Phys. Rev. D **88**, 043007 (2013)
4. J. Abadie et al., Class. Quantum Grav. **27**, 173001 (2010)
5. N. Stergioulas, A. Bauswein, K. Zagkouris, H.T. Janka, Mon. Not. R. Astron. Soc. **418**, 427 (2011)
6. A. Bauswein, H.T. Janka, Phys. Rev. Lett. **108**, 011101 (2012)
7. A. Bauswein, H.T. Janka, K. Hebeler, A. Schwenk, Phys. Rev. D **86**, 063001 (2012)
8. D. Alic, C. Bona-Casas, C. Bona, L. Rezzolla, C. Palenzuela, Phys. Rev. D **85**, 064040 (2012)
9. L. Baiotti, I. Hawke, P.J. Montero, F. Löffler, L. Rezzolla, N. Stergioulas, J.A. Font, E. Seidel, Phys. Rev. D **71**, 024035 (2005)
10. E. Schnetter, S.H. Hawley, I. Hawke, Class. Quantum Grav. **21**, 146588 (2004)
11. J.S. Read, D.L. Benjamin, B.J. Owen, J.L. Friedman, Phys. Rev. D **79**, 124032 (2009)
12. http://www.lorene.obspm.fr
13. J.S. Read, L. Baiotti, J.D.E. Creighton, J.L. Friedman, B. Giacomazzo, K. Kyutoku, C. Markakis, L. Rezzolla, M. Shibata, K. Taniguchi, Phys. Rev. D **88**, 044042 (2013)
14. B.S. Sathyaprakash, B.F. Schutz, Living reviews in relativity (2009)

Part III
Black Holes in Quantum Gravity and String Theory

Chapter 16
Higher Spin AdS/CFT Correspondence and Quantum Gravity Aspects of AdS/CFT

Martin Ammon

Abstract The AdS/CFT correspondence is one of the most fascinating developments in theoretical physics of the past two decades. Using it we can get useful insights into quantum gravity aspects of non-perturbative string theory. Here, we will mainly focus on AdS/CFT dualities relating CFTs in two dimensions to higher spin gravity in three dimensions. These dualities are interesting toy-models of AdS/CFT since explicit computations can be performed on both sides and since higher spin gravity may arise as the tensionless limit of string theory. In particular, we discuss (quantum) gravity aspects of three-dimensional higher spin gravity focussing on black holes with higher spin charge. We show that due to higher spin gauge transformations the notion of an event-horizon and singularities is gauge-dependent. Moreover, we discuss how scalar correlation functions behave in this background and how a concept of spacetime may be deduced from entanglement entropy.

16.1 Introduction

In the past two decades a lot of progress was made in understanding strongly coupled quantum field theories and quantum gravity. Part of this progress is due to the discovery of AdS/CFT duality [1] which is one of the most fascinating new developments within fundamental theoretical physics in the past two decades. AdS/CFT duality is a conjecture relating quantum theories of gravity and ordinary quantum field theories containing no gravitons. The most prominent example of such a duality was discovered by Maldacena [1] and relates $\mathcal{N} = 4$ super-Yang Mills theory and type IIB string theory on $AdS_5 \times S^5$.

In particular, the conjecture states that the Hilbert space of quantum field theory is identical to the Hilbert space of the dual quantum gravity theory and that the partition function of the gravity theory is equal to the generating functional of correlation functions on the quantum field theory side. In other words, we should view ordinary

M. Ammon (✉)
Theoretisch-Physikalisches Institut, Friedrich-Schiller-Universität Jena,
Max-Wien-Platz 1, 07743 Jena, Germany
e-mail: martin.ammon@uni-jena.de

P. Nicolini et al. (eds.), *1st Karl Schwarzschild Meeting on Gravitational Physics*,
Springer Proceedings in Physics 170, DOI 10.1007/978-3-319-20046-0_16

quantum field theories and the dual quantum gravity theories as just two sides of the same coin describing the same physics albeit in seemingly very different degrees of freedom. In particular states of the Hilbert space of the CFT are encoded in the geometry of the dual gravity theory. For example, non-extremal black holes on the gravity side correspond to states of thermal equilibrium on the CFT side [2] with finite temperature given by the Hawking temperature [3] of the black hole.

In recent years a lot of evidence in favour of the AdS/CFT correspondence was found. Besides comparing correlation functions (e.g., of BPS operators) of the CFT side to its gravitational counterpart as well as computing the conformal anomaly on both sides, integrable structures were found within classical string theory on $AdS_5 \times S^5$ as well as within perturbation theory of $\mathcal{N} = 4$ super Yang-Mills theory in the large N limit. For a review and more details see the contribution of V. Forini in this volume.

Moreover, the AdS/CFT correspondence provides a powerful toolkit for studying strongly coupled quantum field theories. Much of the usefulness of the duality results from the fact that it is a strong-weak duality: In particular, in a certain limit of the AdS/CFT correspondence, the fields of the quantum field theory are strongly interacting, the ones in the gravitational theory are weakly interacting and thus more mathematically tractable. This fact has been used to study many aspects of nuclear and condensed matter physics by translating problems in those subjects into more mathematically tractable problems in string theory. For concrete applications see e.g., the articles of A. Karch and D.-W. Pang in this volume.

Finally, the duality represents a major advance in our understanding of string theory and quantum gravity. Using the framework of the AdS/CFT correspondence it is possible to study quantum gravity effects within non-perturbative string theory. It is widely believed that AdS/CFT duality should address profound questions concerning quantum gravity albeit in spacetimes with asymptotic AdS boundary conditions. For example, an immediate consequence of AdS/CFT is that the process of forming and evaporating a black hole must be unitary, since in principle such a process can be mapped to the dual gauge theory in which states evolve unitary by the standard field theory Hamiltonian.

However, we are still lacking an understanding of these dualities from first principles, in the sense that we do not know how the spacetime on the gravity side is encoded on the dual CFT side. For example, one may ask how the dual gravitational description of free field theories in arbitrary dimensions or minimal conformal field theories in two spacetime dimensions look like.

Given a conformal field theory as well as its gravitational dual description, we furthermore may ask how to describe local physics, e.g., near the event horizon of a black hole, in asymptotically AdS spacetime in terms of (presumably non-local) gauge-invariant operators on the CFT side. Answers to this question may shed light on what is wrong with Hawking's original semi-classical argument [4] that black hole evaporation leads to mixed states and thus is not unitary. And if the process is unitary, information is not lost and we have to ask ourselves how does the information get out of the black hole. Moreover, in this it is argued that drastic changes have to happen at the event horizon of the black hole—see e.g., the recent discussion of firewalls

[5]. These are questions which may be answered within AdS/CFT thus giving new insights into quantum gravity aspects of string theory.

The original conjecture due to Maldacena [1], relating $\mathcal{N} = 4$ Super-Yang Mills theory and string theory on AdS_5 is too complicated as a starting point for these questions. Thus we use solvable toy models of AdS/CFT correspondence which may be derived from the string theory and which are helpful in clarifying conceptual and technical confusions. One candidate for such a toy model of AdS/CFT, which is discussed in this article, is the duality between higher spin gravity in asymptotically AdS_3 and conformal field theories in two spacetime dimensions.

Higher spin gravity theories [6–8], studied by Vasiliev and collaborators in the past 20 years, contain an infinite set of massless higher spin fields. The classical equations of motion are known explicitly, but are highly non-linear as well as highly non-local and involve some auxiliary fields. Higher spin gravity has attracted attention within the string theory community in recent years since it contains a larger higher spin gauge symmetry which generalises the spin-2 gauge symmetry of ordinary (super-)gravity. Since we expect that string theory contains also such higher spin symmetries which may be dynamically broken, we view higher spin gravity as being in-between Einstein gravity and full-fledged string theory.

16.2 Higher Spin Gravity in AdS Spacetime

Although interacting higher spin theories are inconsistent for asymptotically flat spacetimes due to various no-go theorems, they play an important role in case of asymptotically AdS spacetimes. Here, we concentrate on three-dimensional spacetimes, since only in three dimensions the infinite tower of higher spin fields can be truncated to a finite number of higher spin fields. Hence three-dimensional higher spin theories are a good starting point for the investigation.

Unfortunately, there is no action principle known for higher spin gravity theories. However, as in ordinary spin-2 gravity with negative cosmological constant, we can reformulate it in terms of a Chern-Simons theory. The dynamics of spin-2 gravity is given by the action

$$S = \frac{1}{16\pi G_N} \int_{\mathcal{M}} d^3x \sqrt{-g} \left(R + \frac{2}{l^2} \right) , \tag{16.1}$$

where l is the radius of curvature of AdS_3. The metric may be rewritten in terms of the vielbein $e^a = e^a_\mu dx^\mu$ by $g_{\mu\nu} = e^a_\mu e^b_\nu \eta_{ab}$ and the dual spin connection is given by $\omega^a_\mu = \frac{1}{2}\varepsilon^{abc}\omega_{\mu bc}$. The action (16.1) is equivalent to the Chern-Simons action

$$S = S_{CS}[A] - S_{CS}[\overline{A}] \qquad S_{CS}[A] = \frac{k}{4\pi} \int \mathrm{Tr} \left[A \wedge dA + \frac{2}{3} A \wedge A \wedge A \right] \tag{16.2}$$

on the classical level (up to a boundary term), where we have introduced the $sl(2,\mathbb{R})$ valued gauge fields $A = A_\mu dx^\mu$ and $A = \overline{A}_\mu dx^\mu$ by

$$A_\mu = (\omega^a_\mu + e^a_\mu)\, L_a\,, \qquad \overline{A} = (\omega^a_\mu - e^a_\mu)\, L_a\,. \tag{16.3}$$

Here, L_a with $a \in \{-1, 0, 1\}$, denote the generators of $sl(2,\mathbb{R})$ satisfying $[L_a, L_b] = (a-b)L_{a+b}$, $\mathrm{Tr}[L_a L_b] = \frac{1}{2}\eta_{ab}$. The Chern-Simons level k in (16.2) can be related to Newton's constant G_N by $k = l/(4G_N)$. From now on, we set $l = 1$ for simplicity. AdS_3 in the Poincaré chart with coordinates (ρ, t, x) and metric

$$ds^2 = d\rho^2 + e^{2\rho}(-dt^2 + dx^2) \tag{16.4}$$

may be written in terms of the gauge fields $A, \overline{A} \in sl(2,\mathbb{R})$ as

$$A = A_+\, dx^+ + A_-\, dx^- + L_0\, d\rho, \qquad \overline{A} = \overline{A}_+\, dx^+ + \overline{A}_-\, dx^- - L_0\, d\rho\,, \tag{16.5}$$

where $x^\pm = t \pm x$, with components

$$A_+ = e^\rho L_1, \qquad \overline{A}_- = -e^\rho L_{-1}, \qquad A_- = \overline{A}_+ = 0\,. \tag{16.6}$$

It is straightforward to check that AdS_3 satisfies the equations of motion of the Chern-Simons action (16.2), given by $F = dA + A\wedge A = 0$, $\overline{F} = d\overline{A} + \overline{A}\wedge\overline{A} = 0$. Moreover, a spacetime is called asymptotically AdS if the gauge fields $A, \overline{A}$ satisfy

$$\begin{aligned} (A - A_{AdS})\big|_{\rho\to\infty} &= \mathcal{O}(1)\,, \qquad A_- = 0\,, \\ (\overline{A} - \overline{A}_{AdS})\big|_{\rho\to\infty} &= \mathcal{O}(1)\,, \qquad \overline{A}_+ = 0\,. \end{aligned} \tag{16.7}$$

The Chern-Simons action (16.2) is invariant under gauge transformations

$$\begin{aligned} A &\mapsto \mathscr{A} = g^{-1}\, A\, g + g^{-1}\, dg \\ \overline{A} &\mapsto \overline{\mathscr{A}} = \tilde{g}\, \overline{A}\, \tilde{g}^{-1} - d\tilde{g}\, \tilde{g}^{-1}\,, \end{aligned} \tag{16.8}$$

where g and $\tilde{g}$ are functions of spacetime coordinates and are valued in $SL(2,\mathbb{R})$. It was shown in [9] that hence the asymptotic symmetry algebra of spin-2 gravity with a negative cosmological constant consists of two copies of the Virasoro algebra.

So far we discussed only spin-2 gravity. However, the generalisation to higher spin gravity turns out to be straightforward in the frame-like notation using the gauge fields A and $\overline{A}$. Let us first consider spin-2 gravity coupled to a single spin-3 field. In this case the gauge algebra $sl(2,\mathbb{R}) \oplus sl(2,\mathbb{R})$ has to be replaced by $sl(3,\mathbb{R}) \oplus sl(3,\mathbb{R})$. Let us still denoted the generators of the principal subalgebra $sl(2,\mathbb{R})$ by L_m, with $m \in \{0, \pm 1\}$, while the remaining five generators are denoted by W_m with $m \in \{0, \pm 1, \pm 2\}$.

The action of the Chern-Simons theory is still given by (16.2) however, the corresponding action (16.1) in the metric-like formulation for this higher spin gravity is not known, although it is simple to determine the components of the metric and the spin-3 field from the (Lie-algebra valued) vielbein $e = e_\mu dx^\mu$ using

$$g_{\mu\nu} = \frac{1}{2}\,\mathrm{Tr}(e_\mu e_\nu)\,, \qquad \phi_{\mu\nu\rho} = \frac{1}{6}\mathrm{Tr}(e_{(\mu} e_\nu e_{\rho)})\,. \tag{16.9}$$

The gauge transformations are still given by (16.8) where the functions g and $\tilde{g}$ are now valued in $SL(3, \mathbb{R})$. Note that some of the gauge-transformations (namely $g, \tilde{g} \in SL(2, \mathbb{R}) \subset SL(3, \mathbb{R})$) correspond again to ordinary diffeomorphisms and frame rotations. The remaining gauge transformations of the Chern-Simons action correspond to higher spin gauge transformations and may change the causal structure of the spacetime as we discuss for black holes. This immediately raises the question what the correct notion of spacetime is in higher spin gravity.

Moreover, the asymptotically AdS boundary conditions are still given by (16.7). Analysing the asymptotic symmetries left unspecified by the boundary condition, two copies of the W-algebra W_3 were found in [10].

Generalisations: It is straightforward to generalise to higher spin gravity theories with spin s up to N or up to ∞ as indicated in the following table:

spin content	gauge algebra	asymptotic symmetry algebra
$s = 2, 3$	$sl(3,\ ) \oplus sl(3,\ )$	$W_3 \oplus W_3$
$s = 2, \ldots, N$	$sl(N,\ ) \oplus sl(N,\ )$	$W_N \oplus W_N$
$s = 2, \ldots, \infty$	$hs(\lambda) \oplus hs(\lambda)$	$W_\infty(\lambda) \oplus W_\infty(\lambda)$

The corresponding asymptotic symmetry algebras were worked out in [11–13], respectively.

Vasiliev higher spin gravity contains besides higher spin fields also a dynamical scalar field, as well as some auxiliary fields. The equations of motion will not be displayed here. Setting the scalar field constant, Vasiliev higher spin theory may be rewritten as a Chern-Simons theory with Lie-algebra $hs(\lambda) \oplus hs(\lambda)$. Moreover, neglecting the backreaction of the scalar field to the metric it is possible [14] to write down an equation of motion for the scalar field coupled to the Chern-Simons gauge fields A and $\overline{A}$.

Higher spin AdS/CFT correspondence: Consider the following two dimensional field theory, a W_N minimal model CFT, which may be represented as a coset theory with a large central charge limit as follows:

$$\frac{su(N)_k \oplus su(N)_1}{su(N)_{k+1}} \qquad k, N \to \infty, \quad \lambda = \frac{N}{k+N} \text{ fixed.} \tag{16.10}$$

According to a conjecture by Gaberdiel and Gopakumar [15] (see [16] for a refined version) this two dimensional CFT is related to three-dimensional Vasiliev gravity. This three dimensional higher spin gravity theory contains the W-algebra $W_\infty(\lambda) \oplus W_\infty(\lambda)$ as (infinite) asymptotic symmetry algebra hence matching the CFT symmetries. On both sides of the duality we can perform several checks [14, 17–20] giving evidence in favour of the conjecture.

16.3 Black Holes in 3D Higher Spin Gravity

In this section we study black holes[1] with higher spin charges in 3D higher spin gravity. Note that the BTZ black hole [23] is a solution of the higher spin theory since it can be obtained from pure AdS via discrete identifications. Hence a strategy is to use the BTZ solution as a starting point and to turn on higher spin charges.

The construction is again efficiently performed in the language of Chern-Simons theory. First we concentrate on $sl(3,\mathbb{R}) \oplus sl(3,\mathbb{R})$ case. Within this theory, black hole solutions carrying higher spin charges were constructed in [11, 24, 25]. The gauge connections read for a rotating higher spin black hole

$$\begin{aligned} A_+ &= e^{\rho} L_1 - \frac{2\pi}{k}\mathscr{L}\, e^{-\rho} L_{-1} - \frac{\pi}{2k}\mathscr{W}\, e^{-2\rho} W_{-2} \\ A_- &= \mu\left(e^{2\rho} W_2 - \frac{4\pi}{k}\mathscr{L}\, W_0 + \frac{4\pi^2}{k^2}\mathscr{L}^2 e^{-2\rho} W_{-2} + \frac{4\pi}{k}\mathscr{W}\, e^{-\rho} L_{-1}\right) \sim \mu A_+^2 \\ \overline{A}_- &= -\left(e^{\rho} L_{-1} - \frac{2\pi}{k}\overline{\mathscr{L}}\, e^{-\rho} L_1 - \frac{\pi}{2k}\overline{\mathscr{W}}\, e^{-2\rho} W_2\right) \\ \overline{A}_+ &= -\overline{\mu}\left(e^{2\rho} W_{-2} - \frac{4\pi}{k}\overline{\mathscr{L}}\, W_0 + \frac{4\pi^2}{k^2}\overline{\mathscr{L}}^2 e^{-2\rho} W_2 + \frac{4\pi}{k}\overline{\mathscr{W}}\, e^{-\rho} L_1\right) \sim \overline{\mu}\overline{A}_-^2 \end{aligned} \qquad (16.11)$$

For the non-rotating case, which we will consider from now on, $\overline{\mathscr{L}} = \mathscr{L}, \overline{\mathscr{W}} = -\mathscr{W}, \overline{\mu} = -\mu$. $\mathscr{L}$ corresponds to the energy of the black hole, while $\mathscr{W}$ is proportional to the higher spin charge.

The functional dependence of the thermodynamic quantities $\mathscr{L}$, and $\mathscr{W}$ on the inverse temperature β and the chemical potential μ can be determined by assuming that black hole thermodynamics and in particular its first law also hold for higher spin black holes. These requirements lead to the holonomy condition [11, 24] associated with the non-contractible time-like circle, stating that

$$\omega = 2\pi(\tau A_+ - \overline{\tau} A_-) \qquad \overline{\omega} = 2\pi(\tau \overline{A}_+ - \overline{\tau}\overline{A}_-) \qquad (16.12)$$

have the same eigenvalues as the BTZ black hole, namely $\{0, 2\pi i, -2\pi i\}$. In other words e^{ω} and $e^{\overline{\omega}}$ are in the center of $SL(3,\mathbb{R})$.

Using the gauge fields (16.11) it is straightforward to determine the metric of the non-rotating black hole which schematically reads $ds^2 = d\rho^2 - \mathscr{F}(\rho)dt^2 +$

[1] For a more detailed review as well as other interesting classical solutions see [21, 22].

$\mathcal{G}(\rho)d\phi^2$. It turns out that $\mathcal{F}$ and $\mathcal{G}$ are strictly positive, i.e., there is no horizon and no singularity. Hence the gauge connection (16.11) was referred to as wormhole gauge.

Note that this is not a bug of the Chern-Simons approach taken, but rather a feature of higher spin gravity (and presumably also of string theory). Since the higher spin gravity theories possess enlarged higher spin symmetry algebras that act non-trivially and unfamiliarly on the metric, black holes in such theories can be gauge transformed into traversable wormholes as shown by constructing the explicit gauge transformation in [11]. Thus the causal structures can be changed and singularities removed. Hence the very notion of geometry in these theories is not gauge invariant and must be replaced by some other more general concept. It is simultaneously exciting and puzzling to imagine how such a concept might look like.

Generalisations: So far, we considered only Chern-Simons theory based on $sl(3, \mathbb{R})$. However, the construction of higher spin black holes generalises straight-forward to other Chern-Simons theories, such as $sl(N, \mathbb{R})$ and $hs(\lambda)$. In the case of $hs(\lambda)$, the partition function and thus the entropy of the dual thermal CFT can be computed [26] and matches with the gravity calculation – adding more evidence to the conjectured higher spin AdS/CFT correspondence.

Are these really black holes? In order to answer this question we should allow for dynamical scalar fields, and hence study black hole solutions in Vasiliev higher spin theory. In particular, scalar correlation functions are of interested, where both scalar fields are inserted on the opposite conformal boundaries of the higher spin black hole in wormhole gauge. In fact, such a correlation function displays the same characteristic features as if these functions were calculated in an honest black hole background. Hence this calculation performed in [27] confirms the black hole nature of the solutions [24]. Subsequently this was also confirmed by the analogous CFT calculation [28].

Entanglement entropy in higher spin gravity: In the case of higher spin gravity with asymptotic symmetry group $W_N \oplus W_N$, there is another way to determine the entropy by considering instead the entanglement entropy of the dual CFT. The conventional Ryu-Takayanagi prescription [29] for entanglement entropy clearly requires modification in higher spin gravity. The idea of a proper distance is no longer meaningful since the line element is no longer gauge invariant under higher spin transformations.

In [30, 31] the dual gravitational description of entanglement entropy in terms of a Wilson line is a particular infinite dimensional representation of the Chern-Simons theory. The calculation of the expectation value may be rephrased in terms of a field $U(s) \in SL(n, \mathbb{R})$ living on the curve associated with the Wilson line. The conjugated momentum $P(s)$ associated to the field $U(s)$ encodes information about the representation of the Wilson line.

It is tempting to think of this proposal as defining a new, generalized notion of geometry for higher spin gravity theories, one that takes as fundamental the notion of the entanglement of the dual field theory.

16.4 Conclusion

We reviewed three dimensional higher spin gravity in AdS spacetimes and its relation to two-dimensional CFTs. In particular, we discussed black hole solutions with non-vanishing higher spin charge and its properties. A necessary condition for consistent thermodynamics of the higher spin black hole is the holonomy condition which may be viewed as a gauge-invariant characterisation of black holes.

The line element is no longer gauge invariant under higher spin transformations and thus the causal structure may change. This indeed happens and we find gauges where no event horizon and singularity is present. Coupling dynamical matter into higher spin gravity, the black hole nature of the solution is confirmed, even in the gauge where no event horizon and no singularity is present. Finally, the CFT entanglement entropy and its dual description is a promising observable for local bulk physics and to obtain the correct notion of spacetime.

Acknowledgments The work reviewed was done in collaboration with A. Castro, M. Gutperle, N. Iqbal, P. Kraus and E. Perlmutter. M. Ammon is grateful to the organizers of the 2013 Karl-Schwarzschild Meeting for providing a very stimulating conference.

References

1. J.M. Maldacena, Adv. Theor. Math. Phys. **2**, 231–252 (1998). arXiv:hep-th/9711200
2. E. Witten, Adv. Theor. Math. Phys. **2**, 505–532 (1998). arXiv:hep-th/9803131
3. S. Hawking, Commun. Math. Phys. **43**, 199–220 (1975)
4. S. Hawking, Phys. Rev. D **14**, 2460–2473 (1976)
5. A. Almheiri, D. Marolf, J. Polchinski, J. Sully, JHEP **1302**, 062 (2013)
6. E. Fradkin, M.A. Vasiliev, Nucl. Phys. B **291**, 141 (1987)
7. M.A. Vasiliev, Phys. Lett. B **243**, 378–382 (1990)
8. X. Bekaert, S. Cnockaert, C. Iazeolla, M. Vasiliev. arXiv:hep-th/0503128
9. J.D. Brown, M. Henneaux, Commun. Math. Phys. **104**, 207–226 (1986)
10. A. Campoleoni, S. Fredenhagen, S. Pfenninger, S. Theisen, JHEP **1011**, 007 (2010)
11. M. Ammon, M. Gutperle, P. Kraus, E. Perlmutter, JHEP **1110**, 053 (2011)
12. A. Campoleoni, S. Fredenhagen, S. Pfenninger, JHEP **1109**, 113 (2011)
13. M. Henneaux, S.-J. Rey, JHEP **1012**, 007 (2010)
14. M. Ammon, P. Kraus, E. Perlmutter, JHEP **1207**, 113 (2012)
15. M.R. Gaberdiel, R. Gopakumar, Phys. Rev. D **83**, 066007 (2011)
16. E. Perlmutter, T. Prochazka, J. Raeymaekers, JHEP **1305**, 007 (2013)
17. M.R. Gaberdiel, T. Hartman, JHEP **1105**, 031 (2011)
18. M.R. Gaberdiel, R. Gopakumar, T. Hartman, S. Raju, JHEP **1108**, 077 (2011)
19. C.-M. Chang, X. Yin, JHEP **1210**, 024 (2012)
20. E. Hijano, P. Kraus, E. Perlmutter, JHEP **1305**, 163 (2013)
21. A. Castro, R. Gopakumar, M. Gutperle, J. Raeymaekers, JHEP **1202**, 096 (2012)
22. M. Ammon, M. Gutperle, P. Kraus, E. Perlmutter, J. Phys. A **46**, 214001 (2013)
23. M. Banados, C. Teitelboim, J. Zanelli, Phys. Rev. Lett. **69**, 1849–1851 (1992)
24. M. Gutperle, P. Kraus, JHEP **1105**, 022 (2011)
25. A. Castro, E. Hijano, A. Lepage-Jutier, A. Maloney, JHEP **1201**, 031 (2012)
26. M.R. Gaberdiel, T. Hartman, K. Jin, JHEP **1204**, 103 (2012)
27. P. Kraus, E. Perlmutter, JHEP **1302**, 096 (2013)

28. M.R. Gaberdiel, K. Jin, E. Perlmutter, JHEP **1310**, 045 (2013)
29. S. Ryu, T. Takayanagi, Phys. Rev. Lett. **96**, 181602 (2006)
30. M. Ammon, A. Castro, N. Iqbal, JHEP **1310**, 110 (2013)
31. J. de Boer, J.I. Jottar. arXiv:1306.4347

Chapter 17
Black Holes in the Asymptotic Safety Program

Alfio Bonanno

Abstract The Asymptotic Safety scenario in quantum gravity provides a powerful framework to investigate the properties of black holes and spacetime singularities near the Planck scale. A rather robust prediction of this approach is that the final state of the black hole evaporation process is a Planck size remnant. Moreover, if the black hole is formed during a realistic gravitational collapse, quantum gravitational effects tend to suppress the inflation of the mass function at the Cauchy horizon. The scalar singularity in the future evolution of the spacetime is still present although its strength is much weaker than the classical case.

17.1 Introduction

A consistent description of the space-time singularities arising in the interior of a black hole is one the most important issues that should be addressed by a satisfactory theory of quantum gravity. Far from being a speculative question, its origin is instead deeply rooted in modern astrophysics, although the history of this problem has followed a rather peculiar path.

The very existence of black holes, although prefigured by Chandrasekhar's discovery of an upper mass limit for stable white dwarfs, was in the beginning dismissed as "ridiculous and absurd" by Eddington in 1935, and it stayed substantially unnoticed until in 1968 the first quasar was discovered. Since then, the astrophysical evidence for black holes has accumulated up the point that the debate about the existence of these objects belongs to the history of science. For instance we now have a compelling evidence that the compact radio source at the Galactic Centre, Sagittarius A*, is a black hole of about 4 million times the mass of the Sun [1], and for the first time

A. Bonanno (✉)
Istituto Nazionale di Astrofisica, Osservatorio Astrofisico di Catania,
Via S.Sofia 78, 95123 Catania, Italy
e-mail: alfio.bonanno@inaf.it

A. Bonanno
Istituto Nazionale di Fisica Nucleare, Sezione di Catania,
Via S.Sofia 72, 95123 Catania, Italy

P. Nicolini et al. (eds.), *1st Karl Schwarzschild Meeting on Gravitational Physics*,
Springer Proceedings in Physics 170, DOI 10.1007/978-3-319-20046-0_17

in the history of modern astronomy we might also have the possibility to directly observe matter falling into the black hole [2].

What happens then? At variance with the situation in Cosmology, thanks to the no-hair property, the initial value problem for the internal evolution is very well known. Near the outer horizon we have a Kerr-Newman geometry perturbed by a tail of gravitational waves that produces an inward energy flux decaying as an inverse power v^p of advanced time, where $p = 4\ell + 4$ for a multipole of order ℓ [3]. From this point of view, the study of the pre-Planckian layers emerges as a standard, albeit nontrivial, mathematical problem, not very different from following the motion of a fluid up to the onset of turbulence.

The outcome of several studies in this direction is that the final endpoint of the gravitational collapse is represented by a strong spacelike singularity at $r = 0$, *preceded* by a null, much milder singularity at the Cauchy horizon (CH) [4–9]. The properties of these distinct singularities are rather different. In the first case the singularity belongs to the future of the spacetime evolution (remember that the radial coordinate r is timelike inside a black hole), the Weyl curvature Ψ_2 becomes Planckian as $r \sim l_{\rm Pl}$, and diverges as $1/r^3$ as $r \to 0$.

In the second case the mass-inflation singularity occurs at macroscopic values of the radial coordinate for an astrophysical black hole. The Weyl curvature becomes instead Planckian as the combined inflow and outflow of gravitational waves back-scattered from the inner potential barrier accumulates along the early evolution of the CH. Clearly, a consistent theory of quantum gravity should provide a convincing picture of the spacetime which encompasses the complete internal evolution from the CH singularity to the future singularity at $r = 0$.

In recent years the Asymptotic Safety program [10] has a emerged as a possible framework to address the above questions. It implements Weinberg's proposal that quantum gravity might be nonperturbatively renormalizable if its renormalization group (RG) flow possesses a non-vanishing (in dimensionless units) fixed point at which the infinite cutoff limit can by taken [11]. This implies that the dimensionful Newton's constant reduces its strength at higher energies, it is thus *antiscreened*. The key ingredient of approach is the flow equation for the gravitational average action Γ_k, a coarse grained effective action dependent on the characteristic energy scale k where the physics is probed [12]. Since the seminal work of Reuter [13] several investigations [14–37] have accumulated a significant amount of evidence suggesting that Quantum Einstein Gravity does indeed have a non-Gaussian fixed point (NGFP) suitable for the Asymptotic Safety program. In particular the RG flow seems to be confined in a low dimensional ultraviolet critical manifold (see also [38, 39] for recent reviews).

The first model of an asymptotically safe black hole (ASBH) has been discussed in [40] where it has been shown that the causal structure of the RG-improved spacetime is similar to the one of a Reissner-Nordstrom black hole. It has also been argued that the final point of the evaporation process is a Planck size remnant [41] which could possibly be part of the Dark Matter. The possibility of a low-energy phenomenology at LHC have been discussed in [42–47], while the implications of a possible BH remnant for the Dark Matter problem have appeared in [48]. The extension of the

ASBH model to the nonspherical case is studied in [49] while ASBH in higher derivative gravity have been discussed in [50]. In recent times a study of a test scalar field propagating in an ASBH spacetime has recently put forward in [51].

The aim of this paper is to briefly review the construction of an ASBH spacetime, addressing the issue of the cutoff identification and of the resolution of spacetime singularities.

17.2 The Quantum Schwarzschild Black Hole

In its simplest form, according to AS, the the running of G is approximately given by

$$G(k) = \frac{G_0}{1 + \omega\, G_0\, k^2} \tag{17.1}$$

where G_0 denotes the laboratory value of Newton's constant, and ω is a positive number of the order unity. At large distances ($k \to 0$), $G(k)$ approaches G_0, and in the ultraviolet limit ($k \to \infty$), it decreases as $G(k) \propto 1/k^2$. The information about the k-dependence of G can then be exploited as in the case of QED. There the screening of the classic potential $V_{\rm cl}(r) = e^2/4\pi r$ is obtained by replacing e^2 with by the running gauge coupling in the one-loop approximation $e^2(k) = e^2(k_0)[1 - b\ln(k/k_0)]^{-1}$. If the renormalization scale k is identified with the inverse of the distance r we obtain

$$V(r) = -e^2(r_0)[1 - b\ln(r_0/r) + O(e^4)]/4\pi r \tag{17.2}$$

which is the correct Uehling potential obtained with more traditional approaches.

Here the starting point is the classical Schwarzschild metric

$$ds^2 = -f(r)dt^2 + f(r)^{-1}dr^2 + r^2 d\Omega^2 \tag{17.3}$$

with $d\Omega^2 \equiv d\theta^2 + \sin^2\theta d\phi^2$ and the classical lapse function $f(r) = 1 - 2G_0M/r \equiv f_{\rm class}(r)$. The RG improvement is obtained by substituting, in $f_{\rm class}(r)$, G_0 by the r-dependent Newton constant $G(r) \equiv G(k = k(r))$ which obtains from $G(k)$ via an appropriate "cutoff identification" $k = k(r)$. In flat space the natural choice would be $k \propto 1/r$. In [40] it was argued that in the Schwarzschild background the correct choice, in leading order at least, is $k(r) = \xi/d(r)$ where ξ is a constant of the order of unity, and $d(r) \equiv \int_0^r dr' |f_{\rm class}(r')|^{-1/2}$ is the proper distance from a point with coordinate r to the center of the black hole. While the integral defining $d(r)$ can be evaluated exactly, it is sufficient to use the following approximation which becomes exact for both $r \to \infty$ and $r \to 0$:

$$d(r) = \left(\frac{r^3}{r + \gamma\, G_0\, M}\right)^{\frac{1}{2}} \tag{17.4}$$

The resulting $G(r) \equiv G(k = \xi/d(r))$ reads

$$G(r) = \frac{G_0\, r^3}{r^3 + \tilde{\omega}\, G_0\, [r + \gamma G_0 M]} \tag{17.5}$$

where $\widetilde{\omega} \equiv \omega\xi^2$ is a nonuniversal constant which cannot be obtained by renormalization group arguments alone. However, standard perturbative quantization of Einstein gravity leads to

$$V(r) = -G_0 \frac{m_1 m_2}{r}\left[1 - \frac{G_0(m_1 + m_2)}{2c^2 r} - \hat{\omega}\frac{G_0 \hbar}{r^2 c^3}\right] \tag{17.6}$$

at large distances [52], where $\hat{\omega} = 118/15\pi$. We are then led to the identification

$$\widetilde{\omega} = \hat{\omega} = \frac{118}{15\pi}. \tag{17.7}$$

for $\widetilde{\omega}$. In these equations the parameter γ has the value $\gamma = 9/2$ if one sets $k = \xi/d(r)$ as above. It turns out, however, that most of the qualitative properties of the improved metric, in particular all those related to the structure of its horizons, are fairly insensitive to the precise value of γ. In particular, $\gamma = 0$ (corresponding to $k = \xi/r$) and $\gamma = 9/2$ where found [40] to lead to rather similar results throughout. For this reason one can adopt the choice $\gamma = 0$ to simplify the calculation unless one is interested in the behavior near $r = 0$, where $G(r) = r^3/\gamma\widetilde{\omega}G_0 M + O(r^4)$.

The metric of the RG improved Schwarzschild black hole is given by the line element (17.3) with

$$f(r) = 1 - \frac{2G(r)M}{r} \tag{17.8}$$

Let us briefly list its essential features [40]:

(a) There exists a critical mass value

$$M_{\rm cr} = \sqrt{\widetilde{\omega}/G_0} = \sqrt{\widetilde{\omega}}\, m_{\rm Pl} \tag{17.9}$$

such that $f(r)$ has two simple zeros at r_- and $r_+ > r_-$ if $M > M_{\rm cr}$, one double zero at $r_+ = r_- = \sqrt{\widetilde{\omega} G_0}$ if $M = M_{\rm cr}$, and no zero at all if $M < M_{\rm cr}$. For $M > M_{\rm cr}$ the zeros are at

$$r_\pm = G_0 M\, [1 \pm \sqrt{1 - \Omega}] \tag{17.10}$$

with the convenient abbreviation

$$\Omega \equiv \frac{M_{\rm cr}^2}{M^2} = \widetilde{\omega}\left(\frac{m_{\rm Pl}}{M}\right)^2 \tag{17.11}$$

The spacetime has an outer horizon at r_+ and in inner (Cauchy) horizon at r_-. At $M_{\rm cr}$, the black hole is extremal, the two horizons coincide, and the spacetime is free from any horizon if the mass is sufficiently small, $M < M_{\rm cr}$.

(b) The Bekenstein-Hawking temperature $T_{\rm BH} = \kappa/2\pi$ is given by the surface gravity at the outer horizon, $\kappa = \frac{1}{2}f'(r_+)$. Explicitly,

$$T_{\rm BH}(M) = \frac{1}{4\pi G_0 M}\frac{\sqrt{1-\Omega}}{1+\sqrt{1-\Omega}} = \frac{1}{4\pi G_0 M_{\rm cr}}\frac{\sqrt{\Omega(1-\Omega)}}{1+\sqrt{1-\Omega}} = \frac{M_{\rm cr}}{4\pi\widetilde{\omega}}\frac{\sqrt{\Omega(1-\Omega)}}{1+\sqrt{1-\Omega}} \quad (17.12)$$

This temperature vanishes for $M \searrow M_{\rm cr}$, *i.e.* $\Omega \nearrow 1$, thus motivating the interpretation of the improved Schwarzschild metric with $M = M_{\rm cr}$ as describing a "cold" remnant of the evaporation process CITE.

(c) The energy flux from the black hole, its luminosity L, can be estimated using Stefan's law. It is given by $L = \sigma\mathscr{A}(M)T_{\rm BH}(M)^4$ where σ is a constant and $\mathscr{A} \equiv 4\pi r_+^2$ denotes the area of the outer horizon. With (17.10) and (17.12) we obtain

$$L(M) = \frac{\sigma\, M_{\rm cr}^2}{(4\pi)^3\,\widetilde{\omega}^2}\frac{\Omega(1-\Omega)^2}{[1+\sqrt{1-\Omega}]^2} \quad (17.13)$$

For a single massless field with two degrees of freedom one has $\sigma = \pi^2/60$.

(d) The $r = 0$ hypersurface is still singular although less violent than in the standard case as

$$R_{\mu\nu\rho\sigma}R^{\mu\nu\rho\sigma} \sim 1/r^2 \quad (17.14)$$

which has to be contrasted with the standard behavior $R_{\nu\nu\rho\sigma}R^{\mu\nu\rho\sigma} \sim 1/r^6$.

One should not be surprised that the singularity is still present in this approach. The RG-improved Schwarzschild metric is not in general a solution of the Einstein equation, as the Uehling potential in QED is not a solution of the original Maxwell equation, although the screening of the electric charge is very well described by the RG-improved Coulomb potential. Similarly we can argue that, as long as we are in mildly planckian region, the physical predictions of the RG-improved metric are certainly reliable, but near the singularity, in a deep transplanckian region, a more consistent approach should be followed to encode the information on the running of G into the field equations.

17.3 The Quantum Astrophysical Black Hole

We can find a clue to the above issue if we study the singularity occurring in a realistic black hole. In fact in a dynamical situation the actual cutoff might be different from a simple measure of distance. In general it is plausible to expect that, provided its strength is not negligible compared to the energy scale in a typical graviton loop, the field itself can act as a cutoff in the graviton propagator.

An instructive example of a singularity occurring in a dynamic situation which might capture the essential features of the realistic gravitational collapse can be obtained considering charged spherically symmetric spacetime with a fluid of ingoing and outgoing photons [4]. In this case it is convenient to introduce a two-dimensional reduction of the spherisymmetric Einstein equation so that the 4-metric is expressible as

$$ds^2 = g_{AB}(x^C)dx^A dx^B + r^2 d\Omega^2 \ (A, B \ldots = 0.1) \tag{17.15}$$

where (x^0, x^1) are an arbitrary pair of co-ordinates which identify a 2-sphere area $4\pi r$. The gradient of r defines the Schwarzschild function $M(x^A)$ through

$$(\nabla r)^2 = g^{AB}(\partial_A r)(\partial_B r) = 1 - 2M/r. \tag{17.16}$$

The field equations can then be reformulated as two-dimensional covariant equations involving the two scalars $r(x^A)$, $M(x^A)$. The final result is a $(1 + 1)$-dimensional wave equation for M,

$$\Box M = -16\pi^2 r^3 G^2 T_{ab} T^{ab} \tag{17.17}$$

which clarifies the nonlinearity of the problem. Here T_{ab} represent the combined contribution of outgoing and ingoing gravitational waves near the inner horizon, which, for any practical situation, can be modelled as a fluid of photons. The inner horizon is a lightlike hypersurface characterized by infinite external advanced time v but reachable in finite proper time by infalling observers. The presence of any time-dependent disturbance propagating into the hole experience unbounded blueshift near the Cauchy horizon, causing the source term in (17.17) to diverge as $v \to \infty$. In particular, the Weyl curvature behaves as

$$\Psi_2 \sim |uv|^{-p} e^{\kappa_0 (u+v)} \tag{17.18}$$

where u is an internal retarded time, calibrated so that $u = -\infty$ on the event horizon (EH), and κ_0 is the inner surface gravity of the Cauchy horizon of the static black hole spacetime.

Is there any self-regulator effect which occurs when the curvature has reached planckian values in order to stop the exponential grow of the mass function? Clearly if G decreases near the Planck scale, the divergence of the mass function is diminished, as the source term is suppressed in (17.17). It is then clear that, if the actual cutoff in this situation is the curvature itself, a self-regulator effect should be expected in the AS scenario. Thus, in this case the cutoff identification should not be provided by the proper distance as in the static case, but by the field strength itself. In the case of spherical symmetry it has been argued that the right cutoff identification is $k \propto \sqrt{|\Psi_2|}$, as discussed in [53]. Following this approach it has been shown in [53] that during the early evolution of the Cauchy horizon the mass inflation singularity is damped because of quantum gravitational effects. (See also [54] for a similar investigation in the context of a Vaidya model).

17.4 Conclusions

In recent times a number of investigations have further studied the properties of ASBH. In [55, 56] the combined effect of the running of G and Λ has been discussed. In particular in [56, 57] it was reassuring to discover that the main properties of the ASBH can also be obtained by RG-improving the field equations, following the approach outlined in [58, 59].

On the other hand in [60] an effective action which captures the physics of trans-planckian physics can be obtained from an RG-improvement at the level of the action. In particular

$$\mathscr{L}_{\rm eff}^{\rm QEG}(R) = R^2 + bR^2 \cos\left[\frac{\theta''}{2}\log\left(\frac{R}{\mu}\right)\right]\left(\frac{R}{\mu}\right)^{-\theta'} \tag{17.19}$$

where μ is an infrared renormalization scale, θ'' and θ' are the real and imaginary part of the critical exponents of the NGFP and b is a positive constant. The solution of the field equation obtained from (17.19) are characterized by an infinite number of de-Sitter spacetime with a regular interior. One can then imagine that a regular de Sitter core could be a generic feature of the interior of ASBH, although further investigations including higher truncations would be essential to demonstrate the robustness of this result.

Acknowledgments It is a pleasure to thank Piero Nicolini and all the organizers of the KSM 2013 meeting for the warm hospitality and the exciting scientific environment.

References

1. S.S. Doeleman, J. Weintroub, A.E.E. Rogers, R. Plambeck, R. Freund, R.P.J. Tilanus, P. Friberg, L.M. Ziurys, J.M. Moran, B. Corey, K.H. Young, D.L. Smythe, M. Titus, D.P. Marrone, R.J. Cappallo, D.C.J. Bock, G.C. Bower, R. Chamberlin, G.R. Davis, T.P. Krichbaum, J. Lamb, H. Maness, A.E. Niell, A. Roy, P. Strittmatter, D. Werthimer, A.R. Whitney, D. Woody **455**, 78 (2008)
2. S. Gillessen, R. Genzel, T.K. Fritz, E. Quataert, C. Alig, A. Burkert, J. Cuadra, F. Eisenhauer, O. Pfuhl, K. Dodds-Eden, C.F. Gammie, T. Ott **481**, 51 (2012)
3. R.H. Price, Phys. Rev. D **5**, 2419 (1972)
4. E. Poisson, W. Israel, Phys. Rev. D **41**, 1796 (1990)
5. A. Bonanno, S. Droz, W. Israel, S.M. Morsink, Phys. Rev. D **50**, 7372 (1994)
6. A. Bonanno, Phys. Rev. D **53**, 7373 (1996)
7. P.R. Brady, S. Droz, S.M. Morsink, Phys. Rev. D **58**(8), 084034 (1998)
8. L.M. Burko, Phys. Rev. D **66**(2), 024046 (2002)

9. A. Ori, Phys. Rev. Lett. **68**, 2117 (1992)
10. M. Niedermaier, M. Reuter, Living Rev. Relativ. **9**, 5 (2006)
11. S. Weinberg, in *General Relativity: An Einstein centenary survey*, eds. by S.W. Hawking, W. Israel (1979), pp. 790–831
12. C. Wetterich, Phys. Lett. B **301**, 90 (1993)
13. M. Reuter, Phys. Rev. D **57**, 971 (1998). doi:10.1103/PhysRevD.57.971
14. O. Lauscher, M. Reuter, Phys. Rev. D **66**(2), 025026 (2002)
15. O. Lauscher, M. Reuter, Phys. Rev. D **65**(2), 025013 (2002)
16. M. Reuter, F. Saueressig, Phys. Rev. D **65**(6), 065016 (2002)
17. M. Reuter, F. Saueressig, Phys. Rev. D **66**(12), 125001 (2002)
18. W. Souma, Prog. Theoret. Phys. **102**, 181 (1999)
19. A. Bonanno, M. Reuter, J. High Energ. Phys. **2**, 035 (2005)
20. R. Percacci, D. Perini, Phys. Rev. D **67**(8), 081503 (2003)
21. R. Percacci, D. Perini, Phys. Rev. D **68**(4), 044018 (2003)
22. A. Codello, R. Percacci, Phys. Rev. Lett. **97**(22), 221301 (2006)
23. D.F. Litim, Phys. Rev. Lett. **92**(20), 201301 (2004)
24. A. Codello, R. Percacci, C. Rahmede, Ann. Phys. **324**, 414 (2009)
25. P.F. Machado, F. Saueressig, Phys. Rev. D **77**(12), 124045 (2008)
26. M. Reuter, H. Weyer, Phys. Rev. D **79**(10), 105005 (2009)
27. M. Reuter, H. Weyer, Phys. Rev. D **80**(2), 025001 (2009)
28. D. Benedetti, P.F. Machado, F. Saueressig, Nucl. Phys. B **824**, 168 (2010)
29. J.E. Daum, U. Harst, M. Reuter, J. High Energ. Phys. **1**, 84 (2010)
30. E. Manrique, S. Rechenberger, F. Saueressig, Phys. Rev. Lett. **106**(25), 251302 (2011)
31. A. Eichhorn, H. Gies, M.M. Scherer, Phys. Rev. D **80**(10), 104003 (2009)
32. A. Eichhorn, H. Gies, Phys. Rev. D **81**(10), 104010 (2010)
33. K. Groh, F. Saueressig, J. Phys. Math. Gen. **43**, 365403 (2010)
34. D. Benedetti, F. Caravelli, J. High Energ. Phys. **6**, 17 (2012)
35. A. Bonanno, F. Guarnieri, Phys. Rev. D **86**(10), 105027 (2012)
36. D. Benedetti, F. Guarnieri. ArXiv e-prints (2013)
37. A. Nink, M. Reuter, J. High Energ. Phys. **1**, 62 (2013)
38. M. Reuter, F. Saueressig, New J. Phys. **14**(5), 055022 (2012). doi:10.1088/1367-2630/14/5/055022
39. M. Reuter, F. Saueressig, *Lecture Notes in Physics*, vol. 863, eds. by G. Calcagni, L. Papantonopoulos, G. Siopsis, N. Tsamis (Springer, Berlin, 2013). (*Lecture Notes in Physics*, vol. 863 (Springer, Berlin, 2013), p. 185)
40. A. Bonanno, M. Reuter, Phys. Rev. D **62**(4), 043008 (2000)
41. A. Bonanno, M. Reuter, Phys. Rev. D **73**(8), 083005 (2006)
42. T.G. Rizzo, Class. Quantum Gravity **23**, 4263 (2006)
43. B. Koch, M. Bleicher, H. Stöcker, Phys. Lett. B **672**, 71 (2009)
44. M. Bleicher, P. Nicolini, J. Phys. Conf. Ser. **237**(1), 012008 (2010)
45. T. Burschil, B. Koch, Soviet. J. Exp. Theor. Phys. Lett. **92**, 193 (2010)
46. K. Falls, D.F. Litim, A. Raghuraman, Int. J. Mod. Phys. A **27**, 1250019 (2012)
47. J. Mureika, P. Nicolini, E. Spallucci, Phys. Rev. D **85**(10), 106007 (2012)
48. A. Barrau, C. Féron, J. Grain, Astrophys. J. **630**, 1015 (2005)
49. M. Reuter, E. Tuiran, Phys. Rev. D **83**(4), 044041 (2011)
50. Y.F. Cai, D.A. Easson, JCAP **9**, 002 (2010)
51. D.J. Liu, B. Yang, Y.J. Zhai, X.Z. Li, Class. Quantum Gravity **29**(14), 145009 (2012)
52. J.F. Donoghue, Phys. Rev. Lett. **72**, 2996 (1994)
53. A. Bonanno, M. Reuter, Phys. Rev. D **60**(8), 084011 (1999)
54. F. Fayos, R. Torres, Class. Quantum Gravity **28**(10), 105004 (2011)
55. B. Koch, F. Saueressig. ArXiv e-prints (2013)
56. C. Contreras, B. Koch, P. Rioseco, Class. Quantum Gravity **30**(17), 175009 (2013)
57. B. Koch, C. Contreras, P. Rioseco, F. Saueressig. ArXiv e-prints (2013)
58. M. Reuter, H. Weyer, Phys. Rev. D **69**(10), 104022 (2004)
59. M. Reuter, H. Weyer, Phys. Rev. D **70**(12), 124028 (2004)
60. A. Bonanno, Phys. Rev. D **85**(8), 081503 (2012)

Chapter 18
Quantum Black Holes and Effective Quantum Gravity Approaches

Xavier Calmet

One of the most exciting developments in theoretical physics in the last 20 years has been the realization that the scale of quantum gravity could be in the TeV region instead of the usually assumed 10^{19} GeV. Indeed, the strength of gravity can be affected by the size of potential extra-dimensions [1–4] or the quantum fluctuations of a large hidden sector of particles [5]. A dramatic signal of quantum gravity in the TeV region would be the production of small black holes in high energy collisions of particles at colliders. The possibility of creating small black holes at colliders has led to some wonderful theoretical works on the formation of black holes in the collisions of particles.

Long before studying the production of such black holes in the high energy collisions of particles became fashionable, in the 1970s Penrose proved that a closed trapped surface forms when two shockwaves traveling at energies much larger than the Planck scale even when the impact parameter is non-zero. Unfortunately, he never published his work. The result was independently rediscovered by Eardley and Giddings in 2002 [6] when the high energy community started to discuss the formation of black holes at colliders. Earlier estimate of the production cross section had been done using the hoop conjecture. Some did not trust the hoop conjecture, thinking that in the collision of particles the situation was too asymmetrical to trust this conjecture. The paper of Eardley and Giddings settled the issue. Proving the formation of a closed trapped surface is enough to establish gravitational collapse and hence the formation of a black hole. This work was extended by Hsu [7] into the semi-classical region using path integral methods. One could thus claim with confidence that black holes with masses 5 to 20 times the Planck scale, depending on the model of quantum gravity, could form in the collision of particles at the CERN

X. Calmet (✉)
Physics and Astronomy, University of Sussex, Falmer, Brighton, BN1 9QH, UK
e-mail: x.calmet@sussex.ac.uk

P. Nicolini et al. (eds.), *1st Karl Schwarzschild Meeting on Gravitational Physics*, Springer Proceedings in Physics 170, DOI 10.1007/978-3-319-20046-0_18

LHC is the Planck scale was low enough. Early phenomenological studies can be found in [8–14].

However, it is obvious that even if the Planck scale was precisely at 1 TeV not many semi-classical black holes could be produced at the LHC since the center of mass energy of the collisions between the protons was at most of 8 TeV so far [15]. Even with the 14 TeV LHC, not many if any semi-classical black holes will be produced.

We thus focussed on quantum black holes, which are black holes with masses of the order of the Planck mass which could be produced copiously at the LHC or in cosmic ray experiments [16–25]. The current bound derived using LHC data on the fist quantum black hole mass if of the order of 5.3 TeV [26, 27]. Note that this bound is slightly model dependent. However, this is a clear sign that there are no quantum gravitational effects at 1 TeV.

At the time we are writing up this paper, there is actually no sign of any physics beyond the standard model in the TeV region. It thus seems that the hierarchy problem was a red herring; a light Higgs boson has been found, but there is no sign of new physics to stabilize the Higgs boson's mass. This is the second nail in the coffin for fine-tuning problems after the discovery of a small and non-zero cosmological constant without new physics to stabilize it.

Instead of trying to probe the Planck scale directly by producing small black holes directly at colliders, it is useful to think of alternative ways to probe the scale of quantum gravity. Effective field theory techniques are very powerful when we know the symmetries of the low energy action which is the case for the standard model of particle physics coupled to general relativity. Integrating out all quantum gravitational effects, we are left with an effective action which we can use to probe the scale of quantum gravity at low energies. We thus consider:

$$S=\int d^4x\,\sqrt{-g}\left[\left(\frac{1}{2}M^2+\xi H^\dagger H\right)R-\Lambda_C^4+c_1R^2+c_2R_{\mu\nu}R^{\mu\nu}+L_{SM}+O(M_\star^{-2})\right] \tag{18.1}$$

The Higgs boson H has a non-zero vacuum expectation value, $v = 246$ GeV and thus contribute to the value of the Planck scale:

$$(M^2+\xi v^2)=M_P^2\,. \tag{18.2}$$

The parameter ξ is the non-minimal coupling between the Higgs boson and space-time curvature. The three parameters c_1, c_2 and ξ are dimensionless free parameters. The Planck scale M_P is equal to 2.4335×10^{18} GeV and the cosmological constant Λ_C is of order of 10^{-3} eV. The scale of the expansion $M_\star$ is often identified with M_P but there is no necessity for that and experiments are very useful to set limits on higher dimensional operators suppressed by $M_\star$. Submillimeter pendulum tests of Newton's law [28] are used to set limits on c_1 and c_2. In the absence of accidental cancellations between the coefficients of the terms R^2 and $R_{\mu\nu}R^{\mu\nu}$, these coefficients are constrained to be less than 10^{61} [5]. It has been shown that astrophysical observations are unlikely to improve these bounds [29]. The LHC data can be used to set a limit on the value of the Higgs boson non-minimal coupling to spacetime curvature: one finds that $|\xi| > 2.6 \times 10^{15}$ is excluded at the 95 % C.L. [30]. Very little is known about higher dimensional operators. The Kretschmann scalar $K = R^{\mu\nu\rho\sigma} R_{\mu\nu\rho\sigma}$ which can be coupled to the Higgs field via $KH^\dagger H$ has been studied in [31], but it seems that any observable effect requires an anomalously large Wilson coefficient for this operator. Clearly one will have to be very creative to find a way to measure the parameters of this effective action. This is important as these terms are in principle calculable in a theory of quantum gravity and this might be the only possibility to ever probe quantum gravity indirectly.

The standard model is very, maybe even, too successful. At what energy scale can we expect it to break down? In other words, up to what energy scale can one trust the effective theory described above? We know that this effective theory does not describe dark matter, but this could be a hidden sector of particles or maybe even primordial black holes with masses of the order of the Planck mass which would not affect the effective action and our previous conclusions. It has been recently pointed out that if gravity is asymptotically safe, the effective theory (18.1) could offer a description of nature up to arbitrarily energy scale and predict the Higgs boson's mass correctly, i.e., at 126 GeV [32]. Within this framework, it is natural that instead of considering the Higgs boson as a source of the hierarchy problem, one should look at it as a solution to another type of fine-tuning issue, namely that of the initial conditions of our universe. The fine-tuning problematic at the beginning of our universe is very different from the fine-tuning problem in the standard model. The fine-tuning issue in cosmology is really an initial condition problem. Why did our universe start from such very specific initial conditions? It has been shown in [33–38] that the Higgs boson with a non-minimal coupling to the Ricci scalar could play the role of the inflaton and thus address this problem.

However, getting the right number of e-folding requires a fairly large non-minimal coupling of the order of 10^4. This large non-minimal coupling is the source of a potential issue with perturbative unitarity (see, e.g., [39–42] and references therein). Naively, unitarity seems to be violated at an energy scale of M_P/ξ in today's Higgs vacuum, while it would be violated at a scale $M_P/\sqrt{\xi}$ in the inflationary background. The breakdown of perturbative unitarity is a sign of strong dynamics or new physics which kicks in at the scale of the breakdown of perturbative unitarity, thereby restoring unitarity. However, both new physics and strong dynamics could jeopardize the flatness of the scalar potential which is needed to obtain the correct number of e-folding required to explain the flatness of our universe. It was shown in [43] that at least at one-loop the cutting relation is fulfilled which implies that perturbative unitarity is fixed by one-loop corrections. This is an example of the self-healing mechanics discussed in [44]. The implication of this calculation is that the standard model could be valid at least up to the Planck scale, and describe particle physics and inflation in one consistent framework.

Unless quantum gravity is asymptotically free, proving or disproving this remains a calculational challenge as it is a purely non-perturbative problem, the effective theory (18.1) will certainly breakdown at the scale at which quantum gravitational effects become large. The lack of success in finding a consistent theory of quantum gravity may be an indication that gravity does not need to be quantized in the usual sense, or that we are trying to quantize the wrong degrees of freedom. The metric may be something purely classical and emergent. Physics seems to be in a crisis again as in 1900 when Lord Kelvin said "There is nothing new to be discovered in physics now. All that remains is more and more precise measurement". We should just hope that, as at the start of the 20th century, we will experience a new scientific revolution. My point of view is that we may have reached the limit of what can be done within our current theoretical framework. After all, quantum field theory is still based on very classical concepts namely that of point mechanics: we specify the energy of a particle which we split into kinetic and potential energies. The couplings and masses of the standard model are nothing but proportionality constants between the kinetic terms and the potentials for the corresponding particles. Yes, we quantize the classical theory to obtain a quantum field theory, but the underlying ideas and principles are desperately classical. This may be the reason why we have been unable to make progress and to calculate some of the fundamental constants of nature such as the coefficients of our effective action (18.1). Any progress will require some bright idea. We can hope that black holes will give us some clues of how to proceed beyond the current paradigm.

Acknowledgments This work is supported in part by the European Cooperation in Science and Technology (COST) action MP0905 "Black Holes in a Violent Universe" and by the Science and Technology Facilities Council (grant number ST/J000477/1).

References

1. N. Arkani-Hamed, S. Dimopoulos, G.R. Dvali, Phys. Lett. B **429**, 263 (1998). arXiv:hep-ph/9803315
2. I. Antoniadis, N. Arkani-Hamed, S. Dimopoulos, G.R. Dvali, Phys. Lett. B **436**, 257 (1998). arXiv:hep-ph/9804398
3. L. Randall, R. Sundrum, Phys. Rev. Lett. **83**, 3370 (1999). arXiv:hep-ph/9905221
4. M. Gogberashvili, Int. J. Mod. Phys. D **11**, 1635 (2002). arXiv:hep-ph/9812296
5. X. Calmet, S.D.H. Hsu, D. Reeb, Phys. Rev. D **77**, 125015 (2008). arXiv:0803.1836 [hep-th]
6. D.M. Eardley, S.B. Giddings, Phys. Rev. D **66**, 044011 (2002). arXiv:gr-qc/0201034
7. S.D.H. Hsu, Phys. Lett. B **555**, 92 (2003). arXiv:hep-ph/0203154
8. S. Dimopoulos, G.L. Landsberg, Phys. Rev. Lett. **87**, 161602 (2001). arXiv:hep-ph/0106295
9. S.B. Giddings, S.D. Thomas, Phys. Rev. D **65**, 056010 (2002). arXiv:hep-ph/0106219
10. J.L. Feng, A.D. Shapere, Phys. Rev. Lett. **88**, 021303 (2002). arXiv:hep-ph/0109106
11. L.A. Anchordoqui, J.L. Feng, H. Goldberg, A.D. Shapere, Phys. Lett. B **594**, 363 (2004). arXiv:hep-ph/0311365
12. L.A. Anchordoqui, J.L. Feng, H. Goldberg, A.D. Shapere, Phys. Rev. D **65**, 124027 (2002). arXiv:hep-ph/0112247
13. L.A. Anchordoqui, J.L. Feng, H. Goldberg, A.D. Shapere, Phys. Rev. D **68**, 104025 (2003). arXiv:hep-ph/0307228
14. S. Hossenfelder, S. Hofmann, M. Bleicher, H. Stoecker, Phys. Rev. D **66**, 101502 (2002). arXiv:hep-ph/0109085

15. P. Meade, L. Randall, JHEP **0805**, 003 (2008). arXiv:0708.3017 [hep-ph]
16. X. Calmet, W. Gong, S.D.H. Hsu, Phys. Lett. B **668**, 20 (2008). arXiv:0806.4605 [hep-ph]
17. X. Calmet, G. Landsberg, in *Lower Dimensional Quantum Black Holes*, chapter 7, eds. by A.J. Bauer, D.G. Eiffel. Black Holes: Evolution, Theory and Thermodynamics (Nova Publishers, New York, 2012). arXiv:1008.3390 [hep-ph]
18. X. Calmet, D. Fragkakis, N. Gausmann, Eur. Phys. J. C **71**, 1781 (2011). arXiv:1105.1779 [hep-ph]
19. X. Calmet, D. Fragkakis, N. Gausmann, in *Non Thermal Small Black Holes*, chapter 8, eds. by A.J. Bauer, D.G.Eiffel. Black Holes: Evolution, Theory and Thermodynamics (Nova Publishers, New York, 2012). arXiv:1201.4463 [hep-ph]
20. X. Calmet, N. Gausmann, Non-thermal quantum black holes with quantized masses. Int. J. Mod. Phys. A **28**, 1350045 (2013). arXiv:1209.4618 [hep-ph]
21. G.L. Alberghi, L. Bellagamba, X. Calmet, R. Casadio, O. Micu, Eur. Phys. J. C **73**, 2448 (2013). arXiv:1303.3150 [hep-ph]
22. X. Calmet, M. Feliciangeli, Phys. Rev. D **78**, 067702 (2008). arXiv:0806.4304 [hep-ph]
23. X. Calmet, L.I. Caramete, O. Micu, JHEP **1211**, 104 (2012). arXiv:1204.2520 [hep-ph]
24. N. Arsene, X. Calmet, L.I. Caramete, O. Micu. arXiv:1303.4603 [hep-ph]
25. X. Calmet, A review of quantum gravity at the Large Hadron Collider. Mod. Phys. Lett. A **25**, 1553 (2010). arXiv:1005.1805 [hep-ph]
26. G. Aad et al., ATLAS Collaboration. arXiv:1311.2006 [hep-ex]
27. M.V.Savina [CMS Collaboration], Phys. Atom. Nucl. **76**, 1090 (2013) [Yad. Fiz. **76**(9), 11501159 (2013)]
28. C.D. Hoyle, D.J. Kapner, B.R. Heckel, E.G. Adelberger, J.H. Gundlach, U. Schmidt, H.E. Swanson, Phys. Rev. D **70**, 042004 (2004). arXiv:hep-ph/0405262
29. X. Calmet. arXiv:1308.6155 [gr-qc]
30. M. Atkins, X. Calmet. Phys. Rev. Lett. **110**(5), 051301 (2013). arXiv:1211.0281 [hep-ph]
31. R. Onofrio, Eur. Phys. J. C **72**, 2006 (2012). arXiv:1303.5695 [gr-qc]
32. M. Shaposhnikov, C. Wetterich, Phys. Lett. B **683**, 196 (2010). arXiv:0912.0208 [hep-th]
33. F.L. Bezrukov, M. Shaposhnikov, Phys. Lett. B **659**, 703 (2008). arXiv:0710.3755 [hep-th]
34. A.O. Barvinsky, A.Y. Kamenshchik, A.A. Starobinsky, JCAP **0811**, 021 (2008). arXiv:0809.2104 [hep-ph]
35. A.O. Barvinsky, A.Y. Kamenshchik, C. Kiefer, A.A. Starobinsky, C. Steinwachs, JCAP **0912**, 003 (2009). arXiv:0904.1698 [hep-ph]
36. A.O. Barvinsky, A.Y. Kamenshchik, C. Kiefer, A.A. Starobinsky, C. Steinwachs, Eur. Phys. J. C **72**, 2219 (2012). arXiv:0910.1041 [hep-ph]
37. A. De Simone, M.P. Hertzberg, F. Wilczek, Phys. Lett. B **678**, 1 (2009). arXiv:0812.4946 [hep-ph]
38. F. Bezrukov. arXiv:1307.0708 [hep-ph]
39. R.N. Lerner, J. McDonald, JCAP **1004**, 015 (2010). arXiv:0912.5463 [hep-ph]
40. C.P. Burgess, H.M. Lee, M. Trott, JHEP **1007**, 007 (2010). arXiv:1002.2730 [hep-ph]
41. M. Atkins, X. Calmet, Phys. Lett. B **697**, 37 (2011). arXiv:1011.4179 [hep-ph]
42. M.P. Hertzberg, JHEP **1011**, 023 (2010). arXiv:1002.2995 [hep-ph]
43. X. Calmet, R. Casadio. arXiv:1310.7410 [hep-ph]
44. U. Aydemir, M.M. Anber, J.F. Donoghue, Phys. Rev. D **86**, 014025 (2012). arXiv:1203.5153 [hep-ph]

Chapter 19
The Black Hole Uncertainty Principle Correspondence

Bernard J. Carr

Abstract The Black Hole Uncertainty Principle correspondence proposes a connection between the Uncertainty Principle on microscopic scales and black holes on macroscopic scales. This is manifested in a unified expression for the Compton wavelength and Schwarzschild radius. It is a natural consequence of the Generalized Uncertainty Principle, which suggests corrections to the Uncertainty Principle as the energy increases towards the Planck value. It also entails corrections to the event horizon size as the black hole mass falls to the Planck value, leading to the concept of a Generalized Event Horizon. One implication of this is that there could be sub-Planckian black holes with a size of order their Compton wavelength. Loop quantum gravity suggests the existence of black holes with precisely this feature. The correspondence leads to a heuristic derivation of the black hole temperature and suggests how the Hawking formula is modified in the sub-Planckian regime.

19.1 Introduction

A key feature of the microscopic domain is the Heisenberg Uncertainty Principle (HUP) which implies that the uncertainty in the position and momentum of a particle must satisfy $\Delta x > \hbar/(2\Delta p)$. It is well known that one can heuristically understand this result as reflecting the momentum transferred to the particle by the probing photon. Since the momentum of a particle of mass M is bounded by Mc, an immediate implication is that one cannot localize a particle of mass M on a scale less $\hbar/(2Mc)$. An important role is therefore played by the reduced Compton wavelength, $R_C = \hbar/(Mc)$, which can be obtained from the HUP with the substitution $\Delta x \to R$ and $\Delta p \to cM$ but without the factor of 2. In the (M, R) diagram of Fig. 19.1, the region corresponding to $R < R_C$ might be regarded as the "quantum domain" in the sense that the classical description breaks down there. A key feature of the macroscopic domain is the existence of black holes. General relativity implies that a spherically

B.J. Carr (✉)
Department of Physics and Astronomy, Queen Mary University of London,
Mile End Rd, London E1 4NS, UK
e-mail: B.J.Carr@qmul.ac.uk

P. Nicolini et al. (eds.), *1st Karl Schwarzschild Meeting on Gravitational Physics*,
Springer Proceedings in Physics 170, DOI 10.1007/978-3-319-20046-0_19

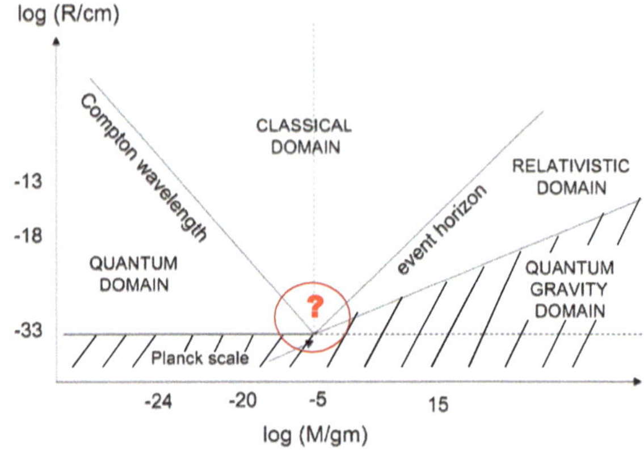

Fig. 19.1 The division of the (M, R) diagram into the classical, quantum, relativistic and quantum gravity domains

symmetric object of mass M forms an event horizon if it falls within its Schwarzschild radius, $R_S = 2GM/c^2$. The region $R < R_S$ might be regarded as the "relativistic domain" in the sense that there is no stable classical configuration in this part of Fig. 19.1.

The Compton and Schwarzschild lines intersect at around the Planck scales, $R_P = \sqrt{\hbar G/c^3} \sim 10^{-33}$ cm, $M_P = \sqrt{\hbar c/G} \sim 10^{-5}$ g, and they divide the (M, R) diagram in Fig. 19.1 into three regimes (quantum, relativistic, classical). There are several other interesting lines in this diagram. The vertical line $M = M_P$ is often assumed to mark the division between elementary particles ($M < M_P$) and black holes ($M > M_P$), because one usually requires a black hole to be larger than its own Compton wavelength. The horizontal line $R = R_P$ is significant because quantum fluctuations in the metric should become important below this. Quantum gravity effects should also be important whenever the density exceeds the Planck value, $\rho_P = c^5/(G^2\hbar) \sim 10^{94}$ g cm^{-3}, corresponding to the sorts of curvature singularities associated with the big bang or the centres of black holes. This implies $R < (M/M_P)^{1/3} R_P$, which is well above the $R = R_P$ line in Fig. 19.1 for $M \gg M_P$, so one might regard the shaded region as specifying the "quantum gravity" domain.

Although the Compton and Schwarzschild boundaries correspond to straight lines in the logarithmic plot of Fig. 19.1, this form presumably breaks down near the Planck point. As one approaches the Planck point from the left, Adler [1–4] and many others have argued that the HUP should be replaced by a Generalized Uncertainty Principle (GUP) of the form

$$\Delta x > \hbar/\Delta p + \alpha R_P^2(\Delta p/\hbar)\,. \tag{19.1}$$

Here α is a dimensionless constant (usually assumed positive) which depends on the particular model and the factor of 2 in the first term has been dropped. A heuristic argument for the second term in (19.1) is that it represents the gravitational effect of the probing photon rather than its momentum effect. This form of the GUP is indicated by the upper curve in Fig. 19.2. Variants of (19.1) can be found in other approaches to quantum gravity, such as non-commutative quantum mechanics or

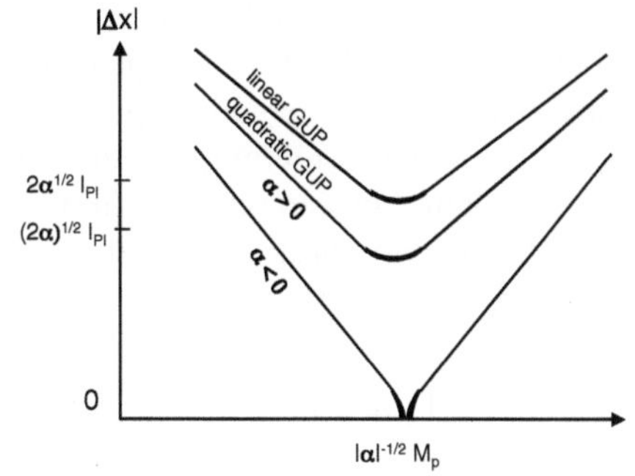

Fig. 19.2 Δx versus Δp for the GUP in its linear (*upper curve*) and quadratic (*lower curve*) forms. For α negative, the smooth minimum is replaced by a cusp at $\Delta x = 0$

general minimum length considerations [5–7]. The GUP can also be derived in loop quantum gravity because of polymer corrections in the structure of spacetime [8, 9] and it is implicit in some approaches to the problem of quantum decoherence [10, 11]. Finally, an expression resembling (19.1) arises in string theory [12–17], although it is usually assumed that the second term cannot correspond to a black hole for $M \gg M_P$ because the string is too elongated to form an horizon.

The second term on the right of (19.1) is much smaller than the first term for $\Delta p \ll M_P c$. Since it can be written as $\alpha G(\Delta p)/c^3$, it roughly corresponds to the Schwarzschild radius for an object of mass $\Delta p/c$. Indeed, if we rewrite (19.1) using the same substitution $\Delta x \to R$ and $\Delta p \to cM$ as before, it becomes

$$R > R'_C = \hbar/(Mc) + \alpha GM/c^2 = \frac{\hbar}{Mc}\left[1 + \alpha(M/M_P)^2\right]. \tag{19.2}$$

The lower limit on R might be regarded as a generalized Compton wavelength, the last term representing a small correction as one approaches the Planck point from the left. However, one can also apply (19.2) for $M \gg M_P$ and it is interesting that in this regime it asymptotes to the Schwarzschild form, apart from a numerical factor [18]. This suggests that there is a different kind of positional uncertainty for an object larger than the Planck mass, related to the existence of black holes. This is not unreasonable since the Compton wavelength is below the Planck scale (and hence meaningless) here and also an outside observer cannot localize an object on a scale smaller than its Schwarzschild radius.

The GUP also has important implications for the black hole horizon size, as can be seen by examining what happens as one approaches the intersect point from the right. In this limit, it is natural to write (19.2) as

$$R > R'_S = \frac{\alpha GM}{c^2}\left[1 + \frac{1}{\alpha}(M_P/M)^2\right] \tag{19.3}$$

and this represents a small perturbation to the Schwarzschild radius for $M \gg M_P$ if one assumes $\alpha = 2$. However, there is no reason for anticipating $\alpha = 2$ in the heuristic derivation of the GUP. Nor is it clear why a more precise calculation (within the context of a specific theory of quantum gravity) would yield this value.

This motivates an alternative approach in which the free constant in (19.2) is associated with the first term rather than the second. After all, the factor of 2 in the expression for the Schwarzschild radius is precise, whereas the coefficient associated with the Compton term is somewhat arbitrary. Thus one might rewrite (19.2) and (19.3) using the expressions

$$R'_C = \frac{\beta\hbar}{Mc}\left[1 + \frac{2}{\beta}(M/M_P)^2\right], \quad R'_S = \frac{2GM}{c^2}\left[1 + \frac{\beta}{2}(M_P/M)^2\right]. \tag{19.4}$$

for some constant β, with the latter being regarded as a Generalized Event Horizon (GEH). The mathematical equivalence of R'_C and R'_S underlies what we have termed the BHUP correspondence.

An important caveat is that (19.1) assumes the two uncertainties add linearly. On the other hand, since they are independent, it might be more natural to assume that they add quadratically:

$$\Delta x > \sqrt{(\hbar/\Delta p)^2 + (\alpha R_P^2 \Delta p/\hbar)^2}\,. \tag{19.5}$$

This corresponds to the lower curve in Fig. 19.2. While the heuristic arguments indicate the form of the two uncertainty terms, they do not specify how one combines them. We refer to (19.1) and (19.5) as the *linear* and *quadratic* forms of the GEP. Adopting the β formalism, as before, then gives a unified expression for generalized Compton wavelength and event horizon size

$$R'_C = R'_S = \sqrt{(\beta\hbar/Mc)^2 + (2GM/c^2)^2}\,, \tag{19.6}$$

leading to the approximations

$$R'_C \approx \frac{\beta\hbar}{Mc}\left[1 + \frac{2}{\beta^2}(M/M_P)^4\right], \quad R'_S \approx \frac{2GM}{c^2}\left[1 + \frac{\beta^2}{8}(M_P/M)^4\right] \tag{19.7}$$

for $M \ll M_P$ and $M \gg M_P$, respectively. These might be compared to the *exact* expressions in the linear case, given by (19.4). As shown below, the horizon size of the black hole solution in loop quantum gravity has precisely the form (19.6).

More generally, the BHUP correspondence might allow any unified expression for $R'_C(M) \equiv R'_S(M)$ which has the asymptotic behaviour $\beta\hbar/(Mc)$ for $M \ll M_P$ and $2GM/c^2$ for $M \gg M_P$. One could envisage many other unified expressions satisfying this condition but they would only be well motivated if based upon some final theory of quantum gravity. One could also consider models with $\alpha < 0$, so that one has a cusp rather than a smooth minimum in Fig. 19.2. Indeed, this may be a feature of loop quantum gravity [8, 9]. It is intriguing that $\alpha < 0$ models could have $G \to 0$ (no gravity) and $\hbar \to 0$ (no quantum discreteness) at the cusp. This relates to models, discussed at this meeting, involving "asymptotic safety" [19] and "world crystals" [20].

19.2 Loop Black Holes

Loop quantum gravity (LQG) is based on a canonical quantization of the Einstein equations written in terms of the Ashtekar variables [21–25]. One important consequence of this is that the area is quantized, with the smallest possible value being

$$a_o \equiv A_{\min}/8\pi = \sqrt{3}\,\gamma\zeta R_P^2/2\,, \tag{19.8}$$

where γ is the Immirzi parameter and ζ is another constant, both being of order 1. The other relevant constant is the dimensionless polymeric parameter δ, which (together with a_0) determines the deviation from classical theory.

One version of LQG gives a black hole solution, known as the loop black hole (LBH) [26–30], which exhibits self-duality and replaces the singularity with another asymptotically flat region. The metric in this solution depends only on the dimensionless parameter $\varepsilon \equiv \delta\gamma$, which must be small, and can be expressed as [26–30]

$$ds^2 = -G(r)c^2dt^2 + dr^2/F(r) + H(r)(d\theta^2 + \sin^2\theta d\phi^2)\,, \tag{19.9}$$

$$H = r^2 + \frac{a_o^2}{r^2},\ G = \frac{(r-r_+)(r-r_-)(r+r_*)^2}{r^4 + a_o^2},\ F = \frac{(r-r_+)(r-r_-)r^4}{(r+r_*)^2(r^4+a_o^2)}.$$

Here $r_+ = 2Gm/c^2$ and $r_- = 2GmP^2/c^2$ are the outer and inner horizons, respectively, and $r_* \equiv \sqrt{r_+ r_-} = 2GmP/c^2$, where m is the black hole mass and

$$P \equiv \frac{\sqrt{1+\varepsilon^2}-1}{\sqrt{1+\varepsilon^2}+1} \tag{19.10}$$

is the polymeric function. For $\varepsilon \ll 1$, we have $P \approx \varepsilon^2/4 \ll 1$, so $r_- \ll r_* \ll r_+$.

In the limit $r \to \infty$, $H(r) \approx r^2$, so r is the usual radial coordinate and $F(r) \approx G(r) \approx 1 - 2GM/(c^2 r)$ where $M = m(1+P)^2$ is the ADM mass. However, the exact expression for $H(r)$ shows that the physical radial coordinate $R = \sqrt{H}$ decreases from ∞ to a minimum $\sqrt{2a_0}$ at $r = \sqrt{a_0}$ and then increases again to ∞ as r decreases from ∞ to 0. In particular, the value of R associated with the event horizon is

$$R_{EH} = \sqrt{H(r_+)} = \sqrt{\left(\frac{2Gm}{c^2}\right)^2 + \left(\frac{a_o c^2}{2Gm}\right)^2}\,. \tag{19.11}$$

Apart from P terms, relating m and M, this is equivalent to (19.6), asymptoting to the Schwarzschild radius for $m \gg M_P$ and the Compton wavelength for $m \ll M_P$ if we put $\beta = \sqrt{3}\gamma\zeta/4$.

The important point is that central singularity of the Schwarzschild solution is replaced with another asymptotic region, so the black hole becomes a wormhole. Metric (19.9) has three other important consequences: (1) it permits the existence of black holes with $m \ll M_P$; (2) it implies a duality between the $m < M_P$ and $m > M_P$ solutions; (3) it involves a unified expression for the Compton and Schwarzschild scales, with expression (19.11) suggesting the quadratic GUP. Further details can be found in [18] and [26–30].

19.3 GUP and Black Hole Thermodynamics

Let us first recall the link between black hole radiation and the HUP [31, 32]. This arises because we can obtain the black hole temperature for $M \gg M_P$ by identifying Δx with the Schwarzschild radius and Δp with a multiple of the black hole temperature:

$$kT = \eta c \Delta p = \frac{\eta \hbar c}{\Delta x} = \frac{\eta \hbar c^3}{2GM} . \tag{19.12}$$

This gives the precise Hawking temperature if we take $\eta = 1/(4\pi)$. The second equality in (19.12) relates to the emitted particle and assumes that Δx and Δp satisfy the HUP. The third equality relates to the black hole and assumes that Δx is the Schwarzschild radius. Both these assumptions require $M \gg M_P$ but the GUP and GEH suggest how they should be modified for $M \ll M_P$.

Adler et al. [1–4] calculate the modification required if Δx is still associated with the Schwarzschild radius but Δp and Δx are related by the linear form of the GUP. In this case, one obtains a temperature

$$T = \frac{\eta M c^2}{\alpha k}\left(1 - \sqrt{1 - \frac{\alpha M_P^2}{M^2}}\right) \approx \frac{\eta \hbar c^3}{2GkM}\left[1 + \frac{\alpha M_P^2}{4M^2}\right] , \tag{19.13}$$

where the last expression applies for $M \gg M_P$ and just represents a small perturbation to the standard Hawking temperature. However, as indicated in Fig. 19.3, the exact expression becomes complex when M falls below $\sqrt{\alpha}\, M_P$, indicating a minimum mass. If we adopt the quadratic form of the relationship between Δp and Δx, the temperature becomes

$$T = \frac{\sqrt{2}\,\eta M c^2}{\alpha k}\left(1 - \sqrt{1 - \frac{\alpha^2}{4}\left(\frac{M_P}{M}\right)^4}\right)^{1/2} \approx \frac{\eta \hbar c^3}{2GkM}\left[1 + \frac{\alpha^2}{32}\left(\frac{M_P}{M}\right)^4\right] , \tag{19.14}$$

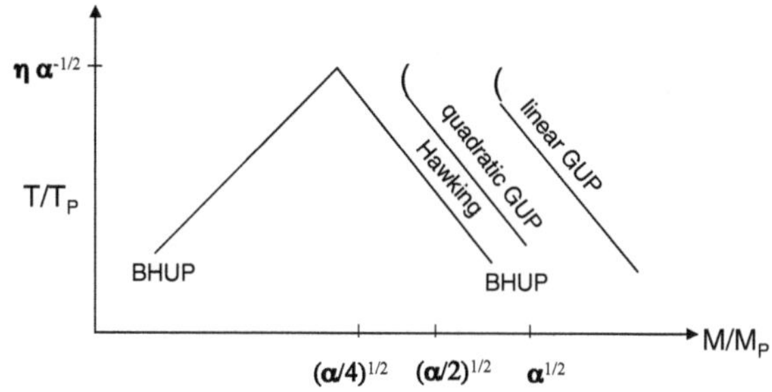

Fig. 19.3 Comparing black hole temperature predicted by Hawking, linear and quadratic GUP, BHUP correspondence

so the deviation from the Hawking prediction is smaller than implied by (19.13) but the exact expression still goes complex for $M < \sqrt{\alpha/2}\, M_P$. In either case, evaporation ceases at about the Planck mass. So the GUP stabilizes the ground state of a black hole just as the HUP stabilizes the ground state of a hydrogen atom.

The BHUP correspondence suggests that the Alder et al. argument must be modified since Δx is given by (19.11) rather than $2GM/c^2$. This makes a major qualitative difference for $M \ll M_P$ because Δx then scales as M^{-1} rather than M and this means that the temperature no longer goes complex. As shown in Fig. 19.3, one obtains the *exact* solution

$$kT = \min\left[\frac{\hbar\eta c^3}{2GM}, \frac{2\eta M c^2}{\alpha}\right]. \tag{19.15}$$

The first expression is the exact Hawking temperature, with no small correction terms. However, one must cross over to the second expression below $M = \sqrt{\alpha/4}\, M_P$ in order to avoid the temperature going above the Planck value T_P. The second expression can be obtained by putting $\Delta x \approx \hbar/(Mc)$ in (19.12). Since this is always less than T_P, the second equality in (19.12) still applies to a good approximation. The different M-dependences for $M < M_P$ and $M > M_P$ arise because there are two different asymptotic spaces in the LBH solution, corresponding to the R and r coordinates, so the quantity Δx needs to be specified more precisely. Putting $r = 2GM/c^2$ implies $(\Delta x)_R/(\Delta x)_r \approx 1$ for $M \gg M_P$ and $(M/M_P)^{-2}$ for $M \ll M_P$.

Note that one can use another argument which gives a different temperature in the sub-Planckian regime. If the temperature is determined by the black hole's surface gravity [31, 32], (19.11) suggests $T \propto GM/R_S'^2 \propto M^3$ rather than M for $M \ll M_P$. The discrepancy arises because the temperature differs in the two asymptotic spaces by a factor $(M/M_P)^2$. The GUP argument only gives the temperature in the same space as the black hole event horizon, which is our space for $M > M_P$ but the other space for $M < M_P$. So the temperature of a sub-Planckian hole scales as M^3 in our space, as predicted by the surface gravity argument, and as M in the other space, as predicted by the GUP argument [18]. Although there is no value of M for which T becomes zero, there are still *effectively* stable relics since the temperature falls below the background radiation density—suppressing evaporation altogether—below some critical mass and such relics might provide the dark matter [30].

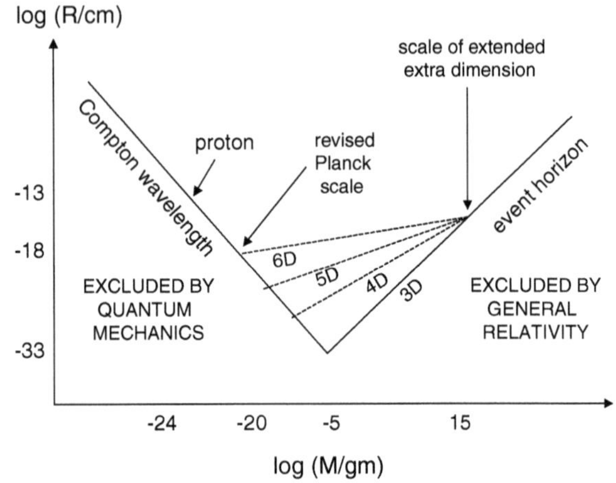

Fig. 19.4 Modification to Fig. 19.1 for various numbers of spatial dimensions with same scale

19.4 Changing the Dimensionality

The black hole boundary in Fig. 19.1 assumes there are three spatial dimensions but many theories suggest that the dimensionality could increase on small scales. Either the extra dimensions are compactified or matter is confined to a brane of finite thickness in the extra dimensions due to warping. In both cases, the extra dimensions are associated with some scale R_C. If there are n extra dimensions and the black holes with mass below $M_C = c^2 R_C/(2G)$ are assumed to be spherically symmetric in the higher dimensional space, then the Schwarzschild radius must be replaced with

$$R_S = R_C \left(\frac{M}{M_C} \right)^{1/(n+1)} \tag{19.16}$$

for $M < M_C$, so the slope of the black hole boundary in Fig. 19.1 becomes shallower, as indicated in Fig. 19.4 for various values of n. The new intersect with the Compton boundary just corresponds to the revised Planck scales. We note that $R_S \propto M^{-1}$ for 2-dimensional holes ($n = -2$). This suggests some link with the idea that physics becomes 2-dimensional (rather than higher dimensional) close to the Planck scale [33, 34], which offers an intriguing alternative interpretation of the BHUP correspondence.

References

1. R.J. Adler, D.I. Santiago, Mod. Phys. Lett. A **14**, 1371 (1999)
2. R.J. Adler, P. Chen, D.I. Santiago, Gen. Rel. Grav. **33**, 2101 (2001)
3. P. Chen, R.J. Adler. arXiv:0205106
4. R.J. Adler. arXiv:1001.1205 [gr-qc]
5. M. Maggiore, Phys. Lett. B **304**, 65 (1993)
6. M. Maggiore, Phys. Lett. B **319**, 83 (1993)
7. M. Maggiore, Phys. Rev. D **49**, 5182 (1994)
8. A. Ashtekar, S. Fiarhurst, J.L. Willis, Class. Quant. Grav. **20**, 1031 (2003)
9. G.M. Hossain, V. Husain, S.S. Seahra, Class. Quant. Grav. **207**, 165013 (2010)
10. B.S. Kay, Class. Quant. Grav. **15**, L89–L98 (1998)
11. B.S. Kay, V. Abyaneh. arXiv:0710.0992 (2007)
12. G. Veneziano, Europhys. Lett. **2**, 199 (1986)
13. E. Witten, Phys. Today April 24 (1996)
14. F. Scardigli, Phys. Lett. B **452**, 39 (1999)
15. D.J. Gross, P.F. Mende, Nuc. Phys. B **303**, 407 (1988)
16. D. Amati, M. Ciafaloni, G. Veneziano, Phys, Lett. B **216**, 41 (1989)
17. T. Yoneya, Mod. Phys. Lett. A **4**, 1587 (1989)
18. B.J. Carr, L. Modesto, I. Prémont-Schwarz. arXiv:1107.0708 [gr-qc]
19. A. Bonnano, M. Reuter, Phys. Rev. D. **73**, 083005 (2006)
20. P. Jizba, H. Kleinert, F. Scardigli, AIP Conf. Proc. **1446**, 181 (2012)
21. C. Rovelli, *Quantum Gravity* (Cambridge University Press, Cambridge, 2004)
22. A. Ashtekar, Class. Quant. Grav. **21**, R53 (2004)
23. T. Thiemann, Lect. Notes Phys. **631**, 41–135 (2003)
24. T. Thiemann, Lect. Notes Phys. **721**, 185–263 (2007)
25. A. Ashtekar, Phys. Rev. Lett. **57**, 2244–2247 (1986)
26. L. Modesto, Phys. Rev. D **70**, 124009 (2004)
27. L. Modesto, Class. Quant. Grav. **23**, 5587–5602 (2006)
28. L. Modesto, Adv. High Energy Phys. **2008**, 459290 (2008)
29. L. Modesto, Int. J. Theor. Phys (2010). arXiv:0811.2196 [gr-qc]
30. L. Modesto, I. Premont-Schwarz, Phys. Rev. D **80**, 064041 (2009)
31. S.W. Hawking, Nature **248**, 30 (1974)
32. S.W. Hawking, Comm. Math. Phys. **43**, 199 (1975)
33. J. Mureika, P. Nicolini, Eur. Phys. J. Plus **128**, 78 (2013)
34. I. Maximiliano, J. Mureika, P. Nicolini, JHEP **1311**, 139 (2013)

Chapter 20
Scattering and Unitarity Methods in Two Dimensions

Valentina Forini, Lorenzo Bianchi and Ben Hoare

Abstract The standard unitarity-cut method is applied to several massive two-dimensional models, including the world-sheet $AdS_5 \times S^5$ superstring, to compute $2 \to 2$ scattering S-matrices at one loop from tree level amplitudes. Evidence is found for the cut-constructibility of supersymmetric models, while for models without supersymmetry (but integrable) the missing rational terms can be interpreted as a shift in the coupling.

20.1 Discussion

Unitarity-based methods, whose use in four dimensions has been crucial for an efficient evaluation of scattering amplitudes [1] in non-abelian gauge theories as well as gravity theories [2], have never really been applied in two dimensions.[1] The aim of our work [6] (we refer the reader to the independent results of [7]) has been to initiate the use of unitarity methods in the perturbative study of the S-matrix for *massive two-dimensional* field theories. Limiting ourselves to the use of *standard* unitarity (therefore placing on shell only two internal lines[2]) we present a formula for the one-loop $2 \to 2$ scattering amplitude built directly from the corresponding on-shell tree-level amplitudes.

As reviewed below, we have applied our method to various models, finding enough evidence to postulate that *supersymmetric, integrable two-dimensional theories should be cut-constructible via standard unitarity methods*. For bosonic theories

[1]For the three-dimensional case see [3–5].
[2]This is nothing but the application of the optical theorem. The case where the loop amplitude issubdivided into more than two pieces is referred to as *generalized* unitarity.

V. Forini (✉) · L. Bianchi · B. Hoare
Institut für Physik und Integrative Research Institute for the Sciences IRIS Adlershof, Humboldt-Universität zu Berlin, Newtonstraße 15, 12489 Berlin, Germany
e-mail: forini@physik.hu-berlin.de

L. Bianchi
e-mail: bianchi@physik.hu-berlin.de

B. Hoare
e-mail: hoare@physik.hu-berlin.de

P. Nicolini et al. (eds.), *1st Karl Schwarzschild Meeting on Gravitational Physics*, Springer Proceedings in Physics 170, DOI 10.1007/978-3-319-20046-0_20

with integrability, we find agreement with perturbation theory up to a finite shift in the coupling.[3] We also successfully apply our method to the light-cone gauge-fixed sigma-model for the $AdS_5 \times S^5$ superstring, where—importantly—standard perturbation theory seems to fail in evaluating the S-matrix beyond the leading order due to regularization issues.

20.2 Two-Particle S-Matrix from Unitarity Cuts at One Loop

In two dimensions, the two-body scattering process of a translational-invariant field theory is described via the four-point amplitude

$$\begin{aligned}\langle \Phi^P(p_3)\Phi^Q(p_4)\,|\mathbb{S}|\,\Phi_M(p_1)\Phi_N(p_2)\rangle &= \mathscr{A}_{MN}^{PQ}(p_1,p_2,p_3,p_4)\\ &\equiv (2\pi)^2\delta^{(2)}(p_1+p_2-p_3-p_4)\,\widetilde{\mathscr{A}}_{MN}^{PQ}(p_1,p_2,p_3,p_4)\,, \quad (20.1)\end{aligned}$$

where $\mathbb{S}$ is the scattering operator, the fields Φ have on-shell momenta p_i (for us, all the particles have equal non-vanishing mass set to unity) and can carry flavor indices. Importantly, the energy-momentum conservation δ-function satisfies

$$\delta^{(2)}(p_1+p_2-p_3-p_4) = J(p_1,p_2)\left(\delta(\mathrm{p}_1-\mathrm{p}_3)\delta(\mathrm{p}_2-\mathrm{p}_4)+\delta(\mathrm{p}_1-\mathrm{p}_4)\delta(\mathrm{p}_2-\mathrm{p}_3)\right), \quad (20.2)$$

which accounts for the fact that in $d = 2$ there is no phase space, and the only thing particles can do is either preserve or exchange their momenta. Above, p is the spatial momentum, the Jacobian $J(p_1,p_2) = 1/(\partial\varepsilon_{\mathrm{p}_1}/\partial\mathrm{p}_1 - \partial\varepsilon_{\mathrm{p}_2}/\partial\mathrm{p}_2)$ depends on the dispersion relation ε_{p} (the on-shell energy associated to p) for the theory at hand, and spatial momenta are assumed to be ordered $\mathrm{p}_1 > \mathrm{p}_2$. The S-matrix elements relevant for the description of the $2 \to 2$ scattering in the two-dimensional case are then defined[4] as

$$S_{MN}^{PQ}(p_1,p_2) \equiv \frac{J(p_1,p_2)}{4\varepsilon_1\varepsilon_2}\widetilde{\mathscr{A}}_{MN}^{PQ}(p_1,p_2,p_1,p_2)\,. \quad (20.3)$$

In applying the standard unitarity rules (derived from the optical theorem) [8] to the one-loop four point amplitude (20.1) one considers *two-particle cuts*, obtained by putting two intermediate lines on-shell. The contributions that follow to the imaginary

[3] It would be interesting to analyze models which are just supersymmetric and not integrable.

[4] Without loss of generality, one can consider in (20.1) the amplitudes associated to the first product of δ-functions $\delta(\mathrm{p}_1-\mathrm{p}_3)\delta(\mathrm{p}_2-\mathrm{p}_4)$. The denominator in (20.3) is required to make contact with the standard definition of the S-matrix in two dimensions.

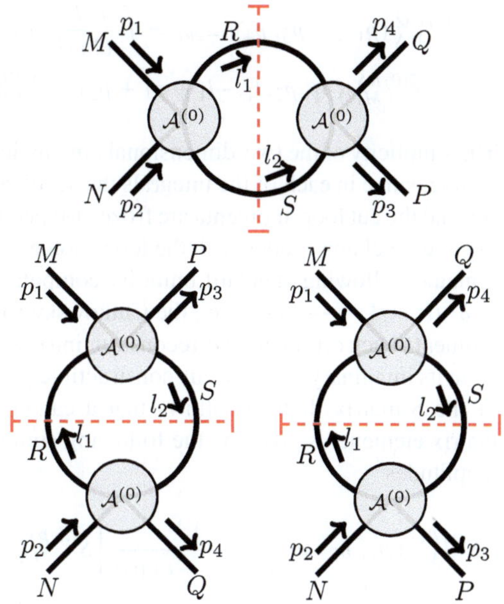

Fig. 20.1 Diagrams representing s-, t- and u-channel cuts contributing to the four-point one-loop amplitude

part of the amplitude are therefore given by the sum of s-, t- and u- channel cuts illustrated in Fig. 20.1, explicitly

$$\mathscr{A}^{(1)}{}^{PQ}_{MN}(p_1, p_2, p_3, p_4)|_{s-cut} = \int \frac{d^2 l_1}{(2\pi)^2} \int \frac{d^2 l_2}{(2\pi)^2}\, i\pi\delta^+(l_1{}^2 - 1)\, i\pi\delta^+(l_2^2 - 1) \times \mathscr{A}^{(0)}{}^{RS}_{MN}(p_1, p_2, l_1, l_2)\mathscr{A}^{(0)}{}^{PQ}_{SR}(l_2, l_1, p_3, p_4) \quad (20.4)$$

$$\mathscr{A}^{(1)}{}^{PQ}_{MN}(p_1, p_2, p_3, p_4)|_{t-cut} = \int \frac{d^2 l_1}{(2\pi)^2} \int \frac{d^2 l_2}{(2\pi)^2}\, i\pi\delta^+(l_1{}^2 - 1)\, i\pi\delta^+(l_2{}^2 - 1) \times \mathscr{A}^{(0)}{}^{SP}_{MR}(p_1, l_1, l_2, p_3)\mathscr{A}^{(0)}{}^{RQ}_{SN}(l_2, p_2, l_1, p_4) \quad (20.5)$$

$$\mathscr{A}^{(1)}{}^{PQ}_{MN}(p_1, p_2, p_3, p_4)|_{u-cut} = \int \frac{d^2 l_1}{(2\pi)^2} \int \frac{d^2 l_2}{(2\pi)^2}\, i\pi\delta^+(l_1{}^2 - 1)\, i\pi\delta^+(l_2{}^2 - 1) \times \mathscr{A}^{(0)}{}^{SQ}_{MR}(p_1, l_1, l_2, p_4)\mathscr{A}^{(0)}{}^{RP}_{SN}(l_2, p_2, l_1, p_3) \quad (20.6)$$

where $\mathscr{A}^{(0)}$ are tree-level amplitudes and a sum over the complete set of intermediate states R, S (all allowed particles for the cut lines) is understood. Notice that tadpole graphs, having no physical two-particle cuts, are by definition ignored in this procedure.

To proceed, in each case one uses (20.1) and the momentum conservation at the vertex involving the momentum p_1 to integrate over l_2, e.g., for the s-channel

$$\mathscr{A}^{(1)}{}^{PQ}_{MN}(p_1, p_2, p_3, p_4)|_{s-cut} = \int \frac{d^2 l_1}{(2\pi)^2}\, i\pi\delta^+(l_1{}^2 - 1)\, i\pi\delta^+((l_1 - p_1 - p_2)^2 - 1)$$
$$\times\, \mathscr{A}^{(0)}{}^{RS}_{MN}(p_1, p_2, l_1, -l_1 + p_1 + p_2)\, \mathscr{A}^{(0)}{}^{PQ}_{SR}(-l_1 + p_1 + p_2, l_1, p_3, p_4)\,, \quad (20.7)$$

The simplicity of the two-dimensional kinematics and of being at one loop plays now its role, since in each of the integrals the set of zeroes of the δ-functions is a discrete set, and the cut loop-momenta are frozen to specific values.[5] This allows us to pull out the tree-level amplitudes with the loop-momenta evaluated at those zeroes.[6] In what remains, following standard unitarity computations [8], we apply the replacement $i\pi\delta^+(l^2 - 1) \longrightarrow \frac{1}{l^2-1}$ (i.e., the Cutkowsky rule in reverse order) which sets loop momenta back off-shell, thus reconstructing scalar bubbles. This allows us to rebuild, from its imaginary part, the cut-constructible piece of the amplitude and, via (20.3),[7] of the S-matrix. It then follows that a candidate expression for the one-loop S-matrix elements is given by the following simple sum of products of two tree-level amplitudes[8]

$$S^{(1)}{}^{PQ}_{MN}(p_1, p_2) = \frac{1}{4(\varepsilon_2\, \mathrm{p}_1 - \varepsilon_1\, \mathrm{p}_2)} \Big[\tilde{S}^{(0)}{}^{RS}_{MN}(p_1, p_2)\tilde{S}^{(0)}{}^{PQ}_{RS}(p_1, p_2)\, I_{p_1+p_2} \quad (20.8)$$
$$+\, \tilde{S}^{(0)}{}^{SP}_{MR}(p_1, p_1)\tilde{S}^{(0)}{}^{RQ}_{SN}(p_1, p_2)\, I_0 + \tilde{S}^{(0)}{}^{SQ}_{MR}(p_1, p_2)\tilde{S}^{(0)}{}^{PR}_{SN}(p_1, p_2)\, I_{p_1-p_2} \Big]$$

where the coefficients are given in terms of the bubble integral

$$I_p = \int \frac{d^2 q}{(2\pi)^2} \frac{1}{(q^2 - 1 + i\varepsilon)((q - p)^2 - 1 + i\varepsilon)} \quad (20.9)$$

and read explicitly

$$I_{p_1+p_2} = \frac{i\pi - \mathrm{arsinh}(\varepsilon_2\, \mathrm{p}_1 - \varepsilon_1\, \mathrm{p}_2)}{4\pi i\, (\varepsilon_2\, \mathrm{p}_1 - \varepsilon_1\, \mathrm{p}_2)}, \qquad I_0 = \frac{1}{4\pi i}, \qquad I_{p_1-p_2} = \frac{\mathrm{arsinh}(\varepsilon_2\, \mathrm{p}_1 - \varepsilon_1\, \mathrm{p}_2)}{4\pi i\, (\varepsilon_2\, \mathrm{p}_1 - \varepsilon_1\, \mathrm{p}_2)}\,.$$

Few important remarks are in order:

(a) Since the unitarity-cut procedure only ensures the correctness of logarithmic terms (in general, of those terms associated to branch-cut singularities, typically

---

[5] At two loops, to constrain completely the four components of the two momenta circulating in the loops one needs four cuts, each one giving an on-shell δ-function. Two-particle cuts at two loops would result in a manifold of conditions for the loop momenta.

[6] This is like using $f(x)\delta(x - x_0) = f(x_0)\delta(x - x_0)$ where $f(x)$ are the tree-level amplitudes in the integrals.

[7] This corresponds to the choice $p_3 = p_1$, $p_4 = p_2$.

[8] In (20.8), $\tilde{S}^{(0)}(p_1, p_2) = 4(\varepsilon_2\, \mathrm{p}_1 - \varepsilon_1\, \mathrm{p}_2)S^{(0)}(p_1, p_2)$ and the denominator on the right-hand side comes from the Jacobian $J(p_1, p_2)$ assuming a standard relativistic dispersion relation (for the theories we consider, at one-loop this is indeed the case).

logarithms or polilogarithms), the proposal (20.8) and its fermionic generalization [6] crucially need to be tested on known examples.[9]

(b) The t-channel cut requires a prescription, since if one first uses the δ-function identity (20.2) to fix, for example, $p_1 = p_3$ and $p_2 = p_4$ the corresponding integral is ill-defined. To avoid this ambiguity we follow the prescription that we should only impose the δ-function identity at the end.[10] Furthermore, if we choose the alternative solution of the conservation δ-function in (20.5), namely $\ell_2 = \ell_1 + p_4 - p_2$, the coefficient of $I(0)$ in (20.8) would be different, which leads to the consistency condition on the tree-level S-matrix[11]

$$\tilde{S}^{(0)}{}^{SP}_{MR}(p_1, p_1)\,\tilde{S}^{(0)}{}^{RQ}_{SN}(p_1, p_2) = \tilde{S}^{(0)}{}^{PS}_{MR}(p_1, p_2)\,\tilde{S}^{(0)}{}^{QR}_{SN}(p_2, p_2)\,. \quad (20.10)$$

We have checked this for the tree-level S-matrices of all the field theory models treated below.

(c) As they only involve the scalar bubble integral in two dimensions, the result (20.8) following from our procedure is inherently *finite*. No additional regularization is required and the result can be compared directly with the $2 \to 2$ particle S-matrix (following from the finite or renormalized four-point amplitude) found using standard perturbation theory. Of course, this need not be the case for the original bubble integrals before cutting—due to factors of loop-momentum in the numerators. These divergences, along with those coming from tadpole graphs, which we did not consider, should be taken into account for the renormalization of the theory. We have not investigated this issue, since all the theories we consider below are either UV-finite or renormalizable.

To explore the validity of the procedure outlined we have considered both relativistic and non-relativistic (world-sheet field theory for the $\mathrm{AdS}_5 \times S^5$ superstring) models.

20.3 Relativistic models

In the *relativistic, bosonic case*, we looked at a class of generalized sine-Gordon models [9, 10], theories defined by a gauged WZW model for a coset G/H plus a potential, whose classical integrability can be demonstrated through the existence of a Lax connection. Considering the coset $G/H = \mathrm{SO}(n+1)/\mathrm{SO}(n)$, where asymptotic excitations are a free $\mathrm{SO}(n)$ vector with unit mass (which is the case we have considered in our general procedure), this class includes the sine-Gordon ($n = 1$) and complex sine-Gordon ($n = 2$) models, for which the exact S-matrices are known

---

[9]Because its bubble integral I_0 can only contribute to rational terms, the t-channel contribution has been neglected in [7], where all rational terms were determined from symmetry considerations.

[10]In some sense this is natural as, in general dimensions, quantum field theory amplitudes have the form (20.1), while the δ-function identity (20.2) is specific to two dimensions.

[11]See [6] for the generalization to the case which includes fermions.

[11, 12]. In all the cases the one-loop S-matrix got via unitarity cuts agrees–up to a term proportional to the tree-level S-matrix which can be interpreted as a scheme-dependent shift in the coupling[12]—with the one known from perturbation theory. Importantly, the latter includes one-loop corrections coming from a gauge-fixing procedure which integrates out unphysical fields [14] and results in contributions to the one-loop S-matrix which restore various properties of integrability.

As for *relativistic, supersymmetric models*, which have checked the procedure on theories obtained as Pohlmeyer reduction of the Green-Schwarz action for the Type IIB superstring on $AdS_5 \times S^5$ [15, 16], $AdS_3 \times S^3$ [17] and $AdS_2 \times S^2$ [15] which can be seen as supersymmetric generalizations of the bosonic models considered above.[13] These reduced theories are all classically integrable, demonstrated by the existence of a Lax connection, and conjectured to be UV-finite [19]. The tree-level and one-loop S-matrices for these theories were computed in [20, 21], while the exact S-matrices have been conjectured using integrability techniques in [22] for the reduced $AdS_2 \times S^2$ model, [21] for the reduced $AdS_3 \times S^3$ model and [23] for the reduced $AdS_5 \times S^5$ model. In all the cases considered, the agreement is *exact* and no additional shift of the coupling is needed. The presence of the supersymmetry, albeit deformed, may provide an explanation for this, with shifts arising from bosonic loops cancelled by shifts from fermionic loops. Importantly, in the reduced $AdS_3 \times S^3$ standard perturbative computation a contribution coming from a one-loop correction needs to be added so that the S-matrix satisfies the Yang-Baxter equation. It is this S-matrix that the unitarity technique matches. This is then another example of how unitarity methods applied to a classically integrable theory seem to provide a *quantum integrable* result. This seems to suggest a relationship between integrable quantization and unitarity techniques which would be interesting to investigate further.

20.4 $AdS_5 \times S^5$ superstring world-sheet theory

We have finally considered the case of the light-cone gauge-fixed superstring on $AdS_5 \times S^5$ and its world-sheet S-matrix.[14] Assuming the quantum integrability of the full world-sheet theory and using the global symmetries the *exact* world-sheet

---

[12]In the sine-Gordon case the agreement is exact. For $n \geq 2$ the shift in the coupling is by the dual Coxeter number of the group $G = SO(n)$, a structure appears regularly in the quantization of WZW and gauged WZW models, where k is the quantized level (see for example [13]).

[13]The reduced $AdS_2 \times S^2$ theory is in fact given by the $\mathcal{N} = 2$ supersymmetric sine-Gordon model and hence is supersymmetric. The reduced $AdS_3 \times S^3$ and $AdS_5 \times S^5$ theories have a non-local $\mathcal{N} = 4$ and $\mathcal{N} = 8$ supersymmetry respectively, which manifests as a q-deformation of the S-matrix symmetry algebra. Furthermore, we have also checked that the unitarity-cutting procedure matches the perturbative result at one-loop in the $\mathcal{N} = 1$ supersymmetric sine-Gordon model [18].

[14]Notice that this is a *non-relativistic* model, as seen quantizing it perturbatively and noticing that the choice of a flat Minkowski worldsheet metric is incompatible with Virasoro constraints (see for example [24]).

S-matrix has been uniquely determined [25] up to an overall phase, or dressing factor [26]. The determination of the latter exploited the non-relativistic generalization of the crossing symmetry [27, 28] as well as perturbative data both from the string and gauge theory sides [29, 30]. Relaxing the level-matching condition and taking the limit of infinite light-cone momentum (decompactification limit), the world-sheet theory becomes a massive field theory defined on a plane, with well-defined asymptotic states and S-matrix. The scattering of the world-sheet excitations has been studied at tree-level in [31], while one-loop [32] and two-loop [33] results have been carried out only in the simpler near-flat-space limit [34] where interactions are at most quartic in the fields. These studies have also explicitly shown some consequences of the integrability of the model, such as the factorization of the many-body S-matrix and the absence of particle production in the scattering processes [35].

The tree level matrix element were evaluated in [31] in the generalized uniform light-cone gauge (showing therefore an explicit dependence on the parameter a labeling different light-cone gauge choices [36]) at leading order in perturbation theory, where the small parameter is the inverse of the string tension $g = \frac{\sqrt{\lambda}}{2\pi}$. After having explicitly verified that the tree-level matrix elements above verify the sermonic generalization of the consistency relation (20.10), we could safely use them as input of our procedure and get the one-loop S-matrix for the light-cone gauge-fixed sigma model.[15] As a first result, an overall phase could be resumed at the one-loop order, which show the expected gauge dependence [38]. As mentioned above, because of the complicated structure of interactions of the light-cone gauge-fixed sigma model, the perturbative S-matrix is known beyond the leading order [32, 33] only in the kinematic truncation known as near-flat-space limit [34]. Therefore, to test the validity of the unitarity method, we needed to compare our one-loop result to the corresponding limit of the exact world-sheet S-matrix. This was achieved by considering the matrix elements derived in [25] for a single SU(2|2) sector together with the dressing phase, here needed at next-to-leading order in the $1/\sqrt{\lambda}$ expansion.[16]

In comparing the exact S-matrix with the one found via unitarity cuts[17] we found

$$(S_{AB}^{CD})_{\text{exact}} = e^{\frac{i}{4g}\left(([A]+2[B]-[C]-2)\mathrm{p}_1+([B]-2[C]-[D]+2)\mathrm{p}_2\right)}\, e^{\varphi_{a=0}(\mathrm{p}_1,\mathrm{p}_2)}\, (S_{AB}^{CD})_{\text{cut}} + \mathcal{O}(1/g^3)\,. \tag{20.11}$$

From (20.11) we see that we have agreement up to a phase whose argument is linear in momenta. This is not surprising, as it simply amounts to moving from the string

[15]Notice that the non-relativistic dispersion relation $\varepsilon(\mathrm{p}) = \sqrt{1+\frac{\lambda}{\pi^2}\sin^2\frac{\mathrm{p}}{2}}$ [25, 37], when expanded in the near-BMN limit $\mathrm{p} \to \zeta\mathrm{p}$, corresponding to the perturbative regime, leads to a relativistic energy $\varepsilon_i = \sqrt{1+\mathrm{p}_i^2}$.

[16]In the comparison with the world-sheet calculation all dimensional quantities (such as the spin-chain length and the momenta) should be rescaled via a factor of $\sqrt{\lambda}/(2\pi)$ [31], for us $\mathrm{p} \to \zeta\,\mathrm{p}$.

[17]This is done in the so-called constant-J gauge $a = 0$.

frame to the spin-chain frame [39, 40]. As argued already at the tree level [31] such terms should not affect the physical spectrum following from inputting the S-matrix into the asymptotic Bethe equations.

References

1. L.J. Dixon (1996)
2. H. Elvang, Y.t. Huang (2013)
3. W.M. Chen, Y.t. Huang, JHEP **1111**, 057 (2011). doi:10.1007/JHEP11(2011)057
4. S. Caron-Huot, Y.t. Huang, JHEP **1303**, 075 (2013). doi:10.1007/JHEP03(2013)075
5. L. Bianchi, M.S. Bianchi (2013)
6. L. Bianchi, V. Forini, B. Hoare, JHEP **1307**, 088 (2013). doi:10.1007/JHEP07(2013)088
7. O.T. Engelund, R.W. McKeown, R. Roiban (2013)
8. Z. Bern, L.J. Dixon, D.C. Dunbar, D.A. Kosower, Nucl. Phys. B **425**, 217 (1994). doi:10.1016/0550-3213(94)90179-1
9. T.J. Hollowood, J.L. Miramontes, Q.H. Park, Nucl. Phys. B **445**, 451 (1995). doi:10.1016/0550-3213(95)00142-F
10. I. Bakas, Q.H. Park, H.J. Shin, Phys. Lett. B **372**, 45 (1996). doi:10.1016/0370-2693(96)00026-3
11. A.B. Zamolodchikov, A.B. Zamolodchikov, Ann. Phys. **120**, 253 (1979). doi:10.1016/0003-4916(79)90391-9
12. N. Dorey, T.J. Hollowood, Nucl. Phys. B **440**, 215 (1995). doi:10.1016/0550-3213(95)00074-3
13. E. Witten, Commun. Math. Phys. **92**, 455 (1984). doi:10.1007/BF01215276
14. B. Hoare, A. Tseytlin, JHEP **1011**, 111 (2010). doi:10.1007/JHEP11(2010)111
15. M. Grigoriev, A.A. Tseytlin, Nucl. Phys. B **800**, 450 (2008). doi:10.1016/j.nuclphysb.2008.01.006
16. A. Mikhailov, S. Schafer-Nameki, JHEP **0805**, 075 (2008). doi:10.1088/1126-6708/2008/05/075
17. M. Grigoriev, A.A. Tseytlin, Int. J. Mod. Phys. A **23**, 2107 (2008). doi:10.1142/S0217751X08040652
18. R. Shankar, E. Witten, Phys. Rev. D **17**, 2134 (1978). doi:10.1103/PhysRevD.17.2134
19. R. Roiban, A.A. Tseytlin, JHEP **0904**, 078 (2009). doi:10.1088/1126-6708/2009/04/078
20. B. Hoare, A. Tseytlin, JHEP **1002**, 094 (2010). doi:10.1007/JHEP02(2010)094
21. B. Hoare, A. Tseytlin, Nucl. Phys. B **851**, 161 (2011). doi:10.1016/j.nuclphysb.2011.05.016
22. K.i. Kobayashi, T. Uematsu, Phys. Lett. B **275**, 361 (1992). doi:10.1016/0370-2693(92)91603-7
23. B.Hoare, T.J. Hollowood, J.L. Miramontes (2013)
24. J. Callan, G. Curtis, H.K. Lee, T. McLoughlin, J.H. Schwarz, I. Swanson et al. Nucl. Phys. B **673**, 3 (2003). doi:10.1016/j.nuclphysb.2003.09.008
25. N. Beisert, Adv. Theor. Math. Phys. **12**, 945 (2008)
26. G. Arutyunov, S. Frolov, M. Staudacher, JHEP **0410**, 016 (2004). doi:10.1088/1126-6708/2004/10/016
27. R.A. Janik, Phys. Rev. D **73**, 086006 (2006). doi:10.1103/PhysRevD.73.086006
28. D. Volin, J. Phys. A **42**, 372001 (2009). doi:10.1088/1751-8113/42/37/372001
29. N. Beisert, R. Hernandez, E. Lopez, JHEP **0611**, 070 (2006). doi:10.1088/1126-6708/2006/11/070
30. N. Beisert, B. Eden, M. Staudacher, J. Stat. Mech. **0701**, P01021 (2007). doi:10.1088/1742-5468/2007/01/P01021
31. T. Klose, T. McLoughlin, R. Roiban, K. Zarembo, JHEP **0703**, 094 (2007). doi:10.1088/1126-6708/2007/03/094

32. T. Klose, K. Zarembo, JHEP **0702**, 071 (2007). doi:10.1088/1126-6708/2007/02/071
33. T. Klose, T. McLoughlin, J. Minahan, K. Zarembo, JHEP **0708**, 051 (2007). doi:10.1088/1126-6708/2007/08/051
34. J.M. Maldacena, I. Swanson, Phys. Rev. D **76**, 026002 (2007). doi:10.1103/PhysRevD.76.026002
35. V. Giangreco, M. Puletti, T. Klose, O. Ohlsson, Sax. Nucl. Phys. B **792**, 228 (2008). doi:10.1016/j.nuclphysb.2007.09.018
36. G. Arutyunov, S. Frolov, M. Zamaklar, Nucl. Phys. B **778**, 1 (2007). doi:10.1016/j.nuclphysb.2006.12.026
37. N. Beisert, V. Dippel, M. Staudacher, JHEP **0407**, 075 (2004). doi:10.1088/1126-6708/2004/07/075
38. G. Arutyunov, S. Frolov, J. Phys. A **42**, 254003 (2009). doi:10.1088/1751-8113/42/25/254003
39. G. Arutyunov, S. Frolov, M. Zamaklar, JHEP **0704**, 002 (2007). doi:10.1088/1126-6708/2007/04/002
40. C. Ahn, R.I. Nepomechie, Lett. Math. Phys. **99**, 209 (2012). doi:10.1007/s11005-011-0478-9

Chapter 21
Gravity Duals to Non-relativistic Quantum Field Theories

Andreas Karch and Stefan Janiszewski

Abstract We use holography to argue that the zoo of consistent quantum theories of gravity contains many non-relativistic alternatives to standard Einstein gravity. The basic argument will be based on symmetry. Non-relativistic diffeomorphisms are symmetries of a large class of non-relativistic quantum field theories, including the quantum Hall states and the unitary Fermi gas. Based on these symmetries, we argue that generic many-body quantum mechanical systems have a dual holographic description in terms of a modified theory of gravity known as Horava gravity.

21.1 Motivation: Why Non-relativistic Holography

Why study non-relativistic holography? In nature we know the right description for all matter, including solids, is a relativistic quantum field theory (QFT). Ignoring the weak interactions for now, the ionic lattice and electrons making up a solid are described by a local relativistic Lagrangian

$$\mathscr{L} = \mathscr{L}_{QED} + \mathscr{L}_{QCD}. \tag{21.1}$$

The field of condensed matter physics comprises the study of particular (meta)-stable states of this theory with finite baryon and lepton number and to analyze their low energy fluctuations. While in principle correct, this viewpoint is hardly useful in practice. For most condensed matter applications, a more appropriate starting point is a non-relativistic Hamiltonian

$$H = \sum_{Nuclei,A} \frac{p_A^2}{m_A} + \sum_{electrons_i} \frac{p_i^2}{m_e} - \sum_{A,i} \frac{e^2}{|x_i - x_A|} + \sum_{i \neq j} \frac{e^2}{|x_i - x_j|}, \tag{21.2}$$

A. Karch (✉) · S. Janiszewski
Department of Physics, University of Washington, Seattle, WA 98195-1560, Washington
e-mail: akarch@uw.edu

S. Janiszewski
e-mail: stefanjj@uw.edu

P. Nicolini et al. (eds.), *1st Karl Schwarzschild Meeting on Gravitational Physics*, Springer Proceedings in Physics 170, DOI 10.1007/978-3-319-20046-0_21

describing the motion of lattice ions and electrons governed by Coulomb interactions (ignoring spin for now for notational simplicity). This theory can be obtained as a low energy decoupling limit of (21.1). The description in terms of (21.2) is substantially easier than the full relativistic Lagrangian governing QCD and QED. Most degrees of freedom (for example the anti-particles) completely decouple from the dynamics and only the degrees of freedom responsible for the dynamics of the solid remain. While (21.2) can be thought of as a low energy limit of (21.1), and in nature that is how it is realized, it is a consistent quantum theory in its own right. The extra relativistic degrees of freedom are not needed for a self-consistent description. For a relativistic QFT with holographic dual (that is, a relativistic theory which is known to have a completely equivalent description in one higher dimension in terms of a quantum theory of gravity), it is natural to ask whether we can find a holographic dual that directly describes the non-relativistic low energy Hamiltonian. As the non-relativistic Hamiltonian gives a consistent quantum theory in its own right, so should the dual non-relativistic quantum theory of gravity. In fact, it is much easier to write down a consistent non-relativistic Hamiltonian than to write down a relativistic field theory. Neither the issue of anomalies nor the issue of UV divergences arises. Therefore holography implies, or at least strongly suggests, that there should exist a large class of non-relativistic self-consistent quantum theories of gravity dual to generic condensed matter Hamiltonians. And one should be able to construct at least some of them by starting with a known relativistic holographic pair and taking the non-relativistic (NR) decoupling limit.

21.2 Prelude: The Role of Symmetries

As most of our arguments will be based on symmetries, let us quickly review some basic facts about the role of symmetries in physics and establish a few definitions we will use when talking about symmetries. The most fundamental separation of symmetries is into gauge and global symmetries. Gauge symmetries are not really symmetries at all, rather they are redundancies of our chosen description. All physical observables are gauge invariant and so don't see the symmetry. One famous example of this phenomenon in the context of QFTs are Seiberg dualities [5]: two seemingly different field theory Lagrangians based on two completely different gauge groups can be shown to represent one and the same physical particle spectrum and interactions. Global symmetries on the other hand are true symmetries of physical observables. All physical quantities have to furnish a representation of the global symmetry group. In addition, global symmetries imply conservation laws. One famous example is that translation invariance implies momentum conservation. Note that in this distinction between gauge and global symmetries, we do not rely on a classification of whether the symmetry acts locally in space and/or time. The difference is simply how the symmetry acts on physical observables. If we demand that physical reality is invariant under a certain symmetry, by definition this symmetry does not act on observables and so is a redundancy of our description. We introduced this gauge symmetry simply for convenience. We introduced extra degrees of freedom to simplify the dynamics together with a prescription that, in the end, removes them again from our physical spectrum. In contrast, global symmetries are the true symmetries of the system, whether they act globally or locally.

Of special importance to us will be global, spurionic symmetries. Under a spurionic symmetry the Lagrangian is only invariant if couplings, and not just the dynamical fields, transform. Per definition, they contain the "true" global symmetries as a subgroup—one can simply ask what subgroup of the spurionic symmetries leaves a given set of couplings invariant. But unlike gauge symmetries, spurionic symmetries have real physical content. They put strong constraints on the low energy effective action. While in general they do not give rise to new conserved quantities, they can often be used to promote a global charge conservation condition to a local current conservation law. A famous example of a spurionic symmetry are chiral rotations for a massive Dirac fermion. The free Dirac Lagrangian is given by

$$\mathscr{L} = \bar{\psi}(i\partial_\mu \gamma^\mu - M)\psi. \tag{21.3}$$

The massless ($M = 0$) theory is invariant under chiral rotations

$$\psi \to e^{-i\phi\gamma_5/2}\psi \tag{21.4}$$

that independently rotate left and right movers. This symmetry is broken by the mass term. However it can be restored if we treat it as a spurionic symmetry, that is we let the mass parameter itself transform as

$$M \to e^{i\phi} M. \tag{21.5}$$

This spurionic symmetry constrains how the mass term can show up in the low energy effective action, for example in the chiral Lagrangian of QCD.

21.3 Diffeomorphisms in General Relativity and QFT

So how do diffeomorphism fit into our classification of symmetries? Under an infinitesimal diffeomorphism

$$x^\mu \to \tilde{x}^\mu = x^\mu + \xi^\mu \tag{21.6}$$

the metric transforms as

$$\delta g_{\mu\nu} = \xi^\lambda \partial_\lambda g_{\mu\nu} + g_{\lambda\nu}\partial_\mu \xi^\lambda + g_{\mu\lambda}\partial_\nu \xi^\lambda. \tag{21.7}$$

In General Relativity (GR) the transformations from (21.6) and (21.7) are a gauge symmetry. The metric is part of the dynamical fields in the theory. Physical observables have to be invariant under diffeomorphisms, implying that (quantum) gravity has no local observables (as the position x^μ isn't gauge invariant). What the true observables of GR are depends on the asymptotics of spacetime. For example, in asymptotically flat space the observable quantity is the S-matrix, in asymptotically AdS space the boundary correlation functions of a dual CFT. While generic diffeomorphisms are a gauge symmetry there is one exception: diffeomorphisms that do not vanish at infinity. Latter represent a genuine global symmetry of quantum gravity and act on the asymptotic data, be it S-matrix or boundary correlation function.

This has to be contrasted with the role of diffeomorphisms in a relativistic QFT. In a QFT diffeomorphisms are still given by (21.6) and (21.7), but this time the metric is not part of the dynamical fields, rather it should be thought of as a coupling constant. Furthermore, the physical quantities are not required to be invariant under diffeomorphisms—they are no longer a gauge symmetry. Local observables, like correlation functions of local operators, can be defined and measured. Note that diffeomorphisms, despite being a global symmetry, still act locally in space. We should think of the space-time metric $g_{\mu\nu}$ as a set of coupling constants (5 at each space-time point) that transform non-trivially under the global spurionic symmetry. In practice we have all used this fact throughout our career. When solving a simple physics problem, like finding the electromagnetic fields of a point charge, we often change coordinates from Cartesian to Spherical. Under this transformation our differential equations are not invariant—the metric, and hence the Laplacian operator, do change. But they change in a well defined way and we can change coordinates to the system that allows us to find the solution most easily.

Thinking of diffeomorphisms as spurionic global symmetries has two important applications. Spurionic symmetries can be used to constrain the way the coupling constants (here the metric) appear in the low energy effective Lagrangian $\mathscr{L}_{eff}$. These constraints are most powerful in gapped systems, where there are no dynamical fields left in the low energy theory and the dependence on couplings is the *only* information captured in the low energy effective action. As an example, put any gapped relativistic system in a non-trivial curved background geometry. $\mathscr{L}_{eff}$ is only a function of the metric. To the first two orders in a derivative expansion, diffeomorphisms ensure that it is entirely captured by two unknown coefficients:

$$\mathscr{L}_{eff} = \sqrt{g}(2\Lambda + R + \cdots). \tag{21.8}$$

From this simple action we can read off the Casimir stress tensor at every point in space and time up to order $(l/L)^2$, where l is the length scale set by the gap and L the geometric size of the geometry. The second application is that knowing the global spurionic symmetries allows us to deduce the true global symmetries for every particular set of couplings by simply asking which subset of the spurionic symmetry leaves this given set of couplings, which here means the spacetime metric, invariant. For example, when asking what diffeomorphisms leave the trival set of couplings $g_{\mu\nu} = \eta_{\mu\nu}$ invariant, we recover the standard global symmetries under translations, rotations and boosts together with the associated conservation laws.

21.4 Non-relativistic Diffeomorphisms

In order to understand what the spurionic global symmetries of a generic non-relativistic QFT are, let us start with the action of the paradigm of a NR QFT: Schrödinger's equation for the many particle system comprised of free bosons or fermions (in the non-relativistic case, both are given by the same action):

$$S = \int dt d^d x \sqrt{g} \left[\frac{i}{2} \psi^\dagger \overleftrightarrow{\partial}_t \psi - A_0 \psi^\dagger \psi - \frac{g^{ij}}{2m} (\partial_i \psi^\dagger - i A_i \psi^\dagger)(\partial_j \psi + i A_j \psi) \right] \tag{21.9}$$

where we allowed our free particles to propagate on a non-trivial spatial metric g_{ij} and in background electro-magnetic fields described by a vector potential A_i and a scalar potential A_0. As before, g_{ij}, A_0 and A_i are to be considered coupling constants. The least we expect this action to be invariant under (as a spurionic global symmetry) are purely spatial diffeomorphisms as well as the standard "gauge transformations" (which here are really global spurionic symmetries as well) acting on the background potentials. What is somewhat surprising at first sight is that the spurionic symmetries

$$\begin{aligned} \delta A_0 &= -\dot{\alpha} + \xi^k \partial_k A_0 + A_k \dot{\xi}^k \\ \delta A_i &= -\partial_i \alpha + \xi^k \partial_k A_i + A_k \partial_i \xi^k - m g_{ik} \dot{\xi}^k \\ \delta g_{ij} &= +\xi^k \partial_k g_{ij} + g_{ik} \partial_j \xi^k + g_{kj} \partial_i \xi^k \end{aligned} \tag{21.10}$$

include time dependent spatial diffeomorphisms. The importance of the latter can be understood by asking, once again, what subgroup of the spurionic symmetry leaves a particular background invariant. For the trivial background $g_{ij} = \delta_{ij}$, $A_i = A_0 = 0$, we find that in addition to the usual suspects (rotations and translations) the time dependent spatial diffeomorphisms allow us to perform Galilean boosts:

$$\boldsymbol{\xi} = \mathbf{v} t, \quad \alpha = -m \mathbf{v} \cdot \mathbf{x}. \tag{21.11}$$

This is the crucial symmetry which we wanted to implement. While the discussion so far only applies to free non-relativistic QFTs, it is easy to add interactions that preserve this full spurionic symmetry. This, in particular, includes Coulomb interactions and short range 2-particle interactions, which means these symmetry considerations carry over to quantum Hall systems, strongly correlated electron systems, as well as the unitary Fermi gas.

As anticipated, the transformation laws from (21.10) can be obtained from relativistic diffeomorphisms as a decoupling limit. Starting with a free, complex scalar field of mass m coupled to a background space-time metric as well as a background potential C_μ the non-relativistic decoupling limit is a simple two step procedure. First we need to turn on a chemical potential equal to the rest mass, $C_t = mc^2$. This ensures that particles and anti-particles have dispersion relation $E = \sqrt{(cp)^2 + (mc^2)^2} \mp mc^2$ respectively. Second, we can now take the limit that c goes to infinity. In this limit the anti-particles go to infinite E and hence completely decouple, whereas the particles pick up a non-relativistic dispersion relation

$$E = \frac{p^2}{2m} + \cdots . \tag{21.12}$$

In this limit the spurionic diffeomorphism and gauge invariance of the relativistic parent descend exactly to the non-relativistic spurionic symmetries of (21.10). Our goal will be to implement exactly the same decoupling limit in the gravitational dual description.

As in the relativistic case, the spurionic diffeomorphism symmetry has important applications. As shown by Hoyos and Son [2], the symmetries of 21.10 significantly constrain transport in any quantum Hall system. Including the first subleading order in a derivative expansion, the low energy effective action has three input parameters: the filling fraction ν (determining, as usual, the fraction of the Landau level being filled), the Wen-Zee shift κ (which tells us how the density of the system changes when putting the same state on a sphere instead of flat space—a quantity known for example for all the Laughlin states) and a thermodynamic susceptibility, $\varepsilon(B)$. In terms of these quantities we can not just determine, as expected, the Hall conductivity as $\nu/(2\pi)$, but also more interesting transport coefficient. The Hall viscosity is fixed to be $\kappa B/(4\pi)$, as had been known before from other methods. A completely novel prediction based on the spurionic symmetry was the first non-trivial term in the Hall current due to a slowly spatially modulated electric field:

$$j^i = -\frac{1}{B}\left[\frac{\kappa}{4\pi} - m\varepsilon''(B)\right]\varepsilon^{ij}\partial_j\left(\nabla \cdot \mathbf{E}\right). \tag{21.13}$$

This universal formula is in principle accessible to experimental verification, which would be a great victory for this approach.

The symmetries discussed so far aren't the complete set of spurionic symmetries of the free non-relativistic system, as we haven't addressed transformations that act on time yet. In the relativistic case, these were of course part of the diffeomorphism invariance, but here they need to be treated separately. One additional symmetry all the NR theories discussed so far share is time translations, $t \to t + const.$. The free non-relativistic systems however have a larger spurionic symmetry; they are invariant under arbitrary time reparametrizations $t \to t + f(t)$ as long as they are accompanied with a corresponding transformation of the background couplings

$$\delta A_0 = f\dot{A}_0 + \dot{f}A_0, \quad \delta A_i = f\dot{A}_i, \quad \delta g_{ij} = f\dot{g}_{ij} - \dot{f}g_{ij}. \tag{21.14}$$

These spurionic symmetries in particular contain non-relativistic scale transformations and conformal transformations as a subgroup that leaves the trivial background invariant. This spurionic symmetry under time-reparametrizations is only expected in theories that preserve the full conformal or Schrödinger symmetry of the free theory. Unlike for the case of the diffeomorphisms it is much harder to construct interactions that preserve the full NR conformal invariance, but there is at least one known example: the unitary Fermi gas. As shown by Son and Wingate [6], in the unitary Fermi gas the full spurionic symmetry once more gives novel constraints on the low energy effective action, albeit in this case only at a higher order in the derivative expansion.

21.5 Spurionic Symmetries and Holography

We have seen that a large class of non-relativistic quantum field theories are invariant under spurionic symmetries that include gauge transformations on a background gauge field, time-dependent spatial diffeomorphisms and either time translations (for a generic NR QFT) or time reparametrizations (for a NR CFT). Time dependent spatial diffeomorphisms together with time reparametrization invariance are often referred to as foliation preserving diffeomorphism—they maintain a classical notion of causality by respecting a preferred time slicing. In order to use this knowledge to deduce the holographic dual of these theories, we need to remind ourselves how relativistic holography encodes the full spurionic diffeomorphism symmetry of a relativistic QFT. Recall that holography is the statement that gravity in asymptotically AdS space has a dual description in terms of a field theory living on its boundary. One basic piece of evidence for this equivalence is that the symmetries match. Most notably the global $SO(4,2)$ conformal symmetry of a $3+1$ dimensional CFT matches the isometry of the dual AdS_5 space. But for all symmetries to match the bulk has to respect the full global (spurionic) diffeomorphism invariance of the QFT. As the bulk theory is gravitational, diffeomorphisms in the bulk are really a redundancy of the description. One way to remove the redundancy is to gauge fix, e.g. to Fefferman-Graham form $g_{r\mu} = 0$, $g_{rr} = r^{-2}$ where r is the extra holographic direction. The leading near-boundary r^{-2} term in the bulk metric is identified as the field theory metric. As discussed before, the only diffeomorphisms that should be considered as a global symmetry of the gravitational bulk theory are those that do not vanish near the boundary. Here these are radially independent diffeomorphisms. Such radially independent bulk diffeomorphisms maintain the Fefferman Graham form and act on the boundary metric exactly as dictated by (21.6) and (21.7)—they are the bulk manifestation of the spurionic diffeomorphism invariance of the dual field theory.

In order to implement the spurionic symmetry under foliation preserving diffeomorphisms of a NR CFT in the same spirit, the holographic dual needs to be built around a gravitational theory whose defining gauge invariances are not the standard relativistic diffeomorphisms, but rather the foliation preserving diffeomorphisms of the NR CFT (together with a U(1) gauge invariance in the bulk in order to implement the spurionic background gauge transformations). Fortunately such a gravitational theory has recently been proposed: Hořava gravity [1]. Hořava gravity has been introduced as a contender for a UV complete theory of quantum gravity. The different scaling of space and time coordinates under dimensional analysis allows one to construct a power counting renormalizable action. It is this gravitational theory that, by symmetry, is the natural candidate for the holographic dual of a generic NR CFT.

21.6 Implementing the Duality

Turning this basic observation based on symmetries into a concrete proposal was the content of our two publications on this topic [3, 5]. Among other things, we were able to show that (a) Hořava gravity can arise as a consistent limit of a relativistic theory. In fact, the most direct way to get Hořava gravity out of string theory is to use a "vector khronon", that is we set the time component of a gauge field (dual to the chemical potential) equal to a constant before sending the speed of light to infinity. This is the direct implementation of the limit we took on the field theory side. (b) A consistent holographic dictionary can be constructed. The field theory sources transform as they should under the radially independent diffeomorphisms in the bulk. (c) While generic Hořava gravity is dual to a Galilean invariant, non-relativistic QFT, we need to impose an additional redundancy in the bulk, called α invariance, in order to make sure that all sources transform appropriately under conformal transformations. (d) Two point functions of probe scalars give correct two point functions. We believe that this duality has still much to teach us, both about non-relativistic field theory as well as the richness of quantum gravity.

Acknowledgments This work has been supported in part by the US Department of Energy under contract number DE-FG02-96ER40956. It is our pleasure to thank all the participants, and in particular the organizers, of the 2013 Karl-Schwarzschild Meeting for a very productive and stimulating conference and for the opportunity to present our work. Most of the work reported on in this talk has appeared previously in our two publications [3, 5].

References

1. P. Horava, Quantum gravity at a lifshitz point. Phys. Rev. D **79**(084), 008 (2009). doi:10.1103/PhysRevD.79.084008, arXiv:0901.3775
2. C. Hoyos, D.T. Son, Hall viscosity and electromagnetic response. Phys. Rev. Lett. **108**(066), 805 (2012). doi:10.1103/PhysRevLett.108.066805, arXiv:1109.2651
3. S. Janiszewski, A. Karch, Non-relativistic holography from Horava gravity. JHEP **1302**, 123 (2013). doi:10.1007/JHEP02(2013)123, arXiv:1211.0005
4. S. Janiszewski, A. Karch, String theory embeddings of nonrelativistic field theories and their holographic horava gravity duals. Phys. Rev. Lett. **110**(8), 081,601 (2013). doi:10.1103/PhysRevLett.110.081601, arXiv:1211.0010
5. N. Seiberg, Electric—magnetic duality in supersymmetric non abelian gauge theories. Nucl. Phys. B **435**, 129–146 (1995). doi:10.1016/0550-3213(94)00023-8, arXiv:hep-th/9411149
6. D. Son, M. Wingate, General coordinate invariance and conformal invariance in nonrelativistic physics: Unitary Fermi gas. Ann. Phys. **321**, 197–224 (2006). doi:10.1016/j.aop.2005.11.001, arXiv:cond-mat/0509786

Chapter 22
A 'Regularized' Schwarzschild Solution

Frans R. Klinkhamer

Abstract An exact solution of the vacuum Einstein field equations over a particular nonsimply-connected manifold is presented. This solution is spherically symmetric and has no curvature singularity. It provides a regularization of the Schwarzschild solution with a curvature singularity at the center.

22.1 Introduction

The main topic of this contribution concerns a nonsingular black-hole solution of general relativity, which is closely related to (but not identical with) the standard Schwarzschild solution [1–5]. It was arrived at by a detour which is rather interesting by itself.

That earlier investigation started from the following simple question: *how smooth is space and what quantitative bounds can be set?* In order to get a first partial answer to this question, certain Swiss-cheese-type spacetime models were considered, for which the photon propagation can be calculated in the long-wavelength limit [6]. Specifically, these spacetime models have randomly-positioned identical static defects.

The simplest type of defect is obtained in the following way: start with 3-dimensional Euclidean space, remove the interior of a ball ($r < b$), and, finally, identify antipodal points on the boundary ($r = b$). The corresponding Swiss-cheese-type model then has two parameters: the defect size b and the average distance d between neighbouring defects. The photon propagation over this spacetime model is described by the isotropic modified Maxwell theory with a single Lorentz-violating parameter [6]

$$\widetilde{\kappa}_{\text{tr}} = \pi\, b^3/d^3 \,. \tag{22.1}$$

Note added in proof After the conference, further aspects of this new type of solution have been studied in [18, 19].

F.R. Klinkhamer (✉)
Institut für Theoretische Physik, Karlsruher Institut für Technologie,
76128 Karlsruhe, Germany
e-mail: frans.klinkhamer@kit.edu

P. Nicolini et al. (eds.), *1st Karl Schwarzschild Meeting on Gravitational Physics*,
Springer Proceedings in Physics 170, DOI 10.1007/978-3-319-20046-0_22

From the absence of vacuum-Cherenkov radiation for ultrahigh-energy cosmic rays, the Auger data give the following two-σ upper bound [7]:

$$\widetilde{\kappa}_{\rm tr} \, < \, 6 \times 10^{-20} \, . \tag{22.2}$$

Two remarks are in order:

1. Bound (22.2), with the particular identification (22.1), also holds for values of b and d close to $L_{\rm Planck} \equiv (\hbar \, G_N / c^3)^{1/2} \sim 10^{-35}$ m.
2. The extremely small number on the right-hand side of (22.2) then suggest that, whatever the ultimate theory of quantum gravity may be, this quantum theory must leave practically no defects/wrinkles/ripples on the emerging classical flat spacetime.

All this is quite intriguing, but the single defect naively embedded in standard Minkowski spacetime does *not* satisfy the vacuum Einstein field equations (there are delta-function-type curvature singularities at $r = b$) and the same holds for the corresponding Swiss-cheese-type spacetime models. The task, then, is to find a proper defect solution. It turns out that the construction of this nonsingular defect solution [8, 9] produces an interesting spin-off: a black-hole solution without curvature singularity [10–12].

The outline of the present contribution is as follows. In Sect. 22.2, the relevant topology is discussed. In Sect. 22.3, the nonsingular black-hole solution is presented. In Sect. 22.4, some questions related to physics are raised.

Let us emphasize, right from the start, that the solution discussed in this contribution is a solution of general relativity, no more, no less. The only "new" ingredient is the nontrivial topology.

22.2 Manifold and Coordinates

As mentioned in the Introduction, our goal is to find a *nonsingular* solution of the vacuum Einstein field equations with a parameter $b > 0$ and topology as suggested by the sketch in Fig. 22.1. In this section, we describe the construction of the manifold

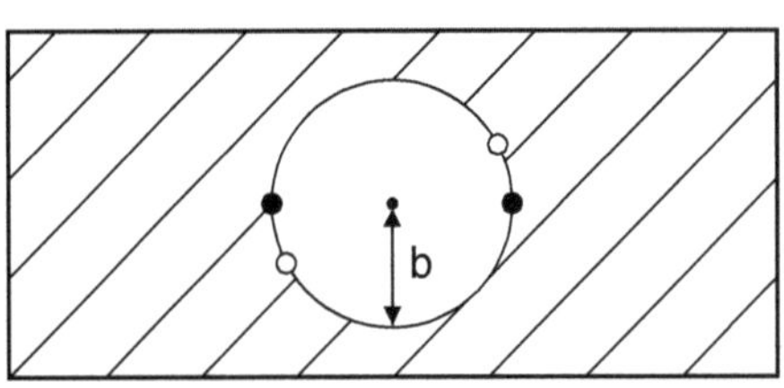

Fig. 22.1 Three-space $\widetilde{\mathcal{M}}_3$ obtained by surgery on $\mathbb{R}^3$: interior of the ball with radius b removed and antipodal points on the boundary of the ball identified (as indicated by open and filled circles)

and identify appropriate coordinates. The solution proper will be given in the next section.

The four-dimensional spacetime manifold considered is

$$\widetilde{\mathscr{M}}_4 = \mathbb{R} \times \widetilde{\mathscr{M}}_3\,, \tag{22.3a}$$

where $\widetilde{\mathscr{M}}_3$ is a noncompact, orientable, nonsimply-connected manifold without boundary. Up to a point, $\widetilde{\mathscr{M}}_3$ is homeomorphic to the 3-dimensional real-projective space,

$$\widetilde{\mathscr{M}}_3 \simeq \mathbb{R}P^3 - \{\text{point}\}\,. \tag{22.3b}$$

Recall that the 3-dimensional real projective space is topologically equivalent to a 3-sphere with antipodal points identified. Here, and in the following, the notation is as follows: $\widetilde{\mathscr{M}}$ with tilde stands for a nonsimply-connected manifold (having a nontrivial first homotopy group, $\pi_1(\widetilde{\mathscr{M}}) \neq 0$) and $\mathscr{M}$ without tilde stands for a simply-connected manifold (having a trivial first homotopy group, $\pi_1(\mathscr{M}) = 0$).

For the explicit construction of $\widetilde{\mathscr{M}}_3$, we perform local surgery on the 3-dimensional Euclidean space $E_3 = \left(\mathbb{R}^3,\, \delta_{mn}\right)$. We use the standard Cartesian and spherical coordinates on $\mathbb{R}^3$,

$$\mathbf{x} = (x^1,\, x^2,\, x^3) = (r\,\sin\theta\,\cos\phi,\; r\,\sin\theta\,\sin\phi,\; r\,\cos\theta)\,, \tag{22.4}$$

with $x^m \in (-\infty,\, +\infty)$, $r \geq 0$, $\theta \in [0,\, \pi]$, and $\phi \in [0,\, 2\pi)$. Now, $\widetilde{\mathscr{M}}_3$ is obtained from $\mathbb{R}^3$ by removing the interior of the ball B_b with radius b and identifying antipodal points on the boundary $S_b \equiv \partial B_b$. With point reflection denoted $P(\mathbf{x}) = -\mathbf{x}$, the 3-space $\widetilde{\mathscr{M}}_3$ is given by

$$\widetilde{\mathscr{M}}_3 = \Big\{\mathbf{x} \in \mathbb{R}^3 \,:\, \Big(|\mathbf{x}| \geq b > 0\Big) \wedge \Big(P(\mathbf{x}) \cong \mathbf{x} \;\text{ for }\; |\mathbf{x}| = b\Big)\Big\}\,, \tag{22.5}$$

where $\cong$ stands for point-wise identification (Fig. 22.1).

The next step is to identify *appropriate coordinates* and an exhaustive discussion of this issue can be found in [8]. The standard coordinates of Euclidean 3-space are unsatisfactory, because a single point may have different coordinates. For example, $(x^1,\, x^2,\, x^3) = (b,\, 0,\, 0)$ and $(x^1,\, x^2,\, x^3) = (-b,\, 0,\, 0)$ correspond to the same point.

A relatively simple covering of $\widetilde{\mathscr{M}}_3$ uses three charts of coordinates, labeled by $n = 1, 2, 3$. Each chart covers and surrounds part of one of the three Cartesian coordinate axes but does not intersect the other two Cartesian coordinate axes. For example, the $n = 1$ coordinate chart covers and surrounds the $|x^1| \geq b$ parts of the x^1 coordinate axis but does not intersect the x^2 and x^3 axes. The domains of the chart-1 coordinates consist of two 'wedges,' on both sides of the defect and pierced by the x^1 axis.

These coordinates are denoted

$$(X_n,\, Y_n,\, Z_n)\,, \quad \text{for} \quad n = 1,\, 2,\, 3\,, \tag{22.6}$$

and have the following ranges:

$$X_1 \in (-\infty,\, \infty)\,, \quad Y_1 \in (0,\, \pi)\,, \quad Z_1 \in (0,\, \pi)\,, \tag{22.7a}$$
$$X_2 \in (0,\, \pi)\,, \quad Y_2 \in (-\infty,\, \infty)\,, \quad Z_2 \in (0,\, \pi)\,, \tag{22.7b}$$
$$X_3 \in (0,\, \pi)\,, \quad Y_3 \in (0,\, \pi)\,, \quad Z_3 \in (-\infty,\, \infty)\,. \tag{22.7c}$$

In each chart, there is one radial-type coordinate with infinite range, one polar-type angular coordinate of finite range, and one azimuthal-type angular coordinate of finite range.

In order to describe the interrelation of the coordinates $(X_n,\, Y_n,\, Z_n)$ in the overlap regions, we express them in terms of the coordinates of the 3-dimensional Euclidean space E_3. For the latter, we use the standard spherical coordinates $(r,\, \theta,\, \phi)$ defined by (22.4) and the nonstandard spherical coordinates $(r,\, \vartheta,\, \varphi)$ defined as follows:

$$(x^1,\, x^2,\, x^3) = (r \sin\vartheta \sin\varphi,\, r\cos\vartheta,\, r\sin\vartheta\cos\varphi)\,, \tag{22.8}$$

with $r \geq 0$, $\vartheta \in [0,\, \pi]$, and $\varphi \in [0,\, 2\pi)$.

Now, the chart-1 and chart-2 coordinates over the appropriate regions (wedges) of $\widetilde{M}_3$ are given by

$$X_1 = \begin{cases} r - b & \text{for} \quad \cos\phi > 0\,, \\ b - r & \text{for} \quad \cos\phi < 0\,, \end{cases} \tag{22.9a}$$

$$Y_1 = \begin{cases} \phi - \pi/2 & \text{for} \quad \pi/2 < \phi < 3\pi/2\,, \\ \phi - 3\pi/2 & \text{for} \quad 3\pi/2 < \phi < 2\pi\,, \\ \phi + \pi/2 & \text{for} \quad 0 \leq \phi < \pi/2\,, \end{cases} \tag{22.9b}$$

$$Z_1 = \begin{cases} \theta & \text{for} \quad \cos\phi > 0\,, \\ \pi - \theta & \text{for} \quad \cos\phi < 0\,, \end{cases} \tag{22.9c}$$

and

$$X_2 = \begin{cases} \phi & \text{for} \quad 0 < \phi < \pi\,, \\ \phi - \pi & \text{for} \quad \pi < \phi < 2\pi\,, \end{cases} \tag{22.10a}$$

$$Y_2 = \begin{cases} r - b & \text{for} \quad 0 < \phi < \pi\,, \\ b - r & \text{for} \quad \pi < \phi < 2\pi\,, \end{cases} \tag{22.10b}$$

$$Z_2 = \begin{cases} \theta & \text{for} \quad 0 < \phi < \pi\,, \\ \pi - \theta & \text{for} \quad \pi < \phi < 2\pi\,. \end{cases} \tag{22.10c}$$

For the $n = 3$ chart, we require coordinates of E_3 that are regular on the Cartesian x^3 axis. These nonstandard spherical coordinates have already been defined in (22.8).

Then, the chart-3 coordinates over the relevant regions (wedges) of $\widetilde{M}_3$ are given by

$$X_3 = \begin{cases} \varphi - \pi/2 & \text{for } \pi/2 < \varphi < 3\pi/2\,, \\ \varphi - 3\pi/2 & \text{for } 3\pi/2 < \varphi < 2\pi\,, \\ \varphi + \pi/2 & \text{for } 0 \le \varphi < \pi/2\,, \end{cases} \tag{22.11a}$$

$$Y_3 = \begin{cases} \vartheta & \text{for } \cos\varphi > 0\,, \\ \pi - \vartheta & \text{for } \cos\varphi < 0\,, \end{cases} \tag{22.11b}$$

$$Z_3 = \begin{cases} r - b & \text{for } \cos\varphi > 0\,, \\ b - r & \text{for } \cos\varphi < 0\,. \end{cases} \tag{22.11c}$$

Having expressed the coordinates $(X_n,\, Y_n,\, Z_n)$ in terms of coordinates of the Euclidean 3-space, it is possible to verify that the $(X_n,\, Y_n,\, Z_n)$ coordinates are invertible and infinitely-differentiable functions of each other in the overlap regions. These coordinates therefore describe a manifold. Moreover, the manifold satisfies [8] the Hausdorff property (two distinct points x and y are always surrounded by two disjoint open sets U and V: $x \in U$, $y \in V$, and $U \cap V = \emptyset$).

22.3 Nonsingular Black-Hole Solution

We will now present a black-hole solution for the topology (22.3) with parameters

$$2M \ge b > 0\,, \tag{22.12}$$

using geometrical units with $G_N = c = 1$. For this new black-hole solution, the curvature singularity will be eliminated by a spacetime defect, i.e., a "hole" in spacetime.

Instead of starting from an *Ansatz* based on Kruskal–Szekeres coordinates [2, 3] as was done in our original article [10], we start from an *Ansatz* based on Painlevé–Gullstrand coordinates [13, 14] (a useful review appears in [15]). Turning to the chart-1 coordinates (22.7a), we arrive at the following line element [12]:

$$ds^2\Big|_{\text{chart-1}} = -dT^2 + \left(\frac{X_1}{\sqrt{b^2 + (X_1)^2}}\, dX_1 + \sqrt{\frac{2M}{\sqrt{b^2 + (X_1)^2}}}\, dT \right)^2 + \left(b^2 + (X_1)^2\right) \left((dZ_1)^2 + \left(\sin Z_1\right)^2 (dY_1)^2\right), \tag{22.13a}$$

for mass parameter $M > 0$ and length parameter $b > 0$. Remark that the surfaces of constant T are intrinsically flat [15]. An advantage of the metric (22.13a) is that it applies not only to the black-hole case $2M \ge b$ (including the special case $2M = b$) but also to the defect case $2M < b$.

The metrics for the $n = 2$ and $n = 3$ charts are obtained from (22.13a) by taking the coordinates $(X_2,\ Y_2,\ Z_2)$ and $(X_3,\ Y_3,\ Z_3)$ instead of $(X_1,\ Y_1,\ Z_1)$. The corresponding metrics are given by the following line elements [12]:

$$ds^2\Big|_{\text{chart-2}} = -dT^2 + \left(\frac{Y_2}{\sqrt{b^2+(Y_2)^2}}\,dY_2 + \sqrt{\frac{2M}{\sqrt{b^2+(Y_2)^2}}}\,dT\right)^2 + \left(b^2+(Y_2)^2\right)\left((dZ_2)^2 + (\sin Z_2)^2\,(dX_2)^2\right), \tag{22.13b}$$

$$ds^2\Big|_{\text{chart-3}} = -dT^2 + \left(\frac{Z_3}{\sqrt{b^2+(Z_3)^2}}\,dZ_3 + \sqrt{\frac{2M}{\sqrt{b^2+(Z_3)^2}}}\,dT\right)^2 + \left(b^2+(Z_3)^2\right)\left((dY_3)^2 + \left(\sin Y_3\right)^2(dX_3)^2\right). \tag{22.13c}$$

The Ricci tensor and Ricci scalar from (22.13) vanish. Hence, the vacuum Einstein field equations are solved. Furthermore, the Kretschmann scalar is found to be given by

$$K \equiv R_{\mu\nu\rho\sigma}\,R^{\mu\nu\rho\sigma} = 48\,\frac{M^2}{\zeta^6}\,, \tag{22.14}$$

with $\zeta^2 = b^2 + (X_1)^2$ for the $n = 1$ chart, and similarly for the other charts. The Kretschmann scalar remains finite because $b > 0$.

Recall that the standard Schwarzschild–Kruskal–Szekeres solution [1–3], with topology

$$\mathscr{M}_{\text{SKS}} = \mathbb{R}^2 \times S^2\,, \tag{22.15}$$

has a physical singularity for $r \to 0$, as shown by the divergence of the Kretschmann scalar,

$$K\Big|_{\text{SKS}} \equiv R_{\mu\nu\rho\sigma}\,R^{\mu\nu\rho\sigma}\Big|_{\text{SKS}} = 48\,\frac{M^2}{r^6}\,. \tag{22.16}$$

The comparison of (22.14) for $b \neq 0$ and (22.16) makes clear that the solution (22.13) over $\mathbb{R} \times \widetilde{\mathscr{M}_3}$ may be considered to be a *regularized* version of the standard Schwarzschild solution over $\mathbb{R}^+ \times \mathbb{R} \times S^2$, with the curvature singularity eliminated by a spacetime defect, i.e., a "hole" in spacetime (Fig. 22.1).

For the moment, specialize to the $n = 1$ chart and transform to Schwarzschild-type coordinates (denoted by a prime). The coordinate T' and the coordinate $X_1' = X_1$ are, respectively, spacelike and timelike for the inner Schwarzschild-type metric [10]. This behavior is analogous to what happens for the standard Schwarzschild solution [5]. Note that the timelike coordinate X_1' of the inner metric ranges from $-\infty$

to $+\infty$, unlike the usual radial coordinate r. Moreover, this timelike coordinate X_1' is part of a topologically nontrivial manifold $\widetilde{\mathscr{M}}_3$. This gives rise to the presence of closed time-like curves (CTCs). These CTCs imply all possible horrors, but, classically, these horrors remain *confined* within the Schwarzschild horizon. Whether or not CTCs in the interior region are physically acceptable depends on the behavior of the matter fields.

The problematic CTCs of the modified Schwarzschild solution (22.13) trace back to the fact that the original singularity was *spacelike*. But it is well-known that the singularity of the standard Reissner–Nordström solution [16, 17] is *timelike* (see, for example, [4, 5]). This suggests, first, to add a small electric charge and, then, to modify the resulting Reissner–Nordström solution in order to arrive at a nonsingular solution. Details can be found in the original article [11] and the review [12].

22.4 Discussion

Apart from the mathematical interest of having a new type of exact solution of the Einstein field equations, these nonsingular solutions (with or without electric charge) may also appear in a physical context.

Start from a nearly flat spacetime with a trivial topology $\mathbb{R}^4$ and a metric approximately equal to the Minkowski metric. Next, arrange for a large amount of matter with total mass $\overline{M}$ and with vanishing net charge $\overline{Q} = 0$ to collapse in a spherically symmetric way. Within the realm of classical Einstein gravity, one expects to end up with part of the singular Schwarzschild solution. But, very close to the final curvature singularity, something else may happen due to quantum effects.

Consider a precursor mass $\overline{\Delta M} \sim \hbar/(b\,c) \ll \overline{M}$ and use typical curvature values from the expressions (22.14) and (22.16) for the Kretschmann scalar. Then, the local spacetime integral of the action density related to the standard Schwarzschild solution differs from that related to (22.13) by an amount $\lesssim \hbar$. As argued by Wheeler [5], the local topology of the manifold may change by a quantum jump if b is sufficiently close to the length scale $L_{\text{Planck}} \equiv (\hbar\, G_N/c^3)^{1/2}$.

This quantum process may then result in a transition from the simply-connected manifold $\mathbb{R}^4$ to the nonsimply-connected manifold $\widetilde{\mathscr{M}}_4$. Hence, *if* the transition amplitude between the different topologies is nonzero for appropriate matter content, quantum mechanics can operate a change between the classical Schwarzschild solution and the classical solution (22.13). In this way, the curvature singularity would be removed. (Note that closed timelike curves can be avoided by adding a small electric charge.)

It is also possible to present an alternative scenario *without* topology change. Now, the spherical collapse of matter is assumed to occur in Minkowski spacetime with a relatively sparse sprinkling of massless static defects (each one given by the solution of Sect. 22.3 with $M = 0$ and $b > 0$). Then, the precursor mass $\overline{\Delta M}$ selects one of the available defect cores and increases its mass ($M = 0 \to \overline{\Delta M} \to \overline{M}$), possibly giving it also a charge by electron-positron pair creation with one charge expelled to infinity. Again, the curvature singularity would be removed.

Many questions remain, the most important of which are the following:

1. Are these regularized Schwarzschild solutions *really* acceptable, both mathematically and physically?
2. Are there perhaps other surprises from this regularization, in addition to restricted elementary flatness (cf. App. D of [12]) and closed time-like curves?
3. Where does the matter go and can the matter be distributed over a thin shell with $\zeta \in [b,\, b + \Delta b)$ for $\Delta b > 0$?
4. Does realistic collapse occur with or without topology change?

These are obviously difficult questions. We can perhaps make further progress by explicitly considering the matter sector.

Acknowledgments It is a pleasure to thank the participants of the Karl Schwarzschild Meeting on Gravitational Physics (Frankfurt Institute for Advanced Studies, July 2013) for interesting discussions and the organizers for making it all happen.

References

1. K. Schwarzschild, Über das Gravitationsfeld eines Massenpunktes nach der Einsteinschen Theorie. Sitzungsberichte der Deutschen Akademie der Wissenschaften zu Berlin, Klasse für Mathematik, Physik, und Technik 189–196 (1916) [scanned version available from http://de.wikisource.org]
2. M.D. Kruskal, Maximal extension of Schwarzschild metric. Phys. Rev. **119**, 1743–1745 (1960)
3. G. Szekeres, On the singularities of a Riemannian manifold. Publ. Math. Debrecen **7**, 285–301 (1960)
4. S.W. Hawking, G.F.R. Ellis, *The Large Scale Structure of Space-Time* (Cambridge University Press, Cambridge, 1973)
5. C.W. Misner, K.S. Thorne, J.A. Wheeler, *Gravitation* (Freeman, New York, 1973)
6. S. Bernadotte, F.R. Klinkhamer, Bounds on length scales of classical spacetime foam models. Phys. Rev. D **75**, 024028 (2007). arXiv:hep-ph/0610216
7. F.R. Klinkhamer, M. Schreck, New two-sided bound on the isotropic Lorentz-violating parameter of modified Maxwell theory. Phys. Rev. D **78**, 085026 (2008). arXiv:0809.3217
8. M. Schwarz, Nontrivial Spacetime Topology, Modified Dispersion Relations, and an $SO(3)$-Skyrme Model. Ph.D. Thesis, KIT, July 2010. Verlag Dr. Hut, München, Germany (2010)
9. F.R. Klinkhamer, C. Rahmede, Nonsingular spacetime defect. Phys. Rev. D **89**, 084064 (2014), arXiv:1303.7219
10. F.R. Klinkhamer, Black-hole solution without curvature singularity. Mod. Phys. Lett. A **28**, 1350136 (2013). arXiv:1304.2305
11. F.R. Klinkhamer, Black-hole solution without curvature singularity and closed timelike curves. Acta Phys. Pol. B **45**, 5–14 (2014), arXiv:1305.2875
12. F.R. Klinkhamer, A new type of nonsingular black-hole solution in general relativity. Mod. Phys. Lett. A **29**, 1430018 (2014), arXiv:1309.7011
13. P. Painlevé, La mécanique classique et la théorie de la relativité. C. R. Acad. Sci. (Paris) **173**, 677–680 (1921)
14. A. Gullstrand, Allgemeine Lösung des statischen Einkörper-problems in der Einsteinschen Gravitationstheorie. Arkiv. Mat. Astron. Fys. **16**, 1–15 (1922)

15. K. Martel, E. Poisson, Regular coordinate systems for Schwarzschild and other spherical spacetimes. Am. J. Phys. **69**, 476–480 (2001). arXiv:gr-qc/0001069
16. H. Reissner, Über die Eigengravitation des elektrischen Feldes nach der Einsteinschen Theorie. Ann. der Phys. **50**, 106–120 (1916)
17. G. Nordström, On the energy of the gravitational field in Einstein's theory. Proc. Acad. Sci. Amst. **26**, 1201–1208 (1918)
18. F.R. Klinkhamer, Skyrmion spacetime defect. Phys. Rev. D **90**, 024007 (2014), arXiv:1402.7048
19. F.R. Klinkhamer, F. Sorba, Comparison of spacetime defects which are homeomorphic but not diffeomorphic. J. Math. Phys. **55**, 112503 (2014), arXiv:1404.2901

Chapter 23
The Chemistry of Black Holes

Robert B. Mann

Abstract I provide a short overview of black hole thermodynamics in extended phase space, where the cosmological constant is interpreted as thermodynamic pressure. This leads to an understanding of black holes from the viewpoint of chemistry, in terms of concepts such as Van der Waals fluids, reentrant phase transitions, and triple points. I discuss first the motivations for the extended phase space, indicating why the mass of a black hole should be understood as the analogue of thermodynamic enthalpy. I think go on to describe how charged and rotating black holes exhibit novel chemical-type phase behaviour, hitherto unseen.

23.1 Introduction

Black hole thermodynamics has remained a subject of interest for four decades, providing us with interesting clues to the underlying structure of quantum gravity. The parallel with standard thermodynamics is quite striking:

$$\begin{array}{rcl} \text{Energy } E & \leftrightarrow M & \text{Mass} \\ \text{Temperature } T & \leftrightarrow \frac{\hbar\kappa}{2\pi} & \text{Surface Gravity} \\ \text{Entropy } S & \leftrightarrow \frac{A}{4G\hbar} & \text{Horizon Area} \end{array}$$

where on the left-hand side we see the basic thermodynamic quantities of a physical system, and on the right their counterparts in black hole physics. However the first law of thermodynamics

$$dE = TdS + VdP + \text{work terms} \leftrightarrow dM = \frac{\kappa}{8G\pi} dA + \Omega dJ + \Phi dQ$$

R.B. Mann (✉)
Department of Physics and Astronomy, University of Waterloo, Waterloo, ON, Canada
e-mail: rbmann@uwaterloo.ca

P. Nicolini et al. (eds.), *1st Karl Schwarzschild Meeting on Gravitational Physics*, Springer Proceedings in Physics 170, DOI 10.1007/978-3-319-20046-0_23

does not quite capture this same correspondence, insofar as there is no counterpart of the "pressure-volume" term; the ΩdJ and ΦdQ terms are understood as thermodynamic work terms. Where is the PV term in black hole thermodynamics [1]?

Recently there have been interesting new developments that address this question. They lead to a picture in which the mass of a black hole should be interpreted as the enthalpy of spacetime. The basic idea is that the cosmological constant of anti de Sitter (AdS) space-time be considered as a thermodynamic variable [2] analogous to pressure in the first law [1, 3–14]. Out of this idea emerges a much richer panoply of thermodynamic behaviour, for both negative and positive [15] cosmological constants. Some of the more interesting results include the discovery of reentrant phase transitions in rotating [16] and Born-Infeld [10] black holes, the existence of a tri-critical points in rotating black holes analogous to the triple point in water [17], and a complete analogy between 4-dimensional Reissner-Nördstrom AdS black holes and the Van der Waals liquid–gas system [9]; the critical exponents coincide with those of the Van der Waals system and predicted by the mean field theory, significantly modifying previous considerations that emerged from the duality description [18, 19].

The purpose of this article is to provide a brief overview and summary of these interesting new developments in black hole thermodynamics. Henceforth we shall set $c = \hbar = G = 1$.

23.2 Smarr Relations

The motivation for extending the thermodynamic phase space of black holes originated from a consideration of the Smarr relation [3, 4]. Consider a Schwarzschild black hole, whose metric is

$$ds^2 = -f dt^2 + \frac{dr^2}{f} + +r^2 d\Omega^2_{d-2} \tag{23.1}$$

where $f = 1 - \frac{2M}{r^{d-3}}$ and $d\Omega^2_d$ is the line element on S^d. The event horizon is at $f(r_+) = 0$, or $r_+^{d-3} = 2M$. The associated thermodynamic quantities are

$$E = M = \frac{d-2}{16\pi}\omega_{d-2} r_+^{d-3} \quad T = \frac{d-3}{4\pi r_+} \quad S = \frac{\omega_{d-2} r_+^{d-2}}{4} \Longrightarrow (d-3)M = (d-2)TS \tag{23.2}$$

where $\omega_d = \frac{2\pi^{\frac{d+1}{2}}}{\Gamma\left(\frac{d+1}{2}\right)}$. The relation on the right is known as the Smarr relation, originally obtained for $d = 4$ [20].

For a Schwarzschild-AdS black hole, with cosmological constant $\Lambda = -\frac{(d-1)(d-2)}{2\ell^2}$, the metric is still given by (23.1) but with $f = \frac{r^2}{\ell^2} + 1 - \frac{2M}{r^{d-3}}$. The associated thermodynamic quantities M and S remain the same, but now

$$T = \frac{d-3}{4\pi r_+}\left(1 + \frac{d-1}{d-3}\frac{r_+^2}{\ell^2}\right) \Longrightarrow (d-3)M \neq (d-2)TS \tag{23.3}$$

and we see that the Smarr relation is no longer satisfied.

A consideration of these quantities indicates that an appropriately modified Smarr relation could be satisfied. For a homogeneous function $h(x, y)$ and some scaling parameter α, Euler's theorem indicates

$$h(\alpha^p x, \alpha^q y) = \alpha^r h(x, y) \Longrightarrow rh(x, y) = p\frac{\partial h}{\partial x} + q\frac{\partial h}{\partial x} \tag{23.4}$$

Regarding the mass $M = M(A, \Lambda)$ we obtain

$$(d-3)M = (d-2)\frac{\partial M}{\partial A}A - 2\frac{\partial M}{\partial \Lambda}\Lambda \tag{23.5}$$

since these quantities scale as $M = [L]^{d-3}$, $A = [L]^{d-2}$, and $\Lambda = [L]^{-2}$. Since it is straightforward to show that $T = 4\frac{\partial M}{\partial A}$, relation (23.5) suggests that we regard Λ as a thermodynamic variable. Writing

$$P = -\frac{\Lambda}{8\pi} = \frac{(d-1)(d-2)}{16\pi\ell^2} \qquad V = \frac{\omega_{d-2}{r_+}^{d-1}}{d-1} \tag{23.6}$$

where $V = -8\pi\frac{\partial M}{\partial \Lambda}$ is the thermodynamic volume conjugate to P [6], we have

$$(d-3)M = (d-2)TS - 2PV \quad dM = TdS + VdP \tag{23.7}$$

as a modified Smarr relation and extended first law of thermodynamics.

We therefore have the complete thermodynamic correspondence

$$\begin{array}{rcl} \text{Enthalpy} & H \leftrightarrow M & \text{Mass} \\ \text{Temperature} & T \leftrightarrow \frac{\hbar\kappa}{2\pi} & \text{Surface Gravity} \\ \text{Entropy} & S \leftrightarrow \frac{A}{4G\hbar} & \text{Horizon Area} \\ \text{Pressure} & P \leftrightarrow -\frac{\Lambda}{8\pi} & \text{Cosmological Constant} \\ \text{1st law} & dE = TdS + VdP + \cdots \leftrightarrow & \text{1st law} \\ & & dM = \frac{\kappa}{8G\pi}dA + VdP + \cdots \end{array} \tag{23.8}$$

where the black hole work terms are $\sum_i \Omega_i dJ_i + \Phi dQ$ for multiply rotating and charged black holes. If included then the more general Smarr relation

$$\frac{d-3}{d-2}M = TS + \sum_i (\Omega^i - \Omega^i_\infty)J^i - \frac{2}{d-2}PV_h, + \Phi Q \tag{23.9}$$

holds for all possible $\left[\frac{d}{2}\right]$ rotation parameters, where the quantities Ω^i_∞ allow for the possibility of a rotating frame at infinity [21]. The quantity P is regarded as a thermodynamic pressure. From the perspective of cosmology, this is quite natural, since a negative cosmological constant induces a vacuum pressure. The mass M is then understood as a gravitational version of chemical enthalpy: the total energy of a system, which includes its internal energy and the energy required to "make room for it" by displacing its environment. For space times with $\Lambda < 0$, the PV term can be regarded as a displacement of vacuum energy.

The above results can be derived from a geometric argument [1], indicating that the extended Smarr formula (23.9) is on a firm footing.

23.3 Van der Waals Phase Transitions

One of the first things to emerge from treating Λ as a pressure term was the realization that charged black holes behave as Van der Waals fluids [9]. Recall that Van der Waals' equation [22]

$$\left(P + \frac{a}{v^2}\right)(v - b) = kT \Longrightarrow Pv^3 - (kT + bP)v^2 + av - ab = 0 \qquad (23.10)$$

where $v = V/N$ is the specific volume of the fluid, P its pressure, T its temperature, and k is Boltzmann's constant, is a modification of the ideal gas law that approximates the behaviour of real fluids, taking into account the nonzero size of molecules and the attraction between them. Critical points occur at isotherms $T = T_c$, where $P = P(v)$ has an inflection point

$$\frac{\partial P}{\partial v} = 0\,, \quad \frac{\partial^2 P}{\partial v} = 0 \Longrightarrow kT_c = \frac{8a}{27b}\,, \quad v_c = 3b\,, \quad P_c = \frac{a}{27b^2} \qquad (23.11)$$

in turn implying the universal relation $\frac{P_c v_c}{kT_c} = \frac{3}{8}$ for any such fluid.

Remarkably, charged black holes in any dimension obey the same basic relationships. The metric is still given by (23.1), but with $V = 1 - \frac{m}{r^{d-3}} + \frac{q^2}{r^{2(d-3)}} + \frac{r^2}{l^2}$, where the charge $Q = \frac{\sqrt{2(d-2)(d-3)}}{8\pi}\,\omega_{d-2}\, q$. The relation $T = \frac{V'(r_+)}{4\pi}$ and (23.6) yield

$$P = \frac{T}{v} - \frac{(d-3)}{\pi(d-2)v^2} + \frac{q^2(d-3)}{4\pi v^{2(d-2)}\kappa^{2d-5}} \qquad (23.12)$$

which is the equation of state. To ensure the correct relation between the 'geometric quantities' P and T and the physical pressure and temperature [10] I have identified $v = \frac{4r_+ l_P^{d-2}}{d-2} = \frac{r_+ l_P^{d-2}}{\kappa}$ as the specific volume of the corresponding fluid (instead of V), with $l_P^{d-2} = G_d\hbar/c^3$ the Planck length. The critical points are

$$(\kappa v)_c^{2(d-3)} = q^2(d-2)(2d-5) \quad T_c = \frac{(d-3)^2}{\pi\kappa v_c(2d-5)} \quad P_c = \frac{(d-3)^2}{16\pi\kappa^2 {v_c}^2} \qquad (23.13)$$

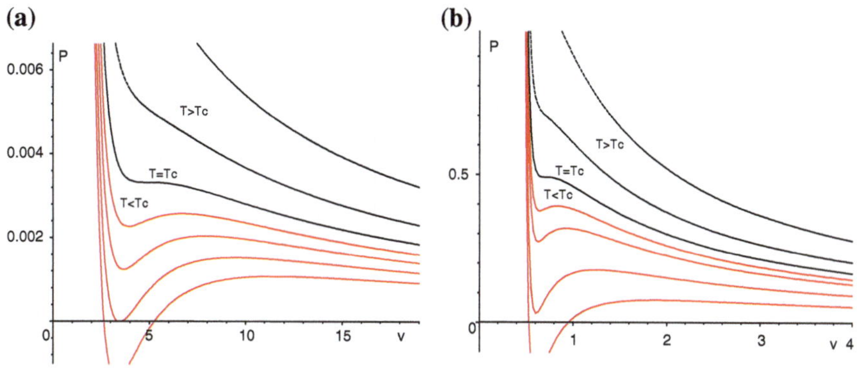

Fig. 23.1 $P - V$ diagrams of charged AdS black holes for $q = 1$. The isotherms decrease in temperature from *top* to *bottom*. The two *upper dark lines* correspond to the "ideal gas" one-phase behaviour for $T > T_c$, the critical isotherm $T = T_c$ is denoted by the *thick solid line*, the *lower* (*red*) *solid lines* correspond to a two-phase state occurring for $T < T_c$. **a** P versus v in $d = 4$ for various values of T. **b** P versus v in $d = 10$ for various values of T

yielding $\frac{P_c v_c}{T_c} = \frac{2d-5}{4d-8}$, which reduces to the $3/8$ result for $d = 4$. It is straightforward to show that the critical exponents (obtained by expanding the equation of state near the critical point in powers of the critical temperature and volume) are the same as those for a Van der Waals fluid; so far no black holes have been found that have different critical exponents. In any dimension, for $T < T_c$ there is a small–large black hole phase transition in the system; Fig. 23.1 illustrates this for $d = 4$ and $d = 10$. Maxwell's equal area law, which states two phases will coexist when the areas above and below a line of constant pressure drawn through a P-v curve are equal, can be used to compute the onset of the small/large black hole phase transition.

23.4 Reentrant Phase Transitions and the Triple Point

Rotating black holes have even more interesting thermodynamic behaviour than their charged counterparts. The general form of the metric [23]

$$ds^2 = -\,W\Big(1+\frac{r^2}{l^2}\Big)d\tau^2 + \frac{2m}{U}\Big(W d\tau - \sum_{i=1}^{N}\frac{a_i\mu_i^2 d\varphi_i}{\Xi_i}\Big)^2 + \frac{U dr^2}{F-2m} \qquad (23.14)$$

$$+\sum_{i=1}^{N}\frac{r^2+a_i^2}{\Xi_i}\mu_i^2 d\varphi_i^2 + \sum_{i=1}^{N+\varepsilon}\frac{r^2+a_i^2}{\Xi_i}d\mu_i^2 - \frac{l^{-2}}{W(1+r^2/l^2)}\Big(\sum_{i=1}^{N+\varepsilon}\frac{r^2+a_i^2}{\Xi_i}\mu_i d\mu_i\Big)^2$$

has N independent angular momenta J_i, described by N rotation parameters a_i, and solves the Einstein equations with cosmological constant $R_{ab} = -\frac{d-1}{l^2}g_{ab}$, where $d = 2N+1+\varepsilon$ and $\varepsilon = \pm 1$ corresponds to even/odd dimensionality. The azimuthal

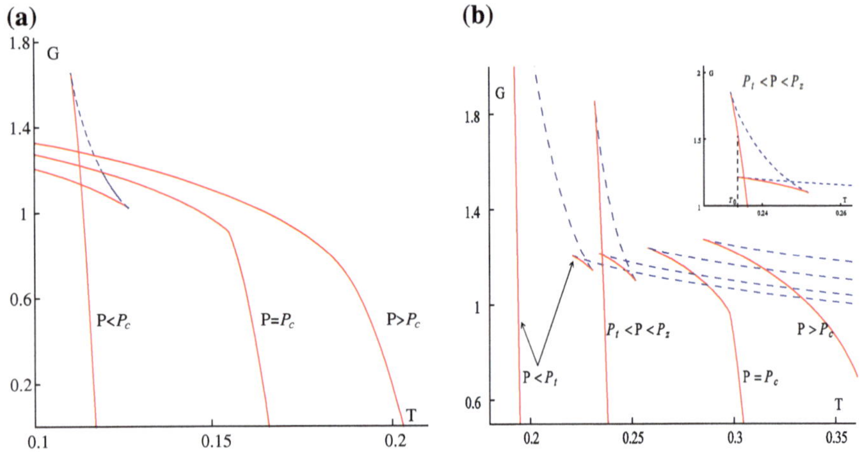

Fig. 23.2 Gibbs free energy for single-spinning black holes. Pressures increase from *left* to *right* and *solid-red/dashed-blue lines* correspond to C_P positive/negative respectively; at their joins C_P diverges. For $P \geq P_c$, the (*lower*) *large* BH *branch* is thermodynamically stable whereas the *upper branch* is unstable, with criticality at $P = P_c$. In $d = 6$, $P \approx 0.0564 \in (P_t, P_z)$ we observe a "zeroth-order phase transition": a discontinuity in the global minimum of G at $T = T_0 \approx 0.2339 \in (T_t, T_z)$ (denoted by the *vertical line* in the *inset*) signifying the onset of an reentrant phase transition. For $P < P_t$ only one branch of stable large BHs exists. **a** Gibbs free energy in $d = 5$ for various values of P and $J = 1$. **b** Gibbs free energy in $d = 6$ for various values of P and $J = 1$.

coordinates μ_i obey $\sum_{i=1}^{N+\varepsilon} \mu_i^2 = 1$, $\Xi_i = 1 - \frac{a_i^2}{l^2}$, $W = \sum_{i=1}^{N+\varepsilon} \frac{\mu_i^2}{\Xi_i}$, and

$$U = r^{\varepsilon} \sum_{i=1}^{N+\varepsilon} \frac{\mu_i^2}{r^2 + a_i^2} \prod_j^N (r^2 + a_j^2) \qquad F = r^{\varepsilon-2}\left(1 + \frac{r^2}{l^2}\right) \prod_{i=1}^N (r^2 + a_i^2) \tag{23.15}$$

Consider first the $d = 5$ case with one rotation parameter J_1, which can be normalized to unity without loss of generality. The behaviour of the Gibbs free energy $G = M - TS$ is illustrated in Fig. 23.2a: as the pressure decreases a characteristic swallowtail emerges, indicating a small/large black hole transition. For fixed P in this range as temperature decreases, G increases and the large black hole becomes smaller, until the crossover at which point a small black hole globally minimizes G. This kind of behaviour is qualitatively the same as that for all Reissner-Nordstrom-AdS black holes, with only the height and width of the swallowtail changing with dimensionality. The $C_P < 0$ line indicates a local thermodynamic instability where the Gibbs energy G is not a local minimum; at the cusps of the curves C_P diverges.

Novel behaviour is manifest for $d > 5$; the behaviour of G for $d = 6$ (and $J_2 = 0$) is plotted in Fig. 23.2b. For $P > P_t$ there is a standard first order small/large black hole phase transition below P_c. Above this value, the upper branch corresponds to small unstable black holes with $C_P < 0$, whereas the lower branch describes stable large black holes with $C_P > 0$.

However three separate phases of black holes emerge for for $P_c > P_z = P > P_t$: intermediate black holes (on the left), small (middle), and large (on the right). This holds for $T \in (T_t, T_z)$, $P \in (P_t, P_z)$. A standard first order phase transition separates the small and large black holes, but the intermediate and small ones are separated by a finite jump in G, which in this range has a discontinuous global minimum, as shown in the inset in Fig. 23.2b. This is the *reentrant phase transition* [16], first observed in a nicotine/water mixture [24], and since seen in multicomponent fluid systems, gels, ferroelectrics, liquid crystals, and binary gases [25]; similar behaviour occurs for Born-Infeld black holes [10]. For $T < T_t$ only one large black hole phase exists. The associated P-T diagram is plotted in Fig. 23.3a.

For $J_1 >> J_2 \neq 0$ the situation completely changes: a new branch of (locally) stable tiny cold black holes appears. Both the unstable branch of tiny hot black holes and the $J_2 = 0$ 'no black hole region' disappear, and the situation becomes quite similar to what happens when a small charge is added to a Schwarzschild black hole [18]. The zeroth-order phase transition is 'replaced' by a 'solid/liquid'-like phase transition of small to large black holes.

Once J_2 becomes sufficiently large, a new phenomenon occurs [17]: a triple point and a second critical point emerge from the coexistence line at $P_{tr} = P_{c_2} \approx 0.09577$ and $T_{tr} = T_{c_2} \approx 0.30039$. As J_2/J_1 increases, the triple point moves away from the second critical point (the values of T_t and P_t decrease), and a small/intermediate/large black hole phase transition occurs, resembling a solid/liquid/gas phase transition. The

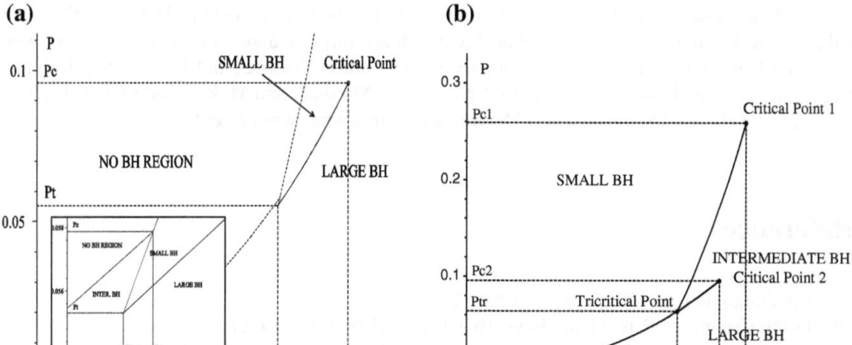

Fig. 23.3 Reentrant phases and triple points. The *left/right* P-T diagrams respectively depict reentrant phase behaviour and a triple (tricritical) point. *Solid black lines* indicate first order phase transitions between the various small/intermediate/large black hole types. The *red solid line* (*inset* at *left*) indicates the 'coexistence line' of small and intermediate black holes, separated by a finite gap in G, indicating a reentrant phase transition. It commences from (T_z, P_z) and terminates at (P_t, T_t). The "No BH region" is to the *left* of the *dashed oblique curve*, containing the (T_z, P_z) point. The diagram at *right* is analogous to the solid/liquid/gas phase diagram; note that the solid-liquid coexistence line does not extend to infinity but rather terminates, similar to the "liquid/gas" coexistence line, in a critical point. Both diagrams are in $d = 6$ dimensions. **a** Reentrant Phase behaviour for $J_2 = 0$. **b** Triple point for $J_2 = 0.05J_1$

situation is illustrated in Fig. 23.3b. The first critical point simultaneously moves towards the triple point (T_{c_1} and P_{c_1} decrease), and at $J_2 = 0.08121 J_1$ both critical points occur at the same pressure $P_{c_1} = P_{c2} \approx 0.0953$, whereas $T_{c_1} \approx 0.2486 < T_{c_2} \approx 0.2997$. Increasing J_2/J_1 even further, the first critical point moves closer and closer to the triple point. For $J_2 \approx 0.0985 J_1$ the two merge at $P_{tr} = P_{c_1} \approx 0.049$; for larger J_2/J_1 only the second critical point remains.

23.5 Summary

The physics associated with extended black hole thermodynamics is only just beginning to be explored. There are many other issues to consider, including a full analysis of rotating black holes in higher dimensions, the significance of $\Lambda > 0$, and higher curvature theories such as Lovelock theory though some work has been done on these latter aspects [15, 26].

One of the most interesting avenues for future research are the implications of the extended phase space for gauge-gravity duality. Since neither the existence of the reentrant phase transition nor of the triple point depends on a variable $\Lambda \sim P$, for any fixed P within the allowed range, these phenomena will take place. Consequently there will be a corresponding reentrant phase transition in the AdS/CFT context within the allowed range of N in the dual $SU(N)$ gauge theory.

Acknowledgments I am grateful to N. Altimirano, B. Dolan, S. Gunasekaran, D. Kastor, D. Kubiznak, Z. Sherkatghanad, and J. Traschen for the interesting collaborations that led to the work described here, which was supported in part by the Natural Sciences and Engineering Research Council of Canada. I am also grateful to P. Nicolini, J. Mureika, and M. Kaminski for their efforts in organizing the Karl Schwarzschild Meeting where this work was presented.

References

1. B.P Dolan, inTech (2012). arXiv:1209.1272
2. J.D.E. Creighton, R.B. Mann, Phys. Rev. D **52**, 4569–4587 (1995)
3. M.M. Caldarelli, G. Cognola, D. Klemm, Class. Quant. Grav. **17**, 399–420 (2000)
4. D. Kastor, S. Ray, J. Traschen. Class. Quant. Grav. **26**, 195011 (2009)
5. B.P. Dolan, Class. Quant. Grav. **28**, 125020 (2011)
6. M. Cvetic, G.W. Gibbons, D. Kubiznak, C.N. Pope, Phys. Rev. D **84**, 024037 (2011)
7. A. Larranaga, A. Cardenas, J. Korean Phys. Soc. **60**, 987–992 (2012)
8. G.W. Gibbons, AIP Conf. Proc. **1460**, 90–100 (2012)
9. D. Kubiznak, R.B. Mann, JHEP **1207**, 033 (2012)
10. S. Gunasekaran, R.B. Mann, D. Kubiznak, JHEP **1211**, 110 (2012)
11. A. Belhaj, M. Chabab, H. El Moumni, M.B. Sedra, Chin. Phys. Lett. **29**, 100401 (2012)
12. H. Lu, Y. Pang, C.N. Pope, J.F. Vazquez-Poritz. Phys. Rev. D **86**, 044011 (2012)
13. A. Smailagic, E. Spallucci. J. Grav. **2013**, 525696 (2013)
14. S.H. Hendi, M.H. Vahidinia. Phys. Rev. D **88**(8), 084045 (2013)
15. B.P. Dolan et al., Phys. Rev. D **87**, 104017 (2013)
16. N. Altamirano, D. Kubiznak, R.B. Mann, Phys. Rev. D **88**, 101502 (2013)

17. N. Altamirano, D. Kubiznak, R.B. Mann, Z. Sherkatghanad. Class. Quant. Grav. **31**, 042001 (2014)
18. A. Chamblin, R. Emparan, C.V. Johnson, R.C. Myers, Phys. Rev. D **60**, 064018 (1999)
19. A. Chamblin, R. Emparan, C.V. Johnson, R.C. Myers, Phys. Rev. D **60**, 104026 (1999)
20. L. Smarr, Phys. Rev. Lett. **30**, 71–73 (1973)
21. G.W. Gibbons, M.J. Perry, C.N. Pope, Class. Quant. Grav. **22**, 1503–1526 (2005)
22. N. Goldenfeld. *Lectures on Phase Transitions and the Renormalization Group* (Westview Press, New York, 1992)
23. G.W. Gibbons, H. Lu, D.N. Page, C.N. Pope, Phys. Rev. Lett. **93**, 171102 (2004)
24. C. Hudson, Z. Phys. Chem. **47**, 113 (1904)
25. T. Narayanan, A. Kumar, Phys. Rep. **249**, 135 (1994)
26. Rong-Gen Cai, Li-Ming Cao, Li Li, Run-Qiu Yang, JHEP **1309**, 005 (2013)

Chapter 24
Black Holes in Supergravity

Kellogg S. Stelle

Abstract A brief review is given of the use of duality symmetries to form orbits of supergravity black-hole solutions and their relation to extremal (i.e. BPS) solutions at the limits of such orbits. An important technique in this analysis uses a timelike dimensional reduction and exchanges the stationary black-hole problem for a nonlinear sigma-model problem. Families of BPS solutions are characterized by nilpotent orbits under the duality symmetries, based upon a tri-graded or penta-graded decomposition of the corresponding duality group algebra.

24.1 Introduction

Aside from the general mathematical interest in classifying black hole solutions of any kind, the study of families of such solutions is also of current interest because it touches other important issues in theoretical physics. For example, the classification of BPS and non-BPS black holes forms part of a more general study of branes in supergravity and superstring theory. Branes and their intersections, as well as their worldvolume modes and attached string modes, are key elements in phenomenological approaches to the marriage of string theory with particle physics phenomenology. The related study of nonsingular and horizon-free BPS gravitational solitons is also central to the "fuzzball" proposal of BPS solutions as candidate black-hole quantum microstates.

The search for supergravity solutions with assumed Killing symmetries can be recast as a Kaluza-Klein problem [1–4]. To see this, consider a 4D theory with a nonlinear bosonic symmetry G_4 (e.g. the "duality" symmetry E_7 for maximal $N = 8$ supergravity). Scalar fields take their values in a target space $\Phi_4 = G_4/H_4$, where H_4 is the corresponding linearly realized subgroup, generally the maximal compact subgroup of G_4 (e.g. $SU(8) \subset E_7$ for $N = 8$ SG). The search will be constrained by the following considerations:

K.S. Stelle (✉)
The Blackett Laboratory, Imperial College London,
Prince Consort Road, London SW7 2AZ, UK
e-mail: k.stelle@imperial.ac.uk

P. Nicolini et al. (eds.), *1st Karl Schwarzschild Meeting on Gravitational Physics*,
Springer Proceedings in Physics 170, DOI 10.1007/978-3-319-20046-0_24

- We assume that a solution spacetime is asymptotically flat or asymptotically Taub-NUT and that there is a 'radial' function r which is divergent in the asymptotic region, $g^{\mu\nu}\partial_\mu r \partial_\nu r \sim 1 + \mathscr{O}(r^{-1})$.
- Searching for stationary solutions amounts to assuming that a solution possesses a timelike Killing vector field $\kappa_\mu(x)$. Lie derivatives with respect to κ_μ are assumed to vanish on all fields. The Killing vector κ_μ will be assumed to have $W := -g_{\mu\nu}\kappa^\mu\kappa^\nu \sim 1 + \mathscr{O}(r^{-1})$.
- We also assume asymptotic hypersurface orthogonality, i.e. $\kappa^\nu(\partial_\mu\kappa_\nu - \partial_\nu\kappa_\mu) \sim \mathscr{O}(r^{-2})$. In any vielbein frame, the curvature will then fall off as $R_{abcd} \sim \mathscr{O}(r^{-3})$.

The 3D theory obtained after dimensional reduction with respect to a timelike Killing vector κ_μ will have an Abelian principal bundle structure, with a metric

$$ds^2 = -W(dt + B_i dx^i)^2 + W^{-1}\gamma_{ij} dx^i dx^j \tag{24.1}$$

where t is a coordinate adapted to the timelike Killing vector κ_μ and γ_{ij} is the metric on the 3-dimensional hypersurface $\mathscr{M}_3$ at constant t. If the 4D theory also has Abelian vector fields $\mathscr{A}_\mu$, they similarly reduce to 3D as

$$4\sqrt{4\pi G}\mathscr{A}_\mu dx^\mu = U(dt + B_i dx^i) + A_i dx^i \tag{24.2}$$

The timelike reduced 3D theory will have a G/H^* coset space structure similar to the G/H coset space structure of a 3D theory reduced with a spacelike Killing vector. Thus, for the spacelike reduction of maximal supergravity down to 3D, one obtains an $E_8/SO(16)$ theory from the sequence of dimensional reductions descending from $D = 11$ [5]. The resulting 3D theory has this exceptional symmetry because 3D Abelian vector fields can be dualized to scalars; this also happens for the analogous theory subjected to a timelike reduction to 3D. The resulting 3D theory contains 3D gravity coupled to a G/H^* nonlinear sigma model.

Although the numerator group G for a timelike reduction is the same as that obtained in a spacelike reduction, the divisor group H^* for a timelike reduction is a *noncompact* form of the spacelike divisor group H [2]. A consequence of this $H \to H^*$ change and the dualization of vectors is the appearance of *negative-sign* kinetic terms for some 3D scalars.

Consequently, maximal supergravity, after a timelike reduction to 3D and the subsequent dualization of 29 vectors to scalars, has a bosonic sector containing 3D gravity coupled to a $E_8/SO^*(16)$ nonlinear sigma model with 128 scalar fields. As a consequence of the timelike dimensional reduction and vector dualizations, however, the scalars do not all have the same signs for their "kinetic" terms:

- There are 72 positive-sign scalars: 70 descending directly from the 4D theory, one emerging from the 4D metric and one more coming from the $D = 4 \to D = 3$ Kaluza-Klein vector, subsequently dualized to a scalar.
- There are 56 negative-sign scalars: 28 descending directly from the time components of the 28 4D vectors, and another 28 emerging from the 3D vectors obtained

from spatial components of the 28 4D vectors, becoming then negative-sign scalars after dualization.

The sigma-model structure of this timelike reduced maximal theory is $E_8/SO^*(16)$. The $SO^*(16)$ divisor group is not an $SO(p,q)$ group defined via preservation of an indefinite metric. Instead it is constructed starting from the SO(16) Clifford algebra $\{\Gamma^I, \Gamma^J\} = 2\delta^{IJ}$ and then by forming the complex U(8)-covariant oscillators $a_i := \frac{1}{2}(\Gamma_{2i-1} + \mathrm{i}\Gamma_{2i})$ and $a^i \equiv (a_i)^\dagger = \frac{1}{2}(\Gamma_{2i-1} - \mathrm{i}\Gamma_{2i})$. These satisfy the standard fermi oscillator annihilation/creation anticommutation relations

$$\{a_i, a_j\} = \{a^i, a^j\} = 0, \qquad \{a_i, a^j\} = \delta_i{}^j \tag{24.3}$$

The 120 $SO^*(16)$ generators are then formed from the 64 hermitian U(8) generators $a_i{}^j$ plus the $2 \times 28 = 56$ antihermitian combinations of $a_{ij} \pm a^{ij}$. Under $SO^*(16)$, the vector representation and the antichiral spinor are pseudo-real, while the 128-dimensional chiral spinor representation is *real*. This is the representation under which the 72 + 56 scalar fields transform in the $E_8/SO^*(16)$ sigma model.

The 3D classification of extended supergravity stationary solutions *via* timelike reduction generalizes the 3D supergravity systems obtained from spacelike reduction [6]. This also connects with $N = 2$ models with coupled vectors [7] and $N = 4$ models with vectors, where solutions have also been generated using duality symmetries [8, 9]

The process of timelike dimensional reduction down to 3 dimensions together with dualization of all form-fields to scalars produces an Euclidean gravity theory coupled to a G/H^* nonlinear sigma model, $I_\sigma = \int d^3x \sqrt{\gamma}(R(\gamma) - \frac{1}{2}G_{AB}(\phi)\partial_i\phi^A\partial_j\phi^B\gamma^{ij})$, where $G_{AB}(\phi)$ is the G/H^* sigma-model target-space metric and γ_{ij} is the 3D metric. Varying this action produces the 3D field equations

$$\frac{1}{\sqrt{\gamma}}\nabla_i(\sqrt{\gamma}\gamma^{ij}G_{AB}(\phi)\partial_j\phi^B) = 0 \tag{24.4}$$

$$R_{ij}(\gamma) = \tfrac{1}{2}G_{AB}(\phi)\partial_i\phi^A\partial_j\phi^B \tag{24.5}$$

where ∇_i is a doubly covariant derivative (for the 3D space $\mathscr{M}_3$ and for the G/H^* target space).

Now one can make the simplifying assumption that $\phi^A(x) = \phi^A(\sigma(x))$, with a single intermediate map $\sigma(x)$. Subject to this assumption, the field equations become

$$\nabla^2\sigma\frac{d\phi^A}{d\sigma} + \gamma^{ij}\partial_i\sigma\partial_j\sigma[\frac{\partial^2\phi^A}{d\sigma^2} + \Gamma^A_{BC}(G)\frac{d\phi^B}{d\sigma}\frac{d\phi^C}{d\sigma}] = 0 \tag{24.6}$$

$$R_{ij} = \left(\tfrac{1}{2}G_{AB}(\phi)\frac{d\phi^A}{d\sigma}\frac{d\phi^B}{d\sigma}\right)\partial_i\phi^A\partial_j\phi^B \tag{24.7}$$

Now one uses the gravitational Bianchi identity $\nabla^i(R_{ij} - \frac{1}{2}\gamma_{ij}R) \equiv 0$ to obtain $\frac{1}{4}\frac{d}{d\sigma}(G_{AB}(\phi)\frac{d\phi^A}{d\sigma}\frac{d\phi^B}{d\sigma})(\nabla^i\sigma\partial_i\sigma) = 0$. Requiring separation of the $\sigma(x)$ properties from the $\frac{d}{d\sigma}$ properties leads to the conditions

$$\nabla^2\sigma = 0 \tag{24.8}$$

$$\frac{d^2\phi^A}{d\sigma^2} + \Gamma^A_{BC}(G)\frac{d\phi^B}{d\sigma}\frac{d\phi^C}{d\sigma} = 0 \tag{24.9}$$

$$\frac{d}{d\sigma}\left(G_{AB}(\phi)\frac{d\phi^A}{d\sigma}\frac{d\phi^B}{d\sigma}\right) = 0 \tag{24.10}$$

The first equation (24.8) above implies that $\sigma(x)$ is a *harmonic map* from the 3D space $\mathscr{M}_3$ into a curve $\phi^A(\sigma)$ in the G/H^* target space. The second equation (24.9) implies that $\phi^A(\sigma)$ is a *geodesic* in G/H^*. The third equation (24.10) implies that σ is an *affine parameter*. The decomposition of $\phi : \mathscr{M}_3 \to G/H^*$ into a harmonic map $\sigma : \mathscr{M}_3 \to \mathbb{R}$ and a geodesic $\phi : \mathbb{R} \to G/H^*$ is in accordance with a general theorem on harmonic maps [10] according to which the composition of a harmonic map with a totally geodesic one is again harmonic. Such factorization into geodesic and harmonic maps is also characteristic of general higher-dimensional p-brane supergravity solutions [1, 3, 4].

Here is a sketch of the map composition:

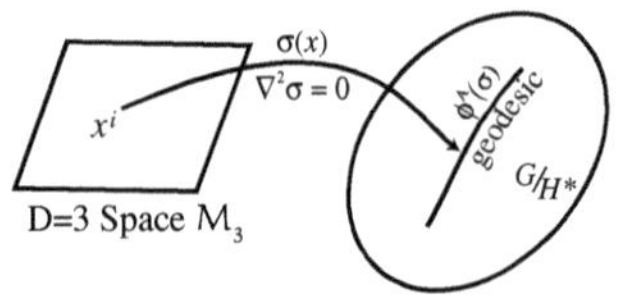

Now define the Komar two-form $K \equiv \partial_\mu \kappa_\nu dx^\mu \wedge dx^\nu$. This is invariant under the action of the timelike isometry and, by the asymptotic hypersurface orthogonality assumption, is asymptotically horizontal. This condition is equivalent to the requirement that the scalar field B dual to the Kaluza-Klein vector arising out of the 4D metric must vanish like $\mathscr{O}(r^{-1})$ as $r \to \infty$. In this case, one can define the Komar mass and NUT charge by (where s^* indicates a pull-back to a section) [11]

$$m \equiv \frac{1}{8\pi}\int_{\partial\mathscr{M}_3} s^* \star K \qquad n \equiv \frac{1}{8\pi}\int_{\partial\mathscr{M}_3} s^* K \tag{24.11}$$

The Maxwell field also defines charges. Using the Maxwell field equation $d\star\mathscr{F} = 0$, where $\mathscr{F} \equiv \delta\mathscr{L}/\delta F$ is a linear combination of the two-form field strengths F depending on the 4D scalar fields, and using the Bianchi identity $dF = 0$, one obtains conserved electric and magnetic charges:

$$q \equiv \frac{1}{2\pi}\int_{\partial\mathscr{M}_3} s^* \star \mathscr{F} \qquad p \equiv \frac{1}{2\pi}\int_{\partial\mathscr{M}_3} s^* F\,. \tag{24.12}$$

Now consider these charges from the three-dimensional point of view in order to clarify their transformation properties under the 3D duality group G The three-dimensional theory is described in terms of a coset representative $\mathscr{V} \in G/H^*$. The

Maurer–Cartan form $\mathcal{V}^{-1}d\mathcal{V}$ for $\mathfrak{g}$ decomposes as

$$\mathcal{V}^{-1}d\mathcal{V} = Q + P \ , \qquad Q \equiv Q_\mu dx^\mu \in \mathfrak{h}^* \ , \quad P \equiv P_\mu dx^\mu \in \mathfrak{g} \ominus \mathfrak{h}^* \ . \tag{24.13}$$

Then the three-dimensional scalar-field equation of motion can be rewritten as $d \star \mathcal{V} P \mathcal{V}^{-1} = 0$, so the $\mathfrak{g}$-valued "Noether current" is $\star\mathcal{V} P \mathcal{V}^{-1}$. Since the three-dimensional theory is Euclidean, one cannot properly speak of a conserved charge. Nevertheless, since $\star\mathcal{V} P \mathcal{V}^{-1}$ is d-closed, the integral of this 2-form over a given homology cycle does not depend on the particular representative of that cycle.

As a result, for stationary solutions, the integral of this three-dimensional 2-form current, taken over any spacelike closed surface $\partial\mathcal{M}_3$ containing in its interior all the singularities and topologically non-trivial subspaces of a solution, defines a $\mathfrak{g} \ominus \mathfrak{h}^*$-valued Noether-charge matrix $\mathcal{C}$:

$$\mathcal{C} \equiv \frac{1}{4\pi} \int_{\partial\mathcal{M}_3} \star\mathcal{V} P \mathcal{V}^{-1} \tag{24.14}$$

This transforms in the adjoint representation of the duality group G in accordance with the standard non-linear action of G on $\mathcal{V} \in G/H^*$. For asymptotically-flat solutions, $\mathcal{V}$ can be arranged to tend asymptotically at infinity to the identity matrix; the charge matrix $\mathcal{C}$ in that case is simply given by the asymptotic value of the one-form P:

$$P = \mathcal{C} \frac{dr}{r^2} + \mathcal{O}(r^{-2}) \ . \tag{24.15}$$

Now follow the evolution of the duality group G down a couple of steps in dimensional reduction. In $D = 5$, maximal supergravity has the maximally noncompact duality group $E_{6,6}$, with the 42 $D = 5$ scalar fields taking their values in the coset space $E_{6,6}/USp(8)$, while the 1-form (i.e. vector) fields transform in the **27** of $E_{6,6}$.

Proceeding on down to 4D, the 27 $D = 5$ vectors produce new scalars upon dimensional reduction, and one also gets a new Kaluza-Klein scalar emerging from the $D = 5$ metric, making up the total of 70 scalars in the 4D theory. These take their values in $E_{7,7}/SU(8)$, while the 4D vector field strengths transform in the **56** of $E_{7,7}$. The new KK scalar corresponds to a $\mathfrak{gl}_1$ *grading generator* of $E_{7,7}$, leading to a tri-graded decomposition of the $E_{7,7}$ algebra as follows:

$$\mathfrak{e}_{7,7} \simeq \overline{\mathbf{27}}^{(-2)} \oplus (\mathfrak{gl}_1 \oplus \mathfrak{e}_{6,6})^{(0)} \oplus \mathbf{27}^{(2)} \tag{24.16}$$

where the superscripts indicate the $\mathfrak{gl}_1$ grading.

Continuing on down to 3D *via* a timelike reduction, one encounters a new phenomenon: 3D vectors can now be dualized to scalars. This is already clear in the timelike reduction of pure 4D GR to 3D, where one obtains a two-scalar system taking values in $SL(2,\mathbb{R})/SO(2)$, where $SL(2,\mathbb{R})$ is the Ehlers group [12]. Its generators can be written

$$\gamma \boldsymbol{h} \oplus \varepsilon \boldsymbol{e} \oplus \varphi \boldsymbol{f} = \begin{pmatrix} \gamma & \varepsilon \\ \varphi & -\gamma \end{pmatrix} \tag{24.17}$$

and its Lie algebra is $[\boldsymbol{h}, \boldsymbol{e}] = 2\boldsymbol{e}, \quad [\boldsymbol{h}, \boldsymbol{f}] = -2\boldsymbol{f}, \quad [\boldsymbol{e}, \boldsymbol{f}] = \boldsymbol{h}$.

Accordingly, in reducing from 4D to 3D a supergravity theory with 4D symmetry group G_4, with corresponding Lie algebra $\mathfrak{g}_4$ and with vectors transforming in the $\mathfrak{l}_4$ representation of $\mathfrak{g}_4$, one obtains a penta-graded structure for the 3D Lie algebra $\mathfrak{g}$, with the Ehlers $\boldsymbol{h}$ now acting as the grading generator $\mathbf{1}^{(0)}$:

$$\mathfrak{g} \simeq \mathbf{1}^{(-2)} \oplus \overline{\mathfrak{l}_4}^{(-1)} \oplus (\mathbf{1} \oplus \mathfrak{g}_4)^{(0)} \oplus \mathfrak{l}_4^{(+1)} \oplus \mathbf{1}^{(2)} \tag{24.18}$$

For example, in 3D maximal supergravity one obtains in this way $\mathfrak{e}_{8,8}$:

$$\mathfrak{e}_{8,8} \simeq \mathbf{1}^{(-2)} \oplus \overline{\mathbf{56}}^{(-1)} \oplus (\mathbf{1} \oplus \mathfrak{e}_{7,7})^{(0)} \oplus \mathbf{56}^{(+1)} \oplus \mathbf{1}^{(2)} \quad \text{(248 generators)} \tag{24.19}$$

Now apply this to the decomposition of the coset-space structure for the 3D scalar fields and the charge matrix $\mathscr{C}$. In 4D, the scalars are associated to the coset generators $\mathfrak{g}_4 \ominus \mathfrak{h}_4$, where $\mathfrak{h}_4$ is the Lie algebra of the 4D divisor group H_4. The representation carried by the 4D electric and magnetic charges q and p is $\mathfrak{l}_4$. Then the 3D scalars and the charge matrix $\mathscr{C}$ can be decomposed into three irreducible representations with respect to $\mathfrak{so}(2) \oplus \mathfrak{h}_4$ according to

$$\mathfrak{g} \ominus \mathfrak{h}^* \cong \big(\mathfrak{sl}(2, \mathbb{R}) \ominus \mathfrak{so}(2)\big) \oplus \mathfrak{l}_4 \oplus \big(\mathfrak{g}_4 \ominus \mathfrak{h}_4\big) \tag{24.20}$$

The metric induced by the $\mathfrak{g}$ algebra's Cartan-Killing metric on this coset space is positive definite for the first and last terms, but is negative definite for λ_4. One associates the $\mathfrak{sl}(2, \mathbb{R}) \ominus \mathfrak{so}(2)$ components with the Komar mass and the Komar NUT charge, while the $\mathfrak{l}_4$ components are associated with the electromagnetic charges. The remaining $\mathfrak{g}_4 \ominus \mathfrak{h}_4$ charges belong to the Noether current of the 4D theory.

Breitenlohner et al. [2] proved that if G is simple, all the non-extremal single-black-hole solutions of a given theory lie on the H^* orbit of a Kerr solution. Moreover, all *static* solutions regular outside the horizon with a charge matrix satisfying $\operatorname{Tr} \mathscr{C}^2 > 0$ lie on the H^*-orbit of a Schwarzschild solution. (Turning on and off angular momentum requires consideration of the $D = 2$ duality group generalizing the Geroch ${A_1}^1$ group.)

Using Weyl coordinates, where the 4D metric takes the form

$$ds^2 = f(x, \rho)^{-1}[e^2 k(x, \rho)(dx^2 + d\rho^2) + \rho^2 d\phi^2] + f(x, \rho)(dt + A(x, \rho)d\phi)^2 \,, \tag{24.21}$$

the coset representative $\mathscr{V}$ associated to the Schwarzschild solution with mass m can be written in terms of the non-compact generator $\boldsymbol{h}$ of the Ehlers $\mathfrak{sl}(2, \mathbb{R})$ only, i.e.

$$\mathscr{V} = \exp\left(\frac{1}{2} \ln \frac{r - m}{r + m} \boldsymbol{h}\right) \quad \rightarrow \quad \mathscr{C} = m\boldsymbol{h} \tag{24.22}$$

For the maximal $N = 8$ theory with symmetry $E_{8(8)}$ (and also for the exceptional 'magic' $N = 2$ supergravity [13] with symmetry $E_{8(-24)}$), one has $h = \text{diag}[2, 1, 0, -1, -2]$, so

$$\boldsymbol{h}^5 = 5\boldsymbol{h}^3 - 4\boldsymbol{h} \tag{24.23}$$

Consequently, the charge matrix $\mathscr{C}$ satisfies in all cases the characteristic equation

$$\mathscr{C}^5 = 5c^2\mathscr{C}^3 - 4c^4\mathscr{C} \tag{24.24}$$

where $c^2 \equiv \frac{1}{\text{Tr}\,\boldsymbol{h}^2}\text{Tr}\,\mathscr{C}^2$ is the *extremality parameter* ($c^2 = 0$ for extremal static solutions; $c^2 = m^2$ for Schwarzschild). Moreover, for all but the two exceptional E_8 cases, a stronger constraint is actually satisfied by the charge matrix $\mathscr{C}$:

$$\mathscr{C}^3 = c^2\mathscr{C} \tag{24.25}$$

The characteristic equation selects acceptable orbits of solutions, i.e. orbits not exclusively containing solutions with naked singularities. It determines $\mathscr{C}$ in terms of the mass and NUT charge and the 4D electromagnetic charges.

The parameter c^2 is the same as the (target space velocity)^2 of the harmonic-map discussion: $c^2 = v^2$. The Maxwell-Einstein theory is the simplest example with an indefinite-signature sigma-model metric, for the scalar-field target space $G/H^* = \text{SU}(2,1)/\text{S}(\text{U}(1,1) \times \text{U}(1))$. The Maxwell-Einstein charge matrix is

$$\mathscr{C}_{\text{ME}} = \begin{pmatrix} m & n & -z/\sqrt{2} \\ n & -m & \mathrm{i}z/\sqrt{2} \\ \bar{z}/\sqrt{2} & \mathrm{i}\bar{z}/\sqrt{2} & 0 \end{pmatrix} \in su(2,2) \ominus u(1,1) \tag{24.26}$$

where $z = q + \mathrm{i}p$ is the complex electromagnetic charge. The Maxwell-Einstein extremality parameter is $c^2 = m^2 + n^2 - z\bar{z}$. Solutions fall into three categories: $c^2 > 0$ nonextremal, $c^2 = 0$ extremal and $c^2 < 0$ hyperextremal. The hyperextremal solutions have naked singularities, while the nonextremal and extremal solutions have their singularities cloaked by horizons.

Extremal solutions have $c^2 = 0$, implying that the charge matrix $\mathscr{C}$ becomes *nilpotent*: $\mathscr{C}^5 = 0$ in the E_8 cases and $\mathscr{C}^3 = 0$ otherwise.

For $\mathscr{N}$ extended supergravity theories, one finds $H^* \cong \text{Spin}^*(2\mathscr{N}) \times H_0$ and the charge matrix $\mathscr{C}$ transforms as a Weyl spinor of $\text{Spin}^*(2\mathscr{N})$ also valued in a representation of $\mathfrak{h}_0$ (where $\mathfrak{h}_0$ acts on the matter content of reducible $N = 4$ theories). As in the $\text{SO}^*(16)$ case considered earlier, one defines the $\text{Spin}^*(2\mathscr{N})$ fermionic oscillators

$$a_i := \frac{1}{2}\left(\Gamma_{2i-1} + \mathrm{i}\Gamma_{2i}\right) \qquad a^i \equiv (a_i)^\dagger = \frac{1}{2}\left(\Gamma_{2i-1} - \mathrm{i}\Gamma_{2i}\right) \tag{24.27}$$

for $i, j, \ldots = 1, \ldots, \mathscr{N}$. These obey standard fermionic annihilation and creation anticommutation relations. Using this annihilation/creation oscillator basis, the charge matrix $\mathscr{C}$ can be represented as a state (where $a_i|0>=0$)

$$|\mathscr{C}> \equiv \left(w + Z_{ij}a^i a^j + \Sigma_{ijkl}a^i a^j a^k a^l + \cdots\right)|0\rangle \tag{24.28}$$

From the requirement that the dilatino fields be left invariant under the unbroken supersymmetry of a BPS solution, one derives a 'Dirac equation' for the charge state vector,

$$\left(\varepsilon^i_\alpha a_i + \Omega_{\alpha\beta}\varepsilon^\beta_i a^i\right)|\mathscr{C}\rangle = 0 \tag{24.29}$$

where $(\varepsilon^i_\alpha, \varepsilon^\alpha_i)$ is the asymptotic (for $r \to \infty$) value of the Killing spinor and $\Omega_{\alpha\beta}$ is a symplectic form on C^{2n} in cases with n/N preserved supersymmetry.

Note that $c^2 = 0 \Longleftrightarrow \langle\mathscr{C}|\mathscr{C}\rangle = 0$ is a *weaker* condition than the supersymmetry Dirac equation. Extremal and BPS are not always synonymous conditions, although they coincide for $\mathcal{N} \leq 5$ pure supergravities. They are not synonymous for $\mathcal{N} = 6$ and 8 or for theories with vector matter coupling.

Earlier analysis of the orbits of the 4D symmetry groups G_4 [14] heavily used the Iwasawa decomposition

$$g = u_{(g,Z)} \exp\left(\ln \lambda_{(g,Z)}\, z\right) b_{(g,Z)} \tag{24.30}$$

with $u_{(g,Z)} \in H_4$ and $b_{(g,Z)} \in \mathfrak{B}_Z$ where $\mathfrak{B}_Z \subset G_4$ is the parabolic subgroup that leaves the charges Z invariant up to a multiplicative factor $\lambda_{(g,Z)}$. This multiplicative factor can be compensated for by 'trombone' transformations combining Weyl scalings with compensating dilational coordinate transformations, leading to a formulation of active symmetry transformations that map solutions into other solutions with *unchanged asymptotic values* of the spacetime metric and scalar fields.

The 4D 'trombone' transformation finds a natural home in the parabolic subgroup of the 3D duality group G. The 3D structure is characterized by the fact that the Iwasawa decomposition *breaks down* for noncompact divisor groups H^*.

The Iwasawa decomposition does, however work "almost everywhere" in the 3D solution space. The places where it fails are precisely the extremal suborbits of the duality group. This has the consequence that G *does not act transitively on its own orbits*. There are G transformations which allow one to send $c^2 \to 0$, thus landing on an extremal (generally BPS) suborbit. However, one cannot then invert the map and return to a generic non-extremal solution from the extremal solution reached on a given G trajectory.

The above framework applies equally to multi-centered as to single-centered solutions [15, 16]. One may start from a general ansatz

$$\mathscr{V}(x) = \mathscr{V}_0 \exp(-\sum_n \mathscr{H}^n(x)\mathscr{C}_n) \tag{24.31}$$

with Lie algebra elements $\mathscr{C}_n \in \mathfrak{g} \ominus \mathfrak{h}^*$ and functions $\mathscr{H}^n(x)$ to be determined by the equations of motion. Defining as above $\mathscr{V}^{-1}d\mathscr{V} = Q+P$ and restricting P to depend linearly on the functions $\mathscr{H}^n(x)$, one finds the requirement $[\mathscr{C}_m, [\mathscr{C}_n, \mathscr{C}_p]] = 0$. The Einstein and scalar equations of motion then reduce to

$$R_{\mu\nu} - \frac{1}{2}g_{\mu\nu}R = \sum_{mn} \partial_\mu \mathscr{H}^m \partial_\nu \mathscr{H}^n \operatorname{Tr} \mathscr{C}_m \mathscr{C}_n \qquad d \star d\mathscr{H}^n = 0 \tag{24.32}$$

Restricting attention to solutions where the 3-space is flat then requires $\operatorname{Tr} \mathscr{C}_m \mathscr{C}_n = 0$. The resulting system generalizes that found in [3, 4]. Solving $[\mathscr{C}_m, [\mathscr{C}_n, \mathscr{C}_p]] = 0 = \operatorname{Tr} \mathscr{C}_m \mathscr{C}_n$ is now reduced to an algebraic problem amenable to the above nilpotent-orbit analysis: non-extremal and extremal stationary solutions can be formed from extremal single-hole constituents.

In summary, what has been developed here is a quite general framework for the analysis of stationary supergravity solutions using duality orbits. The Noether charge matrix $\mathscr{C}$ satisfies a characteristic equation $\mathscr{C}^5 = 5c^2\mathscr{C}^3 - 4c^4\mathscr{C}$ in the maximal E_8 cases and $\mathscr{C}^3 = c^2\mathscr{C}$ in the non-maximal cases, where $c^2 \equiv \frac{1}{\operatorname{Tr} h^2} \operatorname{Tr} \mathscr{C}^2$ is the extremality parameter. Extremal solutions are characterized by $c^2 = 0$, and $\mathscr{C}$ becomes nilpotent ($\mathscr{C}^5 = 0$ or $\mathscr{C}^3 = 0$) on the corresponding extremal suborbits. BPS solutions have a charge matrix $\mathscr{C}$ satisfying an algebraic 'supersymmetry Dirac equation' which encodes the general properties of such solutions. This is a stronger condition than the $c^2 = 0$ extremality condition. The orbits of the 3D duality group G are not always acted upon transitively by G. This is related to the failure of the Iwasawa decomposition for noncompact divisor groups H^*. The Iwasawa failure set corresponds to the extremal suborbits.

References

1. G. Neugebaur, D. Kramer, Ann. Phys. (Leipzig) **24**, 62 (1969)
2. P. Breitenlohner, D. Maison, G.W. Gibbons, Four-dimensional black holes from Kaluza-Klein theories. Commun. Math. Phys. **120**, 295 (1988)
3. G. Clement, D. Gal'tsov, Stationary BPS solutions to dilaton-axion gravity. Phys. Rev. D **54**, 6136 (1996) arXiv:hep-th/9607043
4. D.V. Gal'tsov, O.A. Rytchkov, Generating branes via sigma-models. Phys. Rev. D **58**, 122001 (1998) arXiv:hep-th/9801160
5. E. Cremmer, B. Julia, The $SO(8)$ supergravity. Nucl. Phys. B **159**, 141 (1979)
6. B. de Wit, A.K. Tollsten, H. Nicolai, Locally supersymmetric D = 3 nonlinear sigma models. Nucl. Phys. B **392**, 3 (1993) arXiv:hep-th/9208074
7. P. Meessen, T. Ortin, The Supersymmetric configurations of N = 2, D = 4 supergravity coupled to vector supermultiplets. Nucl. Phys. B **749**, 291 (2006) arXiv:hep-th/0603099
8. M. Cvetič, D. Youm, Dyonic BPS saturated black holes of heterotic string on a six torus. Phys. Rev. D **53**, 584 (1996) arXiv:hep-th/9507090
9. M. Cvetič, A.A. Tseytlin, General class of BPS saturated dyonic black holes as exact superstring solutions. Phys. Lett. B **366**, 95 (1996) arXiv:hep-th/9510097
10. J. Eells Jr, J.H. Sampson, Am. J. Math. **86**, 109 (1964)

11. G. Bossard, H. Nicolai, K.S. Stelle, Universal BPS structure of stationary supergravity solutions. JHEP **0907**, 003 (2009). arXiv:0902.4438 [hep-th]
12. J. Ehlers, Konstruktion und Charakterisierungen von Lösungen der Einsteinschen Gravitationsgleichungen, Dissertation, Hamburg (1957)
13. M. Gunaydin, G. Sierra, P.K. Townsend, Exceptional supergravity theories and the MAGIC square. Phys. Lett. B **133**, 72 (1983)
14. E. Cremmer, H. Lu, C.N. Pope, K.S. Stelle, Spectrum generating symmetries for BPS solitons. Nucl. Phys. B **520**, 132 (1998) arXiv:hep-th/9707207
15. G. Bossard, H. Nicolai, Multi-black holes from nilpotent Lie algebra orbits. Gen. Rel. Grav. **42**, 509 (2010). arXiv:0906.1987 [hep-th]
16. G. Bossard, C. Ruef, Interacting non-BPS black holes. Gen. Rel. Grav. **44**, 21 (2012). arXiv:1106.5806 [hep-th]

Chapter 25
Thermodynamic of Distorted Reissner-Nordström Black Holes in Five-Dimensions

Shohreh Abdolrahimi

Abstract In this paper, we study mechanics and thermodynamics of distorted, five-dimensional, electrically charged (non-extremal) black holes on the example of a static and "axisymmetric" black hole distorted by external, electrically neutral matter. Such a black hole is represented by the derived here solution of the Einstein-Maxwell equations which admits an $\mathbb{R}^1 \times U(1) \times U(1)$ isometry group. We study the properties of this distorted black hole.

25.1 Introduction

This is based on a paper [1], which has been presented as a talk in the Karl Schwarzschild Meeting in Frankfurt. Einstein equations are very complex and describing black hole interaction with external matter and fields usually requires involved numerical computations. To construct exact solutions which would model to some extend the interaction of a black hole with the external matter, Geroch and Hartle [2] proposed to consider static (or stationary), axisymmetric space-times which are not asymptotically flat. Such solutions represent black holes distorted by external matter. Four-dimensional, distorted, axisymmetric, black holes were studied in, e.g., [3–5]. Distorted black holes can show some strange and remarkable properties [6, 7].

In four dimensions, the general static, axisymmetric solution of the vacuum Einstein equations can be written in the form of Weyl solution [8]. Using Weyl form of the metric distorted black hole solutions can be constructed. As it was done in the case of four-dimensions, one can use the generalized Weyl solution [9] to construct distorted black objects by adding the distortion fields to the Newtonian potentials, which define the solution. Here, we shall analyze a solution representing a distorted five-dimensional Reissner-Nordström black hole. This is a static solution of the Einstein-Maxwell equations which has $\mathbb{R}^1 \times U(1) \times U(1)$ isometry group. The construction is based on the gauge transformation of the matrix which is an element

S. Abdolrahimi (✉)
Institut für Physik, Universität Oldenburg, Postfach 2503 D-26111, Oldenburg, Germany
e-mail: shohreh.abdolrahimi@uni-oldenburg.de

P. Nicolini et al. (eds.), *1st Karl Schwarzschild Meeting on Gravitational Physics*, Springer Proceedings in Physics 170, DOI 10.1007/978-3-319-20046-0_25

of the coset target space $SL(2, \mathbb{R}^1)/U(1)$ of the scalar fields which define our model (see, e.g., [10]). We use the following convention of units: $G_{(5)} = c = \hbar = k_B = 1$, where $G_{(5)}$ is the five-dimensional gravitational constant. The space-time signature is +3 and the sign conventions are that adopted in [11].

25.1.1 The Solution

Applying the generating transformation presented in [1] to the distorted five-dimensional Schwarzschild-Tangherlini solution, we derive the solution representing distorted charged black hole,

$$ds^2 = -\frac{4p^2(\eta^2-1)}{\Delta^2} e^{2(\widehat{U}+\widehat{W})} dt^2 + \frac{m\Delta}{2}\left(e^{2(\widehat{V}+\widehat{U}+\widehat{W})}\frac{d\eta^2}{4(\eta^2-1)} + d\hat{\Omega}^2\right), \tag{25.1}$$

$$\Delta = (1+p)(\eta+1) - (1-p)(\eta-1)e^{2(\widehat{U}+\widehat{W})}, \tag{25.2}$$

$$d\hat{\Omega}^2 = \frac{1}{4}\left(e^{2(\widehat{V}+\widehat{U}+\widehat{W})}d\theta^2 + 2(1+\cos\theta)e^{-2\widehat{W}}d\chi^2 + 2(1-\cos\theta)e^{-2\widehat{U}}d\phi^2\right), \tag{25.3}$$

$$\Phi = \frac{\sqrt{3(1-p^2)}}{\Delta}\left(\eta + 1 - (\eta-1)e^{2(\widehat{U}+\widehat{W})}\right), \tag{25.4}$$

where Φ is the electrostatic potential. Here, the functions $\widehat{U}$, $\widehat{W}$, and $\widehat{V}$ are the distortion fields given by the following expressions:

$$\widehat{U}(\eta,\theta) = \sum_{n\geq 0} a_n R^n P_n, \qquad \widehat{W}(\eta,\theta) = \sum_{n\geq 0} b_n R^n P_n, \tag{25.5}$$

$$\widehat{V} = \widehat{V}_1 + \widehat{V}_2, \quad R = (\eta^2 - \sin^2\theta)^{1/2}, \qquad P_n \equiv P_n(\eta\cos\theta/R), \tag{25.6}$$

$$\widehat{V}_1(\eta,\theta) = -\sum_{n\geq 0}\Big\{3(a_n/2 + b_n/2)R^n P_n + (a_n + b_n/2)\sum_{l=0}^{n-1}(\eta - \cos\theta)R^l P_l + (a_n/2 + b_n)\sum_{l=0}^{n-1}(-1)^{n-l}(\eta+\cos\theta)R^l P_l\Big\}, \tag{25.7}$$

$$\widehat{V}_2(\eta,\theta) = \sum_{n,k\geq 1}\frac{nk}{n+k}(a_n a_k + a_n b_k + b_n b_k)R^{n+k}[P_n P_k - P_{n-1}P_{k-1}], \tag{25.8}$$

When the distortion fields $\widehat{U}$, $\widehat{W}$, and $\widehat{V}$ vanish, the solution represents a five-dimensional Reissner-Nordström solution in an empty, asymptotically flat universe. For $p = 1$, this solution represents the distorted five-dimensional Schwarzschild-Tangherlini black hole. If the distortion fields vanish, this solution represents the five-dimensional Reissner-Nordström solution. The parameters m and q are related to the five-dimensional Komar mass of the black hole M and its five-dimensional electric charge Q as follows: $M = 3\pi m/4$ and $Q = 2\sqrt{3}\,q$. In these coordinates, the event (outer) horizon is at $\eta = 1$ and the (inner) Cauchy horizon is at $\eta = -1$, and the space-time singularity is at $\eta = -1/p$. Distortion fields defined by exterior multipole moments correspond to asymptotically flat solutions. To have a regular horizon we shall consider non-asymptotically flat solutions distorted by the external sources only, whose distortion fields are defined by the interior multipole moments. This represents a black hole distorted by external sources. Such fields must be regular and smooth at the horizon. The distortion fields $\widehat{U}$, $\widehat{W}$, and $\widehat{V}$ given above satisfy this condition. If the sources of the distortion fields are included into the solution, then their energy-momentum tensor satisfies the strong energy condition. The strong energy condition implies that $\widehat{U} + \widehat{W} \leq 0$. Then, it follows that, on the "semi-axes" $\theta = 0$ and $\theta = \pi$, at the black hole horizon $\eta = 1$, we have $\sum_{n\geq 0}(\pm 1)^n(a_n + b_n) \leq 0$, where $+1$ corresponds to $\theta = 0$ and -1 corresponds to $\theta = \pi$. One can show that, the space-time curvature invariants diverge in the region where $\Delta = 0$. If $\widehat{U} + \widehat{W} \leq 0$, that is if the sources of the distortion fields satisfy the strong energy condition, then the space-time singularities are located behind the inner (Cauchy) horizon.

In addition, for a regular horizon there should be no conical singularities on the "semi-axes" $\theta = 0$ and $\theta = \pi$, and thus on the horizon. For the metric (25.1)–(25.3) this condition implies (for details see [7]), $\widehat{V} + 2\widehat{U} + \widehat{W}|_{\theta=0} = 0$, for the "semi-axis" $\theta = 0$, and $\widehat{V} + \widehat{U} + 2\widehat{W}|_{\theta=\pi} = 0$, for the "semi-axis" $\theta = \pi$, which can be written in the following form $\sum_{n\geq 0}(a_{2n} - b_{2n}) + 3\sum_{n\geq 0}(a_{2n+1} + b_{2n+1}) = 0$. This condition implies the black hole equilibrium condition.

25.2 Properties of the Distorted Black Hole Solution

In this section we shall discuss the properties of the distorted black hole solution (25.1)–(25.4). The three-dimensional surface of the outer horizon is defined by $t = const$ and $\eta = 1$. The three-dimensional surface of the inner horizon is defined by $t = const$ and $\eta = -1$. One can show that the metrics of the outer and inner horizon surfaces are related to each other by the following transformation, i.e., an exchange between the "semi-axes" and reverse of signs of the multipole moments, $(\theta, \chi, \phi) \longrightarrow (\pi - \theta, \phi, \chi)$, $a_n \longrightarrow -a_n$, $b_n \longrightarrow -b_n$. We shall call this transformation the *duality transformation* between the outer and inner horizons of the distorted black hole. This transformation is exactly the same as the duality transformation between the horizon and the *stretched* singularity surfaces of the distorted five-dimensional Schwarzschild-Tangherlini black hole (see (135)–(137), [7]). The horizon areas of the distorted black hole solution are

$$A_{\pm} = 2\pi^2\sqrt{m^3(1 \pm p)^3}\, e^{\mp\frac{3}{2}\gamma}\,. \quad \gamma = \sum_{n\geq 0} a_{2n} + \sum_{n\geq 0} b_{2n} + \frac{1}{3}\Big(\sum_{n\geq 0} a_{2n+1} - \sum_{n\geq 0} b_{2n+1}\Big). \tag{25.9}$$

One can see that the area product has the same form as that of the Reissner-Nordström black hole:

$$A_+A_- = 4\pi^4 q^3 = \frac{\pi^4}{6\sqrt{3}}Q^3 \,. \tag{25.10}$$

We can define the lower and upper limits for the inner and outer horizon areas of a general distorted Reissner-Nordström black hole, $A_- < \pi^2 Q^{\frac{3}{2}}/(3^{\frac{3}{4}}\sqrt{2}) < A_+$, which can be written in the following form: $A_- < \sqrt{A_- A_+} < A_+$. Thus, the geometric mean of the inner and outer horizon areas of the distorted black hole represents the upper and lower limits of its inner and outer horizon areas, respectively. It is easy to see that, the values of the electrostatic potential at the black hole horizons does not change under the distortion. One can also check that the Smarr formula

$$\pm M = \frac{3}{16\pi}\kappa_\pm A_\pm \pm \frac{\pi}{8}\Phi_\pm Q \,. \tag{25.11}$$

holds for the distorted black hole as well. Here M defines the black hole Komar mass, assuming that the space-time (25.1)–(25.3) can be analytically extended to achieve its asymptotic flatness.

25.3 Mechanics and Thermodynamics of the Distorted Black Hole

In this section, we derive mechanical laws of the distorted black hole and present the corresponding laws of thermodynamics. The zeroth law says that a black hole surface gravity (and accordingly, its temperature) is constant at the black hole horizon. The surface gravity is defined up to an arbitrary constant which depends on the normalization of the time-like Killing vector. However, the normalization does not affect the zeroth law. The surface gravity at the horizons is

$$\kappa_\pm = \frac{2p\, e^{\pm\frac{3}{2}\gamma}}{\sqrt{m(1\pm p)^3}}. \tag{25.12}$$

We see that due to distortion fields, the surface gravity differs from that of the Reissner-Nordström (undistorted) black hole by the factor $e^{\pm\frac{3}{2}\gamma}$. The zeroth law holds for both horizons of our distorted black hole. The corresponding temperature is defined in terms of the surface gravity as $T_\pm = \kappa_\pm/2\pi$. This definition, however, requires a proper normalization of the Killing vector at the spatial infinity. Taking into account the description of black holes within string theory [12], one can view the inner horizon thermodynamics, "temperature" T_-, as the difference of the thermodynamics corresponding to the right- and left-moving excitations of the strings.

Global first law correspond to the total system of the black hole plus the distorting matter. Local first law corresponds to the system of the black hole only. To define a global first law one needs to extend the space-time to achieve its asymptotic flatness. The extension is achieved by requiring that the distortion fields $\widehat{U}$, $\widehat{W}$, and $\widehat{V}$ vanish at the asymptotic infinity and by extending the corresponding space-time manifold. In the extended manifold there exists an electrovacuum region in the interior of the black hole and part of the exterior region where the solution (25.1)–(25.4) is valid. Then, there is a region where the external sources are located. Beyond that region there is asymptotically flat electrovacuum region. Having this extension one can normalize the timelike Killing vector $\boldsymbol{\xi}_{(t)}$ at the spatial infinity as $\boldsymbol{\xi}^2_{(t)} = -1$. As it is done, one naturally finds that the komar mass M of undistorted black hole correspond to the mass of black hole not including the mass of external matter. We derive the global first law of the black hole mechanics,

$$\pm\,\delta M = \frac{\kappa_\pm}{8\pi}\delta A_\pm \pm \frac{\pi}{8}\Phi_\pm \delta Q + M^{loc}_\pm \delta\gamma, \tag{25.13}$$

where the local black hole mass, $M^{loc}_\pm = \pm 3\pi m p/4$, does not depend on the distortion fields. From the global first law of the black hole mechanics, by using the definition of temperature and the black hole entropy, $S_\pm = A_\pm/4$. We derive the global first law of the black hole thermodynamics, $\pm\delta M = T_\pm \delta S_\pm \pm (\pi/8)\Phi_\pm \delta Q + M^{loc}_\pm \delta\gamma$. Here the term $M^{loc}_\pm \delta\gamma$ is interpreted as the work done on the black hole by the variation of the external potential γ due to the distorting matter. If the distortion is adiabatic, $\delta S_\pm = 0$, i.e., such that neither matter nor gravitational waves cross the black hole horizons, and in addition, the black hole charge Q does not change, then the work $M^{loc}_\pm \delta\gamma$ results in the change of the black hole mass δM.

The local first law does not include the distorting matter into consideration of the black hole mechanics. The observers who live near the black hole consider the black hole as an isolated, undistorted object. Thus, assuming that there is no other matter present and the space-time is asymptotically flat, they define its surface gravity $\tilde{\kappa}_+$, the outer horizon area $\tilde{A}_+$, electrostatic potential $\tilde{\Phi}_+$, electric charge $\tilde{Q}$, and the black hole Komar mass $\tilde{M}$. Thus, these observers construct the local first law of the black hole mechanics as that of the Reissner-Nordström (undistorted) black hole,

$$\pm\,\delta\tilde{M} = \frac{\tilde{\kappa}_\pm}{8\pi}\delta\tilde{A}_\pm \pm \frac{\pi}{8}\Phi_\pm \delta Q\,. \tag{25.14}$$

With the definitions of temperature and entropy the local first law of black hole thermodynamics reads $\pm\delta\tilde{M} = \tilde{T}_\pm \delta\tilde{S}_\pm \pm (\pi 8)\Phi_\pm \delta Q$. The measurements of the observers define the black hole area as that which is exactly equal to the black hole area when the presence of the distortion fields is taken into account, i.e.,

$$\tilde{A}_\pm = 2\pi^2\sqrt{\tilde{m}^3(1\pm\tilde{p})^3} = A_\pm = 2\pi^2\sqrt{m^3(1\pm p)^3}\,e^{\mp\frac{3}{2}\gamma}\,, \tag{25.15}$$

where $\tilde{M} = 3\pi\tilde{m}/4$ and $\tilde{p} = \sqrt{\tilde{m}^2 - q^2}/\tilde{m}$. The following relations provide us with the correspondence between the local and the global forms of the first law:

$$\tilde{M} = \frac{M}{2}\left[(1+p)e^{-\gamma} + (1-p)e^{\gamma}\right], \quad \tilde{p} = \frac{(1+p)e^{-\gamma} - (1-p)e^{\gamma}}{(1+p)e^{-\gamma} + (1-p)e^{\gamma}}, \tag{25.16}$$

$$\tilde{\kappa}_{\pm} = \frac{2\tilde{p}}{\sqrt{\tilde{m}(1\pm\tilde{p})^3}} = \frac{\kappa_{\pm}}{2p}\left[(1+p)e^{-\gamma} - (1-p)e^{\gamma}\right], \quad \tilde{\Phi}_{\pm} = \frac{\sqrt{3(1-\tilde{p}^2)}}{1\pm\tilde{p}} = \Phi_{\pm}e^{\gamma}, \tag{25.17}$$

25.4 Conclusion

We have studied the mechanic and thermodynamic of a distorted, five-dimensional Reissner-Nordström black hole solution. The space-time singularities are located behind the black hole's inner (Cauchy) horizon, provided that the sources of the distortion satisfy the strong energy condition. The inner (Cauchy) horizon remains regular if the distortion fields are finite and smooth at the outer horizon. There exists a certain duality transformation between the inner and the outer horizon surfaces which links surface gravity, electrostatic potential, and space-time curvature invariants calculated at the black hole horizons. The product of the inner and outer horizon areas depends only on the black hole's electric charge and the geometric mean of the areas is the upper (lower) limit for the inner (outer) horizon area. The horizon areas, electrostatic potential, and surface gravity satisfy the Smarr formula. We formulated the zeroth and the first laws of mechanics and thermodynamics of the distorted black hole and found a correspondence between the global and local forms of the first law.

Acknowledgments S. A. gratefully acknowledges the Deutsche Forschungsgemeinschaft (DFG) for financial support within the framework of the DFG Research Training group 1620 Models of gravity.

References

1. S. Abdolrahimi, A.A. Shoom, Distorted five-dimensional electrically charged black holes. Phys. Rev. D **89**, 024040 (2014)
2. R. Geroch, J.B. Hartle, Distorted black holes. J. Math. Phys. **23**, 680 (1982)
3. W. Israel, K.A. Khan, Collinear particles and Bondi dipoles in general relativity. Nuovo Cimento **33**, 331 (1964)
4. L.A. Mysak, G. Szekeres, Behavior of the Schwarzschild singularity in superimposed gravitational fields. Can. J. Phys. **44**, 617 (1966)
5. W. Israel, Event horizons in static vacuum space-times. Phys. Rev. **164**, 1776 (1967)
6. A. Abdolrahimi, V.P. Frolov, A.A. Shoom, Interior of a charged distorted black hole. Phys. Rev. D **80**, 024011 (2009). arXiv:0905.0178

7. S. Abdolrahimi, A.A. Shoom, D.N. Page, Distorted 5-dimensional vacuum black hole. Phys. Rev. D **82**, 124039 (2010). arXiv:1009.5971
8. H. Weyl, Zur Gravitationstheorie. Ann. Phys. **54**, 117 (1917)
9. R. Emparan, H.S. Reall, Generalized Weyl solutions. Phys. Rev. D **65**, 084025 (2002)
10. M. Gürses, B.C. Xanthopoulos, Axially symmetric, static self-dual SU(3) gauge fields and stationary Einstein-Maxwell metrics. Phys. Rev. D **26**, 1912 (1982)
11. C.W. Misner, K.S. Thorne, J.A. Wheeler, *Gravitation* (W.H. Freeman and Co., San Francisco, 1973)
12. G. Horowitz, Quantum states of black holes

Chapter 26
What Is the Schwarzschild Radius of a Quantum Mechanical Particle?

Roberto Casadio

Abstract A localised particle in Quantum Mechanics is described by a wave packet in position space, regardless of its energy. However, from the point of view of General Relativity, if the particle's energy density exceeds a certain threshold, it should be a black hole. In order to combine these two pictures, we introduce a horizon wave-function determined by the position wave-function, which yields the probability that the particle is a black hole. The existence of a (fuzzy) minimum mass for black holes naturally follows, and we also show that our construction entails an effective Generalised Uncertainty Principle simply obtained by adding the uncertainties coming from the two wave-functions.

26.1 The Schwarzschild Link

In natural units, with $c = 1$ (and $\hbar = \ell_p \, m_p$), the Newton constant is given by

$$G_N = \ell_p / m_p \, , \tag{26.1}$$

where ℓ_p and m_p are the Planck length and mass, respectively, and converts mass (or energy) into length. This naive observation stands behind Thorne's *hoop conjecture* [1]: A black hole forms when the impact parameter b of two colliding objects is shorter than the *Schwarzschild gravitational radius* of the system, that is for

$$R_H \equiv 2 \, \ell_p \, \frac{E}{m_p} \gtrsim b \, , \tag{26.2}$$

R. Casadio (✉)
Dipartimento di Fisica e Astronomia, Università di Bologna,
40126 Bologna, Italy
e-mail: casadio@bo.infn.it

R. Casadio
Istituto Nazionale di Fisica Nucleare, Sezione di Bologna,
via Irnerio 46, 40126 Bologna, Italy

P. Nicolini et al. (eds.), *1st Karl Schwarzschild Meeting on Gravitational Physics*,
Springer Proceedings in Physics 170, DOI 10.1007/978-3-319-20046-0_26

where E is total energy in the centre-of-mass frame. The emergence of the Schwarzschild radius is indeed easy to understand in a spherically symmetric space-time, where the metric $g_{\mu\nu}$ can be written as

$$\mathrm{d}s^2 = g_{ij}\,\mathrm{d}x^i\,\mathrm{d}x^j + r^2(x^i)\left(\mathrm{d}\theta^2 + \sin^2\theta\,\mathrm{d}\phi^2\right) , \tag{26.3}$$

with $x^i = (x^1, x^2)$ coordinates on surfaces of constant angles θ and ϕ. The location of a trapping horizon, a sphere where the escape velocity equals the speed of light, is then determined by

$$0 = g^{ij}\,\nabla_i r\,\nabla_j r = 1 - \frac{2\,M}{r} , \tag{26.4}$$

where $\nabla_i r$ is the covector perpendicular to surfaces of constant area $\mathscr{A} = 4\,\pi\,r^2$. The active gravitational (or Misner-Sharp) mass M represents the total energy enclosed within a sphere of radius r, and, if we set $x^1 = t$ and $x^2 = r$, is explicitly given by

$$M(t,r) = \frac{4\,\pi\,\ell_\mathrm{p}}{3\,m_\mathrm{p}} \int_0^r \rho(t,\bar{r})\,\bar{r}^2\,\mathrm{d}\bar{r} , \tag{26.5}$$

where $\rho = \rho(x^i)$ is the matter density. It is usually very difficult to follow the dynamics of a given matter distribution and find surfaces satisfying (26.4), but an horizon exists if there are values of r such that $R_\mathrm{H} = 2\,M(t,r) > r$, which is a mathematical reformulation of the hoop conjecture (26.2).

26.2 Horizon Wave-Function

The hoop conjecture was formulated having in mind black holes of astrophysical size [2], for which a classical metric and horizon structure are reasonably safe concepts. However, for elementary particles quantum effects may not be neglected [3]. Consider a spin-less point-like source of mass m, whose Schwarzschild radius is given by R_H in (26.2) with $E = m$. The Heisenberg principle introduces an uncertainty in its spatial localisation, of the order of the Compton-de Broglie length, $\lambda_m \simeq \ell_\mathrm{p}\,m_\mathrm{p}/m$. Assuming quantum physics is a more refined description of reality implies that R_H only makes sense if it is larger than λ_m,

$$R_\mathrm{H} \gtrsim \lambda_m \quad \Rightarrow \quad m \gtrsim m_\mathrm{p} \quad (\text{or } M \gtrsim \ell_\mathrm{p}) . \tag{26.6}$$

Note that this argument employs the flat space Compton length, and it is likely that the particle's self-gravity will affect it. However, we can still assume the condition (26.6) holds as an order of magnitude estimate, hence black holes can only exist with mass (much) larger than the Planck scale.

We are thus facing a deeply conceptual challenge: how can we describe systems containing both quantum mechanical particles and classical horizons? For this purpose, we shall define a horizon wave-function that can be associated with any localised quantum mechanical particle [4, 5], and that will put on quantitative grounds the condition (26.6) that distinguishes black holes from regular particles.

The quantum mechanical state representing an object, which is both *localised in space* and *at rest* in the chosen reference frame, must be described by a wave-function $\psi_S \in L^2(\mathbb{R}^3)$, which can be decomposed into energy eigenstates,

$$|\psi_S\rangle = \sum_E C(E)\, |\psi_E\rangle \ , \tag{26.7}$$

where the sum represents the spectral decomposition in Hamiltonian eigenmodes,

$$\hat{H}\, |\psi_E\rangle = E\, |\psi_E\rangle \ , \tag{26.8}$$

and H can be specified depending on the model we wish to consider. If we also assume the state is *spherically symmetric*, we can introduce a Schwarzschild radius $R_H = R_H(E)$ associated to each component ψ_E of energy E, by inverting (26.2), and define the (unnormalised) *horizon wave-function* as

$$\tilde{\psi}_H(R_H) = C\left(E = m_p\, \frac{R_H}{2\,\ell_p}\right) . \tag{26.9}$$

The normalisation is finally fixed by employing the inner product

$$\langle\, \psi_H \mid \phi_H \,\rangle = 4\,\pi \int_0^\infty \psi_H^*(R_H)\, \phi_H(R_H)\, R_H^2\, dR_H \,. \tag{26.10}$$

We interpret the normalised wave-function ψ_H as yielding the probability that we would detect a horizon of areal radius $r = R_H$ associated with the particle in the quantum state ψ_S. Such a horizon is necessarily "fuzzy", like the particle's position, unless the width of ψ_H is negligibly small. Moreover, the probability density that the particle lies inside its own horizon of radius $r = R_H$ will be given by

$$P_<(r < R_H) = P_S(r < R_H)\, P_H(R_H) \ , \tag{26.11}$$

where $P_S(r < R_H) = 4\,\pi \int_0^{R_H} |\psi_S(r)|^2\, r^2\, dr$ is the probability that the particle is inside the sphere of radius $r = R_H$, and $P_H(R_H) = 4\,\pi\, R_H^2\, |\psi_H(R_H)|^2$ is the probability that the horizon is located on the sphere of radius $r = R_H$. Finally, by integrating (26.11) over all possible values of the radius,

$$P_{BH} = \int_0^\infty P_<(r < R_H)\, dR_H \ , \tag{26.12}$$

the probability that the particle is a black hole will be obtained.

26.2.1 Gaussian Particle

The above construction can be straightforwardly applied to a particle described by the Gaussian wave-function

$$\psi_S(r) = \frac{e^{-\frac{r^2}{2\ell^2}}}{\ell^{3/2}\,\pi^{3/4}}\,, \tag{26.13}$$

where the width $\ell \sim \lambda_m$. This wave-function in position space corresponds to the momentum space wave-function

$$\psi_S(p) = \frac{e^{-\frac{p^2}{2\Delta^2}}}{\Delta^{3/2}\,\pi^{3/4}}\,, \tag{26.14}$$

where $p^2 = \mathbf{p}\cdot\mathbf{p}$ and $\Delta = \hbar/\ell = m_p\,\ell_p/\ell$. For the energy of the particle, we simply assume the relativistic mass-shell relation in flat space, $E^2 = p^2 + m^2$, and we easily obtain the normalised horizon wave-function

$$\psi_H(R_H) = \frac{\ell^{3/2}\,e^{-\frac{\ell^2 R_H^2}{8\ell_p^4}}}{2^{3/2}\,\pi^{3/4}\,\ell_p^3}\,. \tag{26.15}$$

Note that, since $\langle\,\hat{r}^2\,\rangle \simeq \ell^2$ and $\langle\,\hat{R}_H^2\,\rangle \simeq \ell_p^4/\ell^2$, we expect the particle will be inside its own horizon if $\langle\,\hat{r}^2\,\rangle \ll \langle\,\hat{R}_H^2\,\rangle$, which precisely yields the condition (26.6) if $\ell \simeq \lambda_m$. In fact, the probability density (26.11) can now be explicitly computed,

$$P_<(r < R_H) = \frac{\ell^3\,R_H^2}{2\sqrt{\pi}\,\ell_p^6}\,e^{-\frac{\ell^2 R_H^2}{4\ell_p^4}}\left[\mathrm{Erf}\left(\frac{R_H}{\ell}\right) - \frac{2\,R_H}{\sqrt{\pi}\,\ell}\,e^{-\frac{R_H^2}{\ell^2}}\right], \tag{26.16}$$

from which we derive the probability (26.12) for the particle to be a black hole,

$$P_{BH}(\ell) = \frac{2}{\pi}\left[\arctan\left(2\,\frac{\ell_p^2}{\ell^2}\right) + 2\,\frac{\ell^2\,(4 - \ell^4/\ell_p^4)}{\ell_p^2\,(4 + \ell^4/\ell_p^4)^2}\right]. \tag{26.17}$$

In Fig. 26.1, we show the probability (26.17) that the particle is a black hole as a function of the Gaussian width ℓ (in units of ℓ_p). From the plot of P_{BH}, it appears that the particle is most likely a black hole, $P_{BH} \simeq 1$, if $\ell \lesssim \ell_p$. Assuming $\ell = \lambda_m = \ell_p\,m_p/m$, we have thus derived a result in qualitative agreement with the condition (26.6), but from a totally quantum mechanical picture. Strictly speaking, there is no black hole minimum mass in our treatment, but a vanishing probability for a particle of "small" mass (say $m \lesssim m_p/4$, that is $\ell \gtrsim 4\,\ell_p$), to be a black hole.

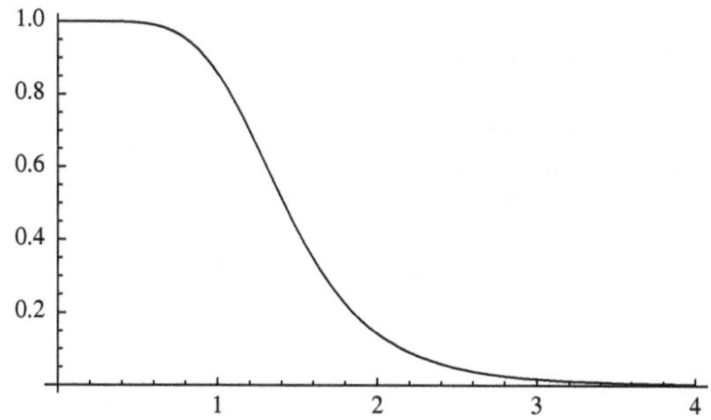

Fig. 26.1 Probability that a particle of width ℓ is a black hole as a function of ℓ/ℓ_p

26.2.2 Generalised Uncertainty Principle

For the Gaussian packet described above, the Heisenberg uncertainty in radial position is given by

$$\langle \Delta r^2 \rangle = 4\pi \int_0^\infty |\psi_S(r)|^2 r^4 \, dr - \left(4\pi \int_0^\infty |\psi_S(r)|^2 r^3 \, dr\right)^2 = \frac{3\pi - 8}{2\pi} \ell^2 , \tag{26.18}$$

and, analogously, the uncertainty in the horizon radius will be given by

$$\langle \Delta R_H^2 \rangle = 4 \frac{3\pi - 8}{2\pi} \frac{\ell_p^4}{\ell^2} . \tag{26.19}$$

Since $\langle \Delta p^2 \rangle = \left(\frac{3\pi-8}{2\pi}\right) m_p^2 \frac{\ell_p^2}{\ell^2} \equiv \Delta p^2$, we can also write

$$\ell^2 = \frac{3\pi - 8}{2\pi} \ell_p^2 \frac{m_p^2}{\Delta p^2} . \tag{26.20}$$

Finally, by combining the uncertainty (26.18) with (26.19) linearly, we find

$$\Delta r \equiv \sqrt{\langle \Delta r^2 \rangle} + \gamma \sqrt{\langle \Delta R_H^2 \rangle} = \frac{3\pi - 8}{2\pi} \ell_p \frac{m_p}{\Delta p} + 2\gamma \, \ell_p \frac{\Delta p}{m_p} , \tag{26.21}$$

where γ is a coefficient of order one, and the result is plotted in Fig. 26.2 (for $\gamma = 1$). This is precisely the kind of result one obtains from the generalised uncertainty principles considered in [6–10], leading to a minimum measurable length

$$\Delta r \geq 2\sqrt{\gamma \frac{3\pi - 8}{\pi}} \ell_p \simeq 1.3 \sqrt{\gamma} \, \ell_p . \tag{26.22}$$

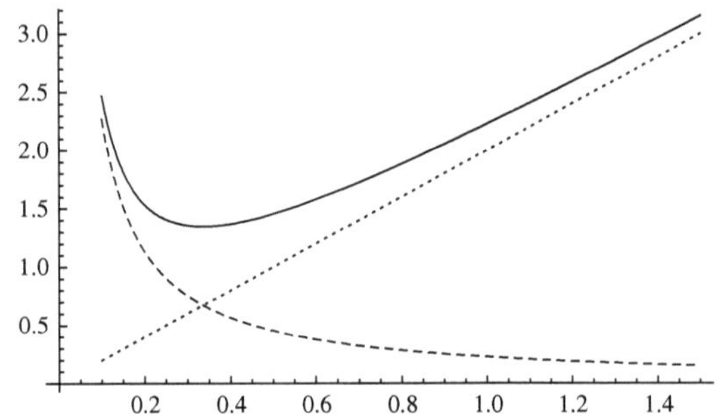

Fig. 26.2 Uncertainty relation (26.21) (*solid line*) as a combination of the Quantum Mechanical uncertainty (*dashed line*) and the uncertainty in horizon radius (*dotted line*) (lengths in units of ℓ_p and momentum in units of m_p)

Of course, one might consider different ways of combining the two uncertainties (26.18) and (26.19), or even avoid this step and just make a direct use of the horizon wave-function. In this respect, the present approach appears more flexible and does *not* require modified commutators for the canonical variables r and p.

26.3 Final Remarks

So far, the idea of the horizon wave-function was just applied to the very simple case of a spinless massive particle, and expected results (existence of a minimum black hole mass and generalised uncertainty relation) were recovered and refined [4, 5]. Next, it should be applied to more realistic systems. For example, one could investigate dispersion relations derived from quantum field theory in curved spacetime, and what a localised state is in the latter context [11]. Regardless of such improvements, the conceptual usefulness of our construction should already be clear, in that it allows us to deal with very quantum mechanical sources, and to do so in a quantitative fashion. For example, one could review the issue of quantum black holes [12, 13] in light of the above formalism, as well as finally tackle the description of black hole formation and dynamical horizons in the gravitational collapse of truly quantum matter [3, 14–16].

References

1. K.S. Thorne, in *Nonspherical Gravitational Collapse: A Short Review*, ed. by J.R. Klauder. Magic Without Magic (San Francisco, 1972), p. 231
2. P.D. D'Eath, P.N. Payne, Phys. Rev. D **46**, 658, 675, 694 (1992)
3. G.L. Alberghi, R. Casadio, O. Micu, A. Orlandi, JHEP **1109**, 023 (2011)
4. R. Casadio, Localised particles and fuzzy horizons: a tool for probing quantum black holes. arXiv:1305.3195 [gr-qc]
5. R. Casadio, F. Scardigli, Horizon wave-function for single localized particles: GUP and quantum black hole decay. arXiv:1306.5298 [gr-qc]
6. M. Maggiore, Phys. Lett. B **319**, 83 (1993)

7. A. Kempf, G. Mangano, R.B. Mann, Phys. Rev. D **52**, 1108 (1995)
8. F. Scardigli, Phys. Lett. B **452**, 39 (1999)
9. F. Scardigli, R. Casadio, Class. Quant. Grav. **20**, 3915 (2003)
10. F. Scardigli, R. Casadio, Int. J. Mod. Phys. D **18**, 319 (2009)
11. T.D. Newton, E.P. Wigner, Rev. Mod. Phys. **3**, 400 (1949)
12. X. Calmet, D. Fragkakis, N. Gausmann, Eur. Phys. J. C **71**, 1781 (2011)
13. X. Calmet, W. Gong, S.D.H. Hsu, Phys. Lett. B **668**, 20 (2008)
14. T. Banks, W. Fischler, A Model for high-energy scattering in quantum gravity. arXiv:hep-th/9906038
15. D.M. Eardley, S.B. Giddings, Phys. Rev. D **66**, 044011 (2002)
16. S.B. Giddings, S.D. Thomas, Phys. Rev. D **65**, 056010 (2002)

Chapter 27
The Background Effective Average Action Approach to Quantum Gravity

Giulio D'Odorico, Alessandro Codello and Carlo Pagani

Abstract We construct a consistent closure for the beta functions of the cosmological and Newton's constants by evaluating the influence that the anomalous dimensions of the fluctuating metric and ghost fields have on their renormalization group flow. In this generalized framework we confirm the presence of an UV attractive non-Gaussian fixed-point, which we find characterized by real critical exponents. Our closure method is general and can be applied systematically to more general truncations of the gravitational effective average action.

27.1 Introduction

As is well known, a quantum field theory of gravity based on the Einstein-Hilbert action is perturbatively non-renormalizable [1]. The Asymptotic Safety scenario [2] is a promising approach which suggests instead that the theory is non-perturbatively renormalizable at a non-Gaussian ultraviolet (UV) fixed-point of the renormalization group (RG) flow. To inquire if a theory is renormalizable in a non-perturbative way one needs non-perturbative tools. One of these tools is the exact functional RG equation satisfied by the (background) effective average action (EAA), first derived in the context of quantum gravity by Reuter [3]. The EAA can be derived from the

G. D'Odorico (✉)
Institute for Mathematics, Astrophysics and Particle Physics, Radboud University, Heyendaalseweg 135, 6525 AJ Nijmegen, The Netherlands
e-mail: G.Dodorico@hef.ru.nl

A. Codello
CP3–Origins and Danish Institute for Advanced Study, University of Southern Denmark, Campusvej 55, 5230 Odense, Denmark
e-mail: codello@cp3-origins.net

C. Pagani · A. Codello
Scuola Internazionale Superiore di Studi Avanzati, via Bonomea 265, 34136 Trieste, Italy
e-mail: cpagani@sissa.it

C. Pagani
Istituto Nazionale di Fisica Nucleare, Sezione di Trieste, 34136 Trieste, Italy

P. Nicolini et al. (eds.), *1st Karl Schwarzschild Meeting on Gravitational Physics*, Springer Proceedings in Physics 170, DOI 10.1007/978-3-319-20046-0_27

path integral by adding to the action a regulator term which suppresses low energy modes and acts effectively as a scale dependent mass term. A standard Legendre transform then gives the EAA, which can be seen as a coarse-grained version of the standard QFT effective action.

So far, various applications of the EAA formalism to the problem of quantum gravity [3, 4] have supported the Asymptotic Safety scenario. Evidence has been found for the existence of a non-Gaussian fixed-point with a finite dimensional UV-critical surface [5–7]. All previous applications of the EAA formalism to quantum gravity were based on the specific closure of the beta functions first proposed in [3]; here we propose a more consistent approach that accounts for the non-trivial influence that the anomalous dimensions of the fluctuating metric and ghost fields have on the RG flow of the cosmological and Newton's constants. Our results show that even in this enlarged framework, these two couplings are characterized by a non-Gaussian fixed point.

The application of these techniques to theories characterized by local symmetries requires overcoming the problem of performing the coarse-graining procedure in a covariant way. A solution to this problem comes from the combination of the EAA and the background field formalisms [8]. The preservation of gauge invariance along the flow comes at the cost of enlarging theory space to include invariants constructed with both background and fluctuating fields. This defines the background EAA (bEAA).

In the construction of the background EAA one introduces, in the path integral, source, gauge-fixing and cutoff terms; in these the background and fluctuating metric do not appear via their sum $g_{\mu\nu}$. As a consequence, the RG flow generates invariants which depend on $\bar{g}_{\mu\nu}$ and $h_{\mu\nu}$ separately.

The background EAA for gravity splits as follows:

$$\Gamma_k[\varphi; \bar{g}] = \bar{\Gamma}_k[\bar{g} + h] + \hat{\Gamma}_k[\varphi; \bar{g}] , \tag{27.1}$$

where $\varphi = (h_{\mu\nu}, \bar{C}_\mu, C^\nu)$ is the fluctuating multiplet comprising the fluctuating metric and the ghost fields. In (27.1) we defined the gauge invariant EAA $\bar{\Gamma}_k[\bar{g}]$ and the remainder EAA $\hat{\Gamma}_k[0; \bar{g}]$. The main virtue of the background EAA for gravity is that it satisfies an exact [3]:

$$\partial_t \Gamma_k[\varphi; \bar{g}] = \frac{1}{2} \mathrm{Tr} \left(\Gamma_k^{(2;0)}[\varphi; \bar{g}] + R_k[\bar{g}] \right)^{-1} \partial_t R_k[\bar{g}] , \tag{27.2}$$

which defines a mathematically consistent RG flow, i.e. UV and IR finite, despite the theory being perturbatively non-renormalizable: this is the non-perturbative tool that is used to inquire if gravity is asymptotically safe.

It is important to note that the RG flow described by (27.2) is driven by the Hessian of the background EAA taken with respect to the fluctuating multiplet φ, thus the RG flow equation for $\bar{\Gamma}_k[\bar{g}]$, resulting from setting $\varphi = 0$ in (27.2), is *not closed* since its rhs depends also on $\hat{\Gamma}_k[\varphi; \bar{g}]$. This fact forces us to consider the flow of the full $\Gamma_k[\varphi, \bar{g}]$ instead of only that of $\bar{\Gamma}_k[\bar{g}]$, which we would like to consider the physically interesting one.

27.2 Consistent Closure

Our truncation ansatz for the gauge invariant part of the EAA will be the RG improved version of the Einstein-Hilbert action:

$$\bar{\Gamma}_k[g] = \frac{1}{16\pi G_k} \int d^d x \sqrt{g}\,(2\Lambda_k - R)\ . \tag{27.3}$$

Wave-function renormalizations of the fluctuating fields,

$$h_{\mu\nu} \to Z_{h,k}^{1/2} h_{\mu\nu} \quad \bar{C}_\mu \to Z_{C,k}^{1/2} \bar{C}_\mu \quad C^\nu \to Z_{C,k}^{1/2} C^\nu\ , \tag{27.4}$$

will also generate the relative anomalous dimensions:

$$\eta_{h,k} = -\partial_t \log Z_{h,k} \qquad \eta_{C,k} = -\partial_t \log Z_{C,k}\ . \tag{27.5}$$

Finally, the remainder functional $\hat{\Gamma}_k[\varphi, \bar{g}]$ will contain the classical background gauge-fixing and ghost actions [4] in the gauge $\alpha = \beta = 1$.

To obtain the beta functions of the physical couplings one computes the Hessian of the background EAA with respect to the fluctuating fields φ, inserts it into the rhs of the RG flow equation (27.2) and then sets $\varphi = 0$. The trace on the rhs of (27.2) can then be expanded in terms of invariants of the background metric using heat kernel techniques in a standard way [4].

After introducing dimensionless cosmological and Newton's constants, $\tilde{\Lambda}_k = k^{-2}\Lambda_k$ and $\tilde{G}_k = k^{d-2} G_k$, one finds the following general, non-closed, system of beta functions:

$$\begin{aligned}
\partial_t \tilde{\Lambda}_k &= -2\tilde{\Lambda}_k \\
&\quad + \left[A_d(\tilde{\Lambda}_k) + C_d(\tilde{\Lambda}_k)\,\eta_{h,k} + E_d(\tilde{\Lambda}_k)\,\eta_{C,k}\right]\tilde{G}_k \\
\partial_t \tilde{G}_k &= (d-2)\tilde{G}_k \\
&\quad + \left[B_d(\tilde{\Lambda}_k) + D_d(\tilde{\Lambda}_k)\,\eta_{h,k} + F_d(\tilde{\Lambda}_k)\,\eta_{C,k}\right]\tilde{G}_k^2\ ,
\end{aligned} \tag{27.6}$$

where $A_d, B_d, C_d, D_d, E_d, F_d$ depend on the cutoff choice.

The first way in which one can close the beta functions (27.6) is the trivial one where one sets $\eta_{h,k} = \eta_{C,k} = 0$. This amounts to a one-loop approximation, which discards the nonperturbative information contained in the functional flow equation.

The second closure method is the "standard" RG improvement adopted in most previous studies [3, 4, 9]:

$$\eta_{h,k} = \frac{\partial_t G_k}{G_k} = 2 - d + \frac{\partial_t \tilde{G}_k}{\tilde{G}_k} \qquad \eta_{C,k} = 0\ . \tag{27.7}$$

Table 27.1 Fixed-points and critical exponents for the various closures of the beta functions of Λ_k and G_k, in $d = 4$

	$\tilde{\Lambda}_*$	$\tilde{G}_*$	$\theta' \pm i\theta''$	$\tilde{\Lambda}_*\tilde{G}_*$	$\eta_{h,*}$	$\eta_{C,*}$
One-loop	0.121	1.172	$-1.868 \pm 1.398i$	0.142	0	0
Reuter [3]	0.193	0.707	$-1.475 \pm 3.043i$	0.137	−2	0
Groh and Saueressig [13]	0.135	0.859	$-1.774 \pm 1.935i$	0.116	−2	−1.8
This work	−0.062	1.617	−4.119, −1.338	−0.100	0.686	−1.356

The beta functions obtained in this way are exactly those first obtained in [3]. In $d = 4$, they have a non-Gaussian fixed-point for the values of $\tilde{\Lambda}_*$ and $\tilde{G}_*$ reported in Table 27.1. The non-Gaussian fixed-point is UV attractive in both directions; thus, within this truncation, quantum gravity is asymptotically safe. The stability matrix has a pair of complex conjugated critical exponents with negative real part. These are also reported in Table 27.1.[1]

The third way to close (27.6) is to separately calculate the anomalous dimensions of the fluctuating metric and ghost fields that enter it. These can be determined as functions of $\tilde{\Lambda}_k$ and $\tilde{G}_k$ that can successively be reinserted in the beta functions. In doing so we make a step further in considering the flow in the enlarged theory space where the background EAA lives.

Our calculations of the anomalous dimensions $\eta_{h,k}$ and $\eta_{C,k}$ have been performed using the diagrammatic techniques presented in [10, 11], where one uses the flow equations for the zero-field proper-vertices of the background EAA to extract the running of the couplings [12].

Both $\eta_{h,k}$ and $\eta_{C,k}$ turn out not to depend on the cutoff operator type (i.e. on the cutoff operator used to separate fast from slow field modes) and have the following general form:

$$\eta_{h,k} = \left[a_d(\tilde{\Lambda}_k) + c_d(\tilde{\Lambda}_k)\,\eta_{h,k} + e_d(\tilde{\Lambda}_k)\,\eta_{C,k}\right]\tilde{G}_k$$
$$\eta_{C,k} = \left[b_d(\tilde{\Lambda}_k) + d_d(\tilde{\Lambda}_k)\,\eta_{h,k} + f_d(\tilde{\Lambda}_k)\,\eta_{C,k}\right]\tilde{G}_k\,. \tag{27.8}$$

Their explicit form is long and will be given in [12]. Equation (27.8) constitutes a linear system for $\eta_{h,k}$ and $\eta_{C,k}$ that can be solved to yield the anomalous dimensions as functions solely of the physical couplings $\tilde{\Lambda}_k$ and $\tilde{G}_k$.

[1]Here we follow the convention of [4] that a negative value for the critical exponent implies that the relative eigendirection is UV attractive.

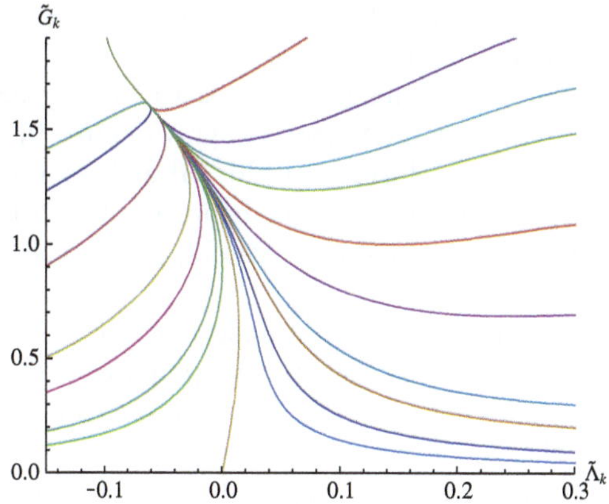

Fig. 27.1 RG flow in $d = 4$ in the $(\tilde{\Lambda}_k, \tilde{G}_k)$ plane for the closure of the beta functions obtained by inserting back in (27.6) the independently computed anomalous dimensions $\eta_{h,k}$ and $\eta_{C,k}$

27.3 Results

The result of the numerical integration of these beta functions, in $d = 4$, is plotted in Fig. 27.1. Note that, despite these new beta functions differ non-trivially from the one-loop and standard RG improved ones, we still find one UV attractive non-Gaussian fixed-point.

This time the critical exponents are real. This is clearly reflected in the fact that the flow next to the non-Gaussian fixed-point is no more spiraling as instead was in the previous cases. Real critical exponents are also suggested by the analysis of [14]. The fixed-point values of the dimensionless couplings and the critical exponents are also given in Table 27.1. Our fixed-point value of the dimensionless cosmological constant is negative and very small; however, the necessary inclusion of matter contributions will in any case change its value.

If we insert the fixed-point values of the cosmological and Newton's constants in the solution of the eta-system, we can determine the fixed-point values for the anomalous dimensions η_{h*} and η_{C*}. The numerical values we find are reported in Table 27.1, together with previous estimates. The anomalous dimension of $h_{\mu\nu}$ results positive, while the anomalous dimension of the ghost fields is negative, as also found in [13, 15].

27.4 Conclusions

We have shown how to account for the non-trivial influence that the anomalous dimensions $\eta_{h,k}$ and $\eta_{C,k}$ of the fluctuating fields have on the RG flow of the cosmological and Newton's constants. We have derived new RG improved beta functions for these couplings which still exhibit a UV attractive non-Gaussian fixed-point, but we have found real critical exponents. These results reinforce the Asymptotic Safety scenario in quantum gravity.

The closure method proposed here is general and can be applied to the beta functions of higher derivative gravity [16–19], to the beta functions obtained using the first-order formalism [20–23] and even to the beta functions present in non-local truncations of the gravitational EAA [24]. The method can also be extended to non-linear sigma models [25–27], membranes [28, 29] or Horava-Liftshitz gravity [30].

It is also important to understand the relation between our closure method and other complementary strategies proposed in the literature. In [31–33] bi-metric truncations were constructed using invariants made with both $g_{\mu\nu}$ and $\bar{g}_{\mu\nu}$; in [34] the problem has been studied using the Vilkovisky-DeWitt formalism; in [35] an analysis similar to ours has been performed, finding too a non-Gaussian fixed point with real critical exponents.

References

1. G. 't Hooft, M.J.G. Veltman, Ann. Poincare Phys. Theor. A **20**, 69 (1974)
2. S. Weinberg, in *General Relativity, An Einstein Centenary Survey*, eds. by S.W. Hawking, W. Israel (Cambridge University Press, 1979)
3. M. Reuter, Phys. Rev. D **57**, 971 (1998). arXiv:hep-th/9605030
4. A. Codello, R. Percacci, C. Rahmede, Ann. Phys. **324**, 414 (2009). arXiv:0805.2909 [hep-th]
5. A. Codello, R. Percacci, C. Rahmede, Int. J. Mod. Phys. A **23**, 143 (2008). arXiv:0705.1769 [hep-th]
6. P.F. Machado, F. Saueressig, Phys. Rev. D **77**, 124045 (2008). arXiv:0712.0445 [hep-th]
7. K. Falls, D.F. Litim, K. Nikolakopoulos, C. Rahmede. arXiv:1301.4191 [hep-th]
8. M. Reuter, C. Wetterich, Nucl. Phys. B **417**, 181 (1994)
9. D. Dou, R. Percacci, Class. Quant. Grav. **15**, 3449 (1998) arXiv:hep-th/9707239
10. A. Codello, Ph.D. thesis, Johannes Gutenberg–Universität, Mainz (2010)
11. A. Codello. arXiv:1304.2059 [hep-th]
12. A. Codello, G. D'Odorico, C. Pagani, in preparation
13. K. Groh, F. Saueressig, J. Phys. A **43**, 365403 (2010). arXiv:1001.5032 [hep-th]
14. D. Benedetti. arXiv:1301.4422 [hep-th]
15. A. Eichhorn, H. Gies, Phys. Rev. D **81**, 104010 (2010). arXiv:1001.5033 [hep-th]
16. A. Codello, R. Percacci, Phys. Rev. Lett. **97**, 221301 (2006). arXiv:hep-th/0607128
17. D. Benedetti, P.F. Machado, F. Saueressig, Mod. Phys. Lett. A **24**, 2233 (2009). arXiv:0901.2984 [hep-th]
18. D. Benedetti, P.F. Machado, F. Saueressig, Nucl. Phys. B **824**, 168 (2010). arXiv:0902.4630 [hep-th]
19. K. Groh, S. Rechenberger, F. Saueressig, O. Zanusso, PoS EPS **-HEP2011** (2011), p. 124. arXiv:1111.1743 [hep-th]
20. J.-E. Daum, M. Reuter, Phys. Lett. B **710**, 215 (2012). arXiv:1012.4280 [hep-th]
21. U. Harst, M. Reuter, JHEP **1205**, 005 (2012). arXiv:1203.2158 [hep-th]
22. P. Dona, R. Percacci, Phys. Rev. D **87**, 045002 (2013). arXiv:1209.3649 [hep-th]
23. J.-E. Daum, M. Reuter. arXiv:1301.5135 [hep-th]
24. A. Satz, A. Codello, F.D. Mazzitelli, Phys. Rev. D **82**, 084011 (2010). arXiv:1006.3808 [hep-th]
25. A. Codello, R. Percacci, Phys. Lett. B **672**, 280 (2009). arXiv:0810.0715 [hep-th]
26. R. Percacci, O. Zanusso, Phys. Rev. D **81**, 065012 (2010). arXiv:0910.0851 [hep-th]

27. R. Flore, A. Wipf, O. Zanusso. arXiv:1207.4499 [hep-th]
28. A. Codello, O. Zanusso, Phys. Rev. D **83**, 125021 (2011). arXiv:1103.1089 [hep-th]
29. A. Codello, N. Tetradis, O. Zanusso. arXiv:1212.4073 [hep-th]
30. S. Rechenberger, F. Saueressig, JHEP **1303**, 010 (2013). arXiv:1212.5114 [hep-th]
31. E. Manrique, M. Reuter, Ann. Phys. **325**, 785 (2010). arXiv:0907.2617 [gr-qc]
32. E. Manrique, M. Reuter, F. Saueressig, Ann. Phys. **326**, 463 (2011). arXiv:1006.0099 [hep-th]
33. E. Manrique, M. Reuter, F. Saueressig, Ann. Phys. **326**, 440 (2011). arXiv:1003.5129 [hep-th]
34. I. Donkin, J. Pawlowski. arXiv:1203.4207 [hep-th]
35. N. Christiansen, D.F. Litim, J.M. Pawlowski, A. Rodigast. arXiv:1209.4038 [hep-th]

Chapter 28
Phase Transitions of Regular Schwarzschild-anti-deSitter Black Holes

Antonia M. Frassino

Abstract We study a solution of the Einstein's equations generated by a self-gravitating, anisotropic, static, non-singular matter fluid. The resulting Schwarzschild like solution is regular and accounts for smearing effects of noncommutative fluctuations of the geometry. We call this solution regular Schwarzschild spacetime. In the presence of an Anti-deSitter cosmological term, the regularized metric offers an extension of the Hawking-Page transition into a van der Waals-like phase diagram. Specifically the regular Schwarzschild-Anti-deSitter geometry undergoes a first order small/large black hole transition similar to the liquid/gas transition of a real fluid. In the present analysis we have considered the cosmological constant as a dynamical quantity and its variation is included in the first law of black hole thermodynamics.

28.1 Regular Schwarzschild-anti-deSitter Spacetime

The regular Schwarzschild anti-deSitter (AdS) metric is a static, spherically symmetric solution of the Einstein's equations with negative cosmological constant $\Lambda = -3/b^2$ and a Gaussian matter source [1–4]. To obtain this metric we replace the vacuum with a Gaussian distribution having variance equivalent to the parameter $\sqrt{\theta}$

$$\rho(r) \equiv \frac{M}{(4\pi\theta)^{3/2}} e^{-r^2/4\theta} . \tag{28.1}$$

A.M. Frassino (✉)
Frankfurt Institute for Advanced Studies, Ruth-MoufangstraßE 1,
60438 Frankfurt am Main, Germany
e-mail: frassino@fias.uni-frankfurt.de

A.M. Frassino
Institut für Theoretische Physik, Johann Wolfgang Goethe-Universität
Frankfurt am Main, Frankfurt am Main, Germany

P. Nicolini et al. (eds.), *1st Karl Schwarzschild Meeting on Gravitational Physics*,
Springer Proceedings in Physics 170, DOI 10.1007/978-3-319-20046-0_28

This type of matter distribution emulates non-commutativity of space-time through the parameter θ that corresponds to the area of the elementary quantum cell, accounting for a natural ultraviolet spacetime cut-off (see [2] and references therein). The resulting energy momentum tensor describes an anisotropic fluid, whose components, fixed by $\nabla_\mu T^{\mu\nu} = 0$ and the condition $g_{00} = -g_{rr}^{-1}$, read

$$T_0^0 = T_r^r = -\rho\,(r) \qquad T_\phi^\phi = T_\theta^\theta = -\rho\,(r) - \frac{r}{2}\frac{\partial\rho\,(r)}{\partial r}\,. \tag{28.2}$$

The spherically symmetric solution of the Einstein's equations with this energy momentum tensor and the cosmological constant Λ is given by the line element

$$ds^2 = -V\,(r)\,dt^2 + \frac{dr^2}{V\,(r)} + r^2 d\Omega^2 \tag{28.3}$$

where $d\Omega^2 = d\vartheta^2 + \sin^2\vartheta\, d\varphi^2$ and

$$V\,(r) = 1 + \frac{r^2}{b^2} - \frac{\omega M}{r}\gamma\left(\frac{3}{2}, \frac{r^2}{4\theta}\right)\,. \tag{28.4}$$

Here $\omega = 2\mathrm{G_N}/\Gamma\,(3/2)$, G_N is the four-dimensional Newton's constant and b is the curvature radius of the AdS space. The function $\gamma\left(\frac{3}{2}, \frac{r^2}{4\theta}\right)$ is the incomplete gamma function defined as $\gamma\,(n, x) \equiv \int_0^x dt\, t^{n-1} e^{-t}$. The line element (28.3) has an event horizon at $r = r_+$, where r_+ is solution of the horizon equation $V\,(r) = 0$. The event horizon radius coincides with the Schwarzschild radius in the limit $\sqrt{\theta}/r_+ \to 0$. The metric (28.3) admits an inner horizon $r_- < r_+$, that coalesces with r_+ in the extremal black hole configuration at $r_0 = r_+ = r_-$. Such a degenerate horizon occurs even without charge or angular momentum.

28.2 Thermodynamics and Equation of state

The temperature associated to the event horizon r_+ can be computed through the formula $T = \frac{1}{4\pi}\, V'\,(r)\big|_{r=r_+}$ and reads

$$T = \frac{1}{4\pi r_+}\left\{1 + \frac{r_+^2}{b^2}\left(3 - r_+\frac{\gamma'\,(r_+)}{\gamma\,(r_+)}\right) - r_+\frac{\gamma'\,(r_+)}{\gamma\,(r_+)}\right\}\,, \tag{28.5}$$

where $\gamma\,(r_+) \equiv \gamma\left(\frac{3}{2}, \frac{r_+^2}{4\theta}\right)$, $\gamma'\,(r_+) = \frac{r_+^2}{4\theta^{3/2}} e^{-r_+^2/4\theta}$ is its derivative with respect to r_+. In contrast to the standard Schwarzschild-anti-deSitter case, extremal solution exists with vanishing Hawking temperature (28.5).

Recently, the idea of including the variation of the cosmological constant in the first law of black hole thermodynamics has been considered [5–7] with interesting consequences: If the cosmological constant Λ behaves like a pressure, we have that for negative cosmological constant the pressure turns to be positive [6, 8], i.e.,

$$\frac{1}{b^2} = -\frac{\Lambda}{3} \equiv \frac{8\pi P}{3}, \tag{28.6}$$

giving rise to several effects (see for example [9–11]). In such a case the equation of state $P(V, T)$ for the regular AdS black hole becomes

$$P = \frac{3\gamma(r_+)}{(3\gamma(r_+) - r_+\gamma'(r_+))} \left\{ \frac{T}{2r_+} - \frac{1}{8\pi r_+^2} + \frac{\gamma'(r_+)}{8\pi r_+ \gamma(r_+)} \right\}. \tag{28.7}$$

Here T is the Hawking temperature of the black hole, i.e. (28.5). Using the equation of state (28.7) it is possible to plot the isotherm functions in a P-V diagram for a regular black hole that resembles the van der Waals pressure-volume diagram (Fig. 28.1).

28.2.1 Gibbs Free Energy

In order to complete the analogy between the regular black hole and a van der Waals gas, we proceed by calculating the Gibbs free energy [6, 7]. This can be done by calculating the action of the Euclidean metric (see for example [12]). Such an action

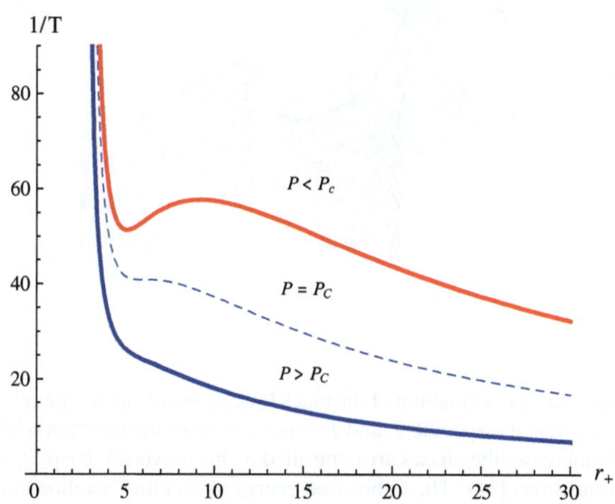

Fig. 28.1 The inverse temperature as function of r_+ (with $\theta = 1$). When $P < P_c$, there are three branches. The *middle* branch is unstable, while the branch with the smaller radii and the one with bigger radii are stable. This graph reproduces the pressure-volume diagram of the van der Waals theory, provided one identifies the black hole thermodynamic variables $\beta \equiv 1/T$, r_+ and P respectively with pressure, volume and temperature of the van der Waals gas

provides the Gibbs free energy via $G = I/\beta$ where β is the period of the imaginary time $\beta \equiv 1/T$. Then, the Gibbs free energy can be expressed as a function of pressure and temperature. The Hawking-Page transition [13] for the standard Schwarzschild-AdS black hole is first order phase transition between a large black hole phase and the purely thermal AdS spacetime. Such a transition takes place when the Gibbs energy changes its sign from positive to negative. In the regular black hole case and considering the cosmological constant as a pressure we find

$$G = \frac{r_+}{12\,G_N}\left[3 - 8P\pi r_+^2 + \frac{r_+\left(3 + 8P\pi r_+^2\right)\gamma'\left(r_+\right)}{\gamma\left(r_+\right)}\right] \tag{28.8}$$

and the Gibbs free energy (28.8) exhibits a characteristic swallowtail behavior (see Fig. 28.2). This usually corresponds to a small black hole/large black hole first-order phase transition [7, 14]. By performing the classical limit for $r \gg \theta$ we get the usual result for a classical uncharged Schwarzschild-AdS black hole that is $G\left(T, P\right) = \left(1/4G_N\right)\left(r_+ - \frac{8\pi}{3} P\, r_+^3\right)$ [7]. Remarkably, in the regular Schwarzschild-AdS black hole case, as in the Reissner-Nordström-AdS (RN-AdS) black hole spacetime, there is a phase transition that occurs at positive Gibbs energy. This fact is visible from the presence of the swallowtail in Fig. 28.2. To investigate this aspect we need to study the sign of the heat capacity. As underlined in [6], the specific heat related to the

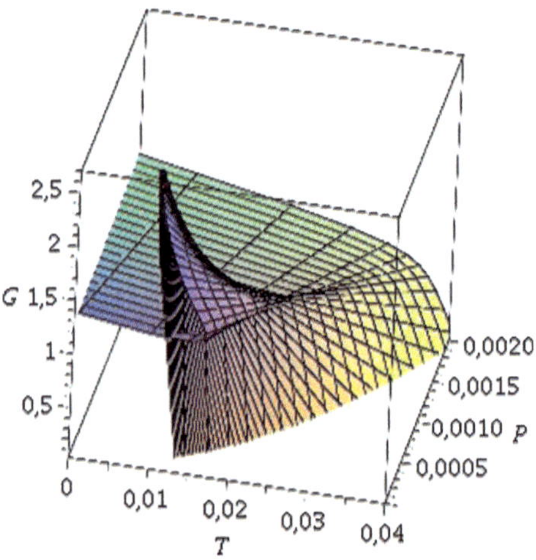

Fig. 28.2 Gibbs free energy as function of the black hole pressure and temperature. The Gibbs free energy G changes its sign at a specific T and P (intersection of the function with the T-P-plane). As in the van der Waals case, the phases are controlled by the universal 'cusp', typical of the theory of discontinuous transitions [14]. The Gibbs free energy shows the "swallowtail" shape, a region where $G(T, P)$ is a multivalued function. This region ends in a point (T_c, P_c). In the region with $P < P_c$ and $T < T_c$ we can see a transition between small black hole/large black hole. Note that r_+ is a function of temperature and pressure via the equation of state (28.7). For large value of P (or T) there is only one branch allowed

black hole is calculated at constant pressure

$$C_p = \left(\frac{\partial H}{\partial T}\right)_P = \left(\frac{\partial H}{\partial r_+}\right)_P \left(\frac{\partial r_+}{\partial T}\right)_P, \tag{28.9}$$

where the enthalpy H is identified with the black hole mass M [6]. Now one can study the phase transitions from the change of the sign of the specific heat: the stability requires that the specific heat at fixed pressure is $C_p \geq 0$ and the specific heat at fixed volume is $C_v \geq 0$. In the case under investigation C_v is always equal to zero because the entropy is only volume dependent. This means that the heat capacity C_v does not diverge at the critical point and its critical exponent is $\alpha = 0$. By studying the sign of the function C_p, we can see that for $P > P_c$ the quantity C_p is always positive and the black hole is stable. In the limit $P \to P_c$ there is a critical value for r_+ for which C_p diverges. For $P < P_c$ there are two discontinuities of the specific heat and the situation is the same as in the Reissner-Nordström-AdS black holes [15]. Thus, in the regular Schwarzschild-AdS case for $P < P_c$ it seems that a different phase transition is allowed because the heat capacity changes again from positive values to negative values. For large r_+ we have the Hawking-Page behavior in which the branch with negative specific heat has lower mass and thus falls in an unstable phase, while the branch with larger mass is locally stable and corresponds to a positive specific heat. Thus, the resulting phase diagram presents a critical point at a critical cosmological constant value in Plank units and a smooth crossover thereafter.

28.2.2 Critical Exponent

We already determined $\alpha = 0$ in the previous section. Now, by defining the variable $t \equiv (T - T_c)/T_c$, we can compute the critical exponent of C_p by evaluating the ratio $\ln(C_p(t))/\ln(t)$ in the limit $t \to 0$. We find that the limit exists and the critical exponent is $\gamma = 1$. This result implies that the heat capacity diverges near the critical point like $C_p \propto |t|^{-1}$. Then using the scaling relations

$$\alpha + 2\beta + \gamma = 2 \tag{28.10}$$

$$\alpha + \beta(1 + \delta) = 2 \tag{28.11}$$

is possible to calculate the other two exponents, i.e., δ that determines the behaviour of the isothermal compressibility of a VdW system and β that describes the behaviour of the difference between of the volume of the gas phase and the liquid phase. For the regular black hole, the scaling relations give $\delta = 3$ and $\beta = 1/2$, result that coincides with the case of charged black holes [7]. These critical exponents are consistent with the Ising mean field values $(\alpha, \beta, \gamma, \delta) = (0, 1/2, 1, 3)$ allowing for an efficient mean field theory description. Since it is believed that the determination of critical exponents define universality classes , i.e., they do not depend on the details of the physical system (exept the number of dimensions), we can say that the phase transitions in the regular Schwarzschild-AdS black holes and in the RN-AdS black holes in four-dimensional spacetime have the same nature.

28.3 Final Remarks

After almost hundred years since the Karl Schwarzschild's exact solution of Einstein's equation, black hole physics is nowadays at the forefront of current research in several branches of theoretical physics. Specific interest has been developed in the thermodynamics of charged black holes in asymptotically AdS spacetime, largely because they admit a gauge duality description via a dual thermal field theory [12]. In recent studies it has been shown that charged Reissner-Nordström AdS black holes exhibit critical behaviour similar to a van der Waals liquid gas phase transition [7]. This analogy become "complete" if the cosmological constant Λ is considered as a dynamical quantity and its variation is included in the first law of black hole thermodynamics [7, 9]. This extended phase space shows new insights with respect to the conventional phase space of a four dimensional black hole in AdS background consisting only of two variables: entropy and temperature. In this work the cosmological constant has been considered as a thermodynamical pressure and its conjugate quantity as a thermodynamical volume. The black hole equation of state (28.7) obtained by considering the regular Schwarzschild-AdS solution shows analogy with the van der Waals liquid-gas system where the parameter θ plays an analogue role of the charge. Note that a detailed description of the not-extended phase space has been presented in [3].

Acknowledgments This work has been supported by the Helmholtz Research School for Quark Matter Studies (H-QM). The author is grateful to P. Nicolini and D. Kubiznak for having carefully read the draft and provided valuable comments.

References

1. P. Nicolini, A. Smailagic, E. Spallucci, Phys. Lett. B **632**, 547 (2006). arXiv:gr-qc/0510112
2. P. Nicolini, Int. J. Mod. Phys. A **24**, 1229 (2009). arXiv:0807.1939 [hep-th]
3. P. Nicolini, G. Torrieri, JHEP **1108**, 097 (2011). arXiv:1105.0188 [gr-qc]
4. A. Smailagic, E. Spallucci, Int. J. Mod. Phys. D **22**, 1350010 (2013). arXiv:1212.5044 [hep-th]
5. D. Kastor, S. Ray, J. Traschen, Class. Quant. Grav. **26**, 195011 (2009). arXiv:0904.2765 [hep-th]
6. B.P. Dolan, Class. Quant. Grav. **28**, 125020 (2011). arXiv:1008.5023 [gr-qc]
7. D. Kubiznak, R.B. Mann, JHEP **1207**, 033 (2012). arXiv:1205.0559 [hep-th]
8. M. Cvetic, G.W. Gibbons, D. Kubiznak, C.N. Pope, Phys. Rev. D **84**, 024037 (2011). arXiv:1012.2888 [hep-th]
9. S. Gunasekaran, R.B. Mann, D. Kubiznak, JHEP **1211**, 110 (2012). arXiv:1208.6251 [hep-th]
10. N. Altamirano, D. Kubiznak, R.B. Mann, Phys. Rev. D **88**, 10 (2013) 101502. arXiv:1306.5756 [hep-th]

11. A.M. Frassino, D. Kubiznak, R.B. Mann, F. Simovic, JHEP **1409**, 080 (2014). arXiv:1406.7015 [hep-th]
12. E. Witten, Adv. Theor. Math. Phys. **2**, 505 (1998). arXiv:hep-th/9803131
13. S.W. Hawking, D.N. Page, Commun. Math. Phys. **87**, 577–588 (1983)
14. A. Chamblin, R. Emparan, C.V. Johnson, R.C. Myers, Phys. Rev. D **60**, 064018 (1999). arXiv:hep-th/9902170
15. C. Niu, Y. Tian, X.-N. Wu, Phys. Rev. D **85**, 024017 (2012). arXiv:1104.3066 [hep-th]

Chapter 29
Vector Fields and Kerr/CFT Correspondence

Amir M. Ghezelbash

Abstract We use the appropriate boundary action for the vector fields near the horizon of near extremal Kerr black hole to calculate the two-point function for the vector fields in Kerr/CFT correspondence. The gauge-independent part of the two-point function is in agreement with what is expected from CFT.

29.1 Introduction

The Kerr/CFT correspondence provides us a new type of holography that may have a connection to the realistic astrophysical objects. In the early days of Kerr/CFT correspondence the discussions were concentrated to the extremal or near extremal case, which are still realistic since the X-ray astrophysical observations support the existence of near extremal rotating black hole candidates [1]. A more general proposal for Kerr/CFT correspondence, which is not limited to extremal rotating black holes, was appeared in [2]. In generic Kerr/CFT correspondence, the conformal symmetry for the generic rotating black holes (also known as hidden conformal symmetry) is revealed from the solution space of the wave function of scalar (or higher spin) fields. The proposal has been applied to several types of generic rotating black holes [3]. In both types of extremal and generic Kerr/CFT correspondence, the results for some physical quantities of rotating black holes, such as the Bekenstein-Hawking entropy and scattering cross section, are in perfect agreement with the corresponding physical quantities in the dual CFT.

Quite recently the authors of [4] found the two-point function of spinor fields in Kerr/CFT correspondence by variation of boundary action for spin-1/2 particles. They determined an appropriate boundary term for the spinors in NHEK geometry and used it to calculate the two-point function of spinors. Moreover they found a

---

Based on preprint arXiv:1308.0096, with H.M. Siahaan, submitted for publication.

A.M. Ghezelbash (✉)
Department of Physics and Engineering Physics, University of Saskatchewan, Saskatoon, SK S7N 5E2, Canada
e-mail: amg142@campus.usask.ca

P. Nicolini et al. (eds.), *1st Karl Schwarzschild Meeting on Gravitational Physics*, Springer Proceedings in Physics 170, DOI 10.1007/978-3-319-20046-0_29

relation between spinors in the four-dimensional bulk and the boundary spinors living in two dimensions. The two-point function of spinor fields is in agreement with the correlation function of a two-dimensional CFT. Inspired by Becker et al. [4], we derive the two-point function for Maxwell fields in Kerr spacetime by varying the corresponding boundary action. Unlike the analysis of spin-1/2 particles that there is no gauge condition, one needs to perform a more careful treatment for the gauge fields where they are subjected to the gauge condition.

29.2 Two-Point Function of Vector Fields

We consider the Maxwell fields in background of Kerr metric in Boyer-Lindquist coordinates,

$$ds^2 = -\frac{\Delta}{\rho^2}\left(dt - a\sin^2\theta d\phi\right)^2 + \frac{\rho^2}{\Delta}dr^2 + \rho^2 d\theta^2 + \frac{\sin^2\theta}{\rho^2}\left(adt - \left(r^2+a^2\right)d\phi\right)^2, \tag{29.1}$$

where $\rho^2 = r^2 + a^2\cos^2\theta$ and $\Delta = r^2 + a^2 - 2Mr$. The action for the Maxwell fields is given by

$$S = \frac{1}{4}\int d^4x\sqrt{-g}\mathbf{F}^*{}_{\mu\nu}\mathbf{F}^{\mu\nu} + c.c. \tag{29.2}$$

The $c.c.$ term should be added in (29.2) to ensure that the action is real valued as we will notice that the Chandrasekhar solutions (29.3)–(29.6) for the Maxwell fields in Kerr spacetime are basically complex quantities. The existence of ∂_t and ∂_ϕ Killing vectors in Kerr geometry leads to write down the dependence of Maxwell fields $\mathbf{A}$ on coordinates t and ϕ as $\mathbf{A} = \begin{pmatrix}\mathbf{A}_t\\ \mathbf{A}_r\\ \mathbf{A}_\theta\\ \mathbf{A}_\phi\end{pmatrix} = e^{-i\omega t + im\phi}\begin{pmatrix}A_t\\ A_r\\ A_\theta\\ A_\phi\end{pmatrix}$, where A_μ's are given by

$$A_t = \frac{ia}{\rho^2\sqrt{2}}\left(\Delta R_+\zeta_+ - R_-\zeta_- - \sin\theta\left(\xi_+S_+ - \xi_-S_-\right)\right), \tag{29.3}$$

$$A_r = \frac{ia}{\sqrt{2}}\left(R_+\zeta_+ + \frac{R_-\zeta_-}{\Delta}\right), \tag{29.4}$$

$$A_\theta = -\frac{1}{\sqrt{2}}\left(\xi_+S_+ + \xi_-S_-\right), \tag{29.5}$$

$$A_\phi = \frac{-i}{\rho^2\sqrt{2}}\left(a^2\sin^2\theta\left(\Delta R_+\zeta_+ - R_-\zeta_-\right) - \sin\theta\left(r^2+a^2\right)\left(\xi_+S_+ - \xi_-S_-\right)\right). \tag{29.6}$$

The functions $\zeta_\pm$ and $\xi_\pm$ are defined by

$$\xi_+ = \frac{1}{2CK}\left(\left(ir\lambda + 2\alpha^2\omega\right)R_- - irC\Delta R_+\right), \tag{29.7}$$

$$\xi_- = \frac{1}{2CK}\left(-\left(ir\lambda - 2\alpha^2\omega\right)\Delta R_+ + irCR_-\right), \tag{29.8}$$

$$\zeta_+ = \frac{1}{2CQ}\left(\left(-\lambda\cos\theta - \frac{2\alpha^2\omega}{a}\right)S_- - CS_+\cos\theta\right), \tag{29.9}$$

$$\zeta_- = \frac{1}{2CQ}\left(\left(\lambda\cos\theta - \frac{2\alpha^2\omega}{a}\right)S_+ + CS_-\cos\theta\right), \tag{29.10}$$

in terms of Teukolsky radial and angular functions $R_\pm$ and $S_\pm$. The functions K, Q and parameters C and α are

$$K = -\left(r^2 + a^2\right)\omega + am, \quad Q = -a\omega\sin\theta + m\left(\sin\theta\right)^{-1}, \tag{29.11}$$

$$|C|^2 = \lambda^2 - 4\alpha^2\omega^2, \; \alpha^2 = a^2 - \frac{am}{\omega}, \tag{29.12}$$

where λ is the separation constant in separated Teukolsky equations for the massless particles with spin 1. We note that in (29.2), $\mathbf{F}_{\mu\nu} = \partial_\mu \mathbf{A}_\nu - \partial_\nu \mathbf{A}_\mu$ and so we can write $S = 2S_0$ where $S_0 = \frac{1}{4}\int d^4x\sqrt{-g}\left(\partial_\mu \mathbf{A}^*_\nu\right)\mathbf{F}^{\mu\nu} + c.c.$. The integrand of S_0 can be written as the difference of a total derivative term and other term which is, in fact, proportional to the Maxwell's equations. Taking a spherical boundary with radius r_B that is the boundary of near-NHEK geometry of Kerr black hole, we can convert the total derivative term to a boundary term, given by

$$S_B = \frac{1}{2}\int d^3x\,\sqrt{-g}\mathbf{A}^*_\nu\mathbf{F}^{r\nu}\Big|_{r=r_B} + c.c., \tag{29.13}$$

where d^3x stands for $dtd\phi d\theta$. The field strength tensor components

$$\mathbf{F}^{r\nu} = g^{rr}g^{\nu\beta}\mathbf{F}_{r\beta} = g^{rr}g^{\nu\beta}\left(\partial_r\mathbf{A}_\beta - \partial_\beta\mathbf{A}_\mathbf{r}\right), \tag{29.14}$$

can be written simply as $\mathbf{F} = \Xi\mathbf{A}$ where $\Xi = g^{rr}\begin{pmatrix} g^{tt}\partial_r & -\left(g^{tt}\partial_t + g^{t\phi}\partial_\phi\right) & 0 & g^{t\phi}\partial_r \\ 0 & 0 & 0 & 0 \\ 0 & -g^{\theta\theta}\partial_\theta & g^{\theta\theta}\partial_r & 0 \\ g^{\phi t}\partial_r & -\left(g^{\phi t}\partial_t + g^{\phi\phi}\partial_\phi\right) & 0 & g^{\phi\phi}\partial_r \end{pmatrix}$.

Using the above expressions, we can rewrite the boundary action (29.13) accordingly as

$$S_B = \frac{1}{2}\int d^3x\,\sqrt{-g}\mathbf{A}^\dagger\Xi\mathbf{A}\Big|_{r=r_B} + c.c., \tag{29.15}$$

where

$$\mathbf{A}^{\dagger}\Xi\mathbf{A} = g^{rr}\left(g^{tt}\left(A_t^*\partial_r A_t - i\omega A_t^* A_r\right) + g^{t\phi}\left(A_t^*\partial_r A_\phi + imA_t^* A_r\right) - g^{\theta\theta}\left(A_\theta^*\partial_\theta A_r - A_\theta^*\partial_r A_\theta\right)\right.$$

$$\left. + g^{\phi t}\left(A_\phi^*\partial_r A_t - i\omega A_\phi^* A_r\right) + g^{\phi\phi}\left(A_\phi^*\partial_r A_\phi + imA_\phi^* A_r\right)\right). \qquad (29.16)$$

To find the explicit form of (29.16), we have to make some approximations to simplify the calculation. The Teukolsky radial functions in the matching region can be read as [5]

$$R_+ = N_+\tau_H^{-1-in/2}\left(\mathscr{A}_+\left(\frac{r}{\tau_H}\right)^{\beta-3/2} + \mathscr{B}_+\left(\frac{r}{\tau_H}\right)^{-\beta-3/2}\right) + \cdots, \qquad (29.17)$$

$$R_- = N_-\tau_H^{1-in/2}\left(\mathscr{A}_-\left(\frac{r}{\tau_H}\right)^{\beta+1/2} + \mathscr{B}_-\left(\frac{r}{\tau_H}\right)^{-\beta+1/2}\right) + \cdots, \qquad (29.18)$$

where β is given by $\beta^2 = \frac{1}{4} + K_l - 2m^2$ and τ_H is the dimensionless Hawking temperature. The parameter K_l is related to separation constant λ by $K_l = \lambda + 2am\omega$ and we consider $K_l \geq 2m^2 - 1/4$ and so β is a real number. The coefficients $\mathscr{A}_\pm$ and $\mathscr{B}_\pm$ are $\mathscr{A}_\pm = \frac{\Gamma(2\beta)\Gamma(1\mp 1-in)}{\Gamma\left(\frac{1}{2}+\beta-i(n-m)\right)\Gamma\left(\frac{1}{2}+\beta\mp 1-im\right)}$, $\mathscr{B}_\pm = \frac{\Gamma(-2\beta)\Gamma(1\mp 1-in)}{\Gamma\left(\frac{1}{2}-\beta-i(n-m)\right)\Gamma\left(\frac{1}{2}-\beta\mp 1-im\right)}$, where $n = \frac{\omega - m\Omega_H}{2\pi T_H}$ and $\Omega_H = \frac{a}{r_+^2+a^2}$ is the angular velocity of the horizon. We notice that since T_H is a very small number and n is a finite number, so $\omega \sim m\Omega_H$. This means we consider only the Maxwell fields with frequency that is around the superradiant bound. The coefficients N_+ and N_- are the normalization constants that their ratio is $\frac{N_-}{N_+} = -\frac{\mathscr{K}_l r_+^2}{n(n+i)}$ where $\mathscr{K}_l = \sqrt{K_l^2 + m^2(m^2 + 1 - 2K_l)}$. In deriving this ratio, we considered the near horizon limit $r \to r_+$. To compact the notation, we rewrite the Maxwell fields in a matrix form as

$$\mathbf{A} = e^{-i\omega t + im\phi}(R_+\mathbf{K}\mathbf{v}_+ + R_-\mathbf{L}\mathbf{v}_-), \qquad (29.19)$$

where the matrices $\mathbf{K}$ and $\mathbf{L}$ are given by

$$\mathbf{K} = \begin{pmatrix} \Delta_B f_1 & 0 & 0 & \Delta_B f_2 \\ f_5 & \kappa_1 & 0 & f_6 \\ \Delta_B f_9 & 0 & \kappa_2 & \Delta_B f_{10} \\ \Delta_B f_{13} & 0 & 0 & \Delta_B f_{14} \end{pmatrix}, \quad \mathbf{L} = \begin{pmatrix} \kappa_3 & f_3 & f_4 & 0 \\ 0 & f_7\Delta_B^{-1} & f_8\Delta_B^{-1} & 0 \\ 0 & f_{11} & f_{12} & 0 \\ 0 & f_{15} & f_{16} & \kappa_4 \end{pmatrix}, \qquad (29.20)$$

and $\mathbf{v}_+ = \left(S_- \; 0 \; 0 \; S_+\right)^T$, $\mathbf{v}_- = \left(0 \; S_- \; S_+ \; 0\right)^T$. For later convenience, we show the first and the second term of (29.19) by $\mathbf{A}_+$ and $\mathbf{A}_-$, respectively. The functions $f_i, i = 1, \ldots, 16$ are given explicitly in [6]. The arbitrary constants $\kappa_i, i = 1, 2, 3, 4$

in (29.20) are introduced to provide the invertibility for matrices **K** and **L**. We set $\kappa_i \to 0$ at the end of calculation wherever κ_i's appear. We may find that there is a relation between the vectors $\mathbf{v}_+$ and $\mathbf{v}_-$ as $\mathbf{v}_- = \chi \mathbf{v}_+$ where the matrix χ is given by $\chi = \begin{pmatrix} 0\,0\,0\,0 \\ 1\,0\,0\,0 \\ 0\,0\,0\,1 \\ 0\,0\,0\,0 \end{pmatrix}$. Denoting the Maxwell fields and the Teukolsky functions on the boundary by $\mathbf{A}^B_\pm$ and $R^B_\pm$ respectively, we get $\frac{\mathbf{A}^B_+}{R^B_+} = e^{-i\omega t + im\phi}\mathbf{K}\mathbf{v}_+$, and $\frac{\mathbf{A}^B_-}{R^B_-} = e^{-i\omega t + im\phi}\mathbf{L}\mathbf{v}_-$. Using the relation between $\mathbf{v}_+$ and $\mathbf{v}_-$, we get

$$\mathbf{A} = (R_+\mathbf{K} + R_-\mathbf{L}\chi)\, e^{-i\omega t + im\phi}\mathbf{v}_+ = \left(R_+ + R_-\mathbf{L}\chi\mathbf{K}^{-1}\right)\frac{(\mathbf{A}^B_+)}{R^B_+}. \qquad (29.21)$$

A very lengthy calculation shows that the integrand $\mathbf{A}^\dagger \Xi \mathbf{A}$ of the boundary action (29.15) is given by

$$\begin{aligned}
\mathbf{A}^\dagger \Xi \mathbf{A} = {} & g^{rr}\frac{\left(\mathbf{A}^B_+\right)^\dagger}{\left(R^B_+\right)^*} R^*_+ \left(R_+(\Pi_1 + \Theta) - Q_+\Pi_2\right)\frac{\mathbf{A}^B_+}{R^B_+} \\
& + g^{rr}\frac{\left(\mathbf{A}^B_+\right)^\dagger}{\left(R^B_+\right)^*} R^*_+ \left(R_-(\Pi_3 + \Theta) - Q_-\Pi_4\right)\mathbf{L}\chi\mathbf{K}^{-1}\frac{\mathbf{A}^B_+}{R^B_+} \\
& + g^{rr}\frac{\left(\mathbf{A}^B_+\right)^\dagger}{\left(R^B_+\right)^*}\left(R_-\mathbf{L}\chi\mathbf{K}^{-1}\right)^\dagger \left(R_+(\Pi_1 + \Theta) - Q_+\Pi_2\right)\frac{\mathbf{A}^B_+}{R^B_+} \\
& + g^{rr}\frac{\left(\mathbf{A}^B_+\right)^\dagger}{\left(R^B_+\right)^*}\left(R_-\mathbf{L}\chi\mathbf{K}^{-1}\right)^\dagger \left(R_-(\Pi_3 + \Theta) - Q_-\Pi_4\right)\mathbf{L}\chi\mathbf{K}^{-1}\frac{\mathbf{A}^B_+}{R^B_+},
\end{aligned} \qquad (29.22)$$

where $\Pi_i, i = 1, \ldots, 4$ and Θ are given explicitly in [6]. Furnished by (29.22) and performing the functional derivative of (29.15) with respect to the real part of the rescaled boundary fields $\mathscr{A}^B_{i+}$, we get the two-point function as

$$\frac{\delta^2 S_B}{\delta\mathscr{A}^B_{i+}\delta\mathscr{A}^B_{j+}} = r_B^{2\beta-4}\frac{\left(R^B_-\right)^* R^B_-}{\left(R^B_+\right)^* R^B_+}\mathscr{Z}^{ij}, \qquad (29.23)$$

where $\mathscr{Z}^{ij} = \int_0^\pi d\theta \sin(\theta)\{\tilde{\theta}^{ij}_4\tilde{\theta}_i\tilde{\theta}^*_j + \tilde{\theta}^{*ij}_4\tilde{\theta}^*_i\tilde{\theta}_j\}_{n.s.}$. The functions $\tilde{\theta}_i$ and $\tilde{\theta}^{ij}_4$ are presented in [6] and n.s. means there is no summation over indices i and j. Moreover we note from the results of [6] that the leading terms in (29.23) correspond to indices i and j to be t and ϕ only. We can simplify (29.23) to

$$\frac{\delta^2 S_B}{\delta\mathscr{A}^B_{i+}\delta\mathscr{A}^B_{j+}} = \mathscr{Y}^{ij} r_B^{2\beta}\left(M^4 + \left(\frac{\mathscr{K}_l M^2}{n\,(n-i)}\right) r_B^{-2\beta}\frac{N_-\mathscr{A}_-}{N_+\mathscr{A}_+}G_R^* \right.$$
$$\left. + \left(\frac{\mathscr{K}_l M^2}{n\,(n+i)}\right) r_B^{-2\beta}\frac{N_-^*\mathscr{A}_-^*}{N_+^*\mathscr{A}_+^*}G_R + \frac{N_-N_-^*}{N_+N_+^*}\frac{\tau_H^{4\beta}}{r_B^{4\beta}}\mathscr{B}_-\mathscr{B}_-^*\right). \quad (29.24)$$

In (29.24), G_R stands for $G_R(n_L, n_R) = -n\,(i+n)\,T_R^{2\beta}\frac{\Gamma(-2\beta)}{\Gamma(2\beta)}\frac{\Gamma\left(\beta+\frac{1}{2}-in_R\right)\Gamma\left(\beta-\frac{1}{2}-in_L\right)}{\Gamma\left(\frac{1}{2}-\beta-in_R\right)\Gamma\left(\frac{3}{2}-\beta-in_L\right)}$, where n_L and n_R are related to m and ω by $m = n_L$, and $n = n_L + n_R$. The first term in bracket in (29.24) clearly is a constant term compared to the other terms. The second term in (29.24) is the complex conjugate of the third term. Moreover, we can ignore the fourth term of (29.24) compared to the other terms, as this term is proportional to $\tau_H^{4\beta}$. Dropping the complex conjugate term in (29.24) according to [4, 7], we find that the field theoretical two-point function (29.23) is equal to $G_R\mathscr{Y}^{ij}$ up to a multiplicative factor that depends on momentum and is not a part of the retarded Green's function. The existence of multiplicative factor has also been found for the field theoretical two-point function of spinors [4]. We note that $G_R(n_L, n_R)$ is in exact agreement with the proposed retarded Green's function for the spin-1 fields in [8]. Using the optical theorem for the obtained retarded Green's function, we get exactly the absorption cross section of spin-1 fields scattered off of the Kerr black hole [9]. Interestingly enough, as we mentioned before, the boundary vector field components that contribute to the leading term of two-point function are only $\mathscr{A}^B_{t+}$ and $\mathscr{A}^B_{\phi+}$. This fact is in agreement with the statement of Kerr/CFT correspondence that the dual boundary theory is a two-dimensional CFT. The two-point function (29.23) (or (29.24)) is a function of ω and m which are the conjugate momenta in t and ϕ directions, respectively. The two-point function (29.23) (or (29.24)) also is in agreement with the correlator of vector operators in the dual boundary CFT to Kerr black hole [6].

29.3 Concluding Remarks

We have obtained the two-point function for the vector fields on the near horizon of near extremal Kerr black holes by varying the appropriate boundary action for the vector fields with respect to the boundary vector fields. One interesting result that emerges from the explicit calculation of the boundary action is that the degrees of freedom of boundary vector fields (which is two) supports the original idea of Kerr/CFT correspondence that the dual theory to the four-dimensional Kerr black hole is a two-dimensional CFT. This is in contrast to the well known AdS$_{d+1}$/CFT$_d$ result that the dimension of bulk theory is exactly one more than the dimension of dual CFT. Moreover the two-point function for the vector fields factorizes in two terms. The first term is not sensitive to the vector indices while the second term depends on vector indices as well as the gauge condition. The structure of the two-point function is exactly in agreement with the correlator of vector operators in a CFT.

Acknowledgments This work was supported by the Natural Sciences and Engineering Research Council of Canada.

References

1. J.E. McClintock, R. Shafee, R. Narayan, R.A. Remillard, S.W. Davis, L.-X. Li, Astrophys. J. **652**, 518 (2006)
2. A. Castro, A. Maloney, A. Strominger, Phys. Rev. D **82**, 024008 (2010)
3. A.M. Ghezelbash, H.M. Siahaan, Class. Quantum Grav. **30**, 135005 (2013)
4. M. Becker, W. Schulgin, JHEP **1204**, 063 (2012)
5. T. Hartman, W. Song, A. Strominger, JHEP **1003**, 118 (2010)
6. A.M. Ghezelbashm, H.M. Siahaan, arXiv:1308.0096, http://arxiv.org/pdf/1308.0096v2.pdf
7. D.T. Son, A.O. Starinets, JHEP **0209**, 042 (2002)
8. B. Chen, C.-S. Chu, JHEP **1005**, 004 (2010)
9. I. Bredberg, T. Hartman, W. Song, A. Strominger, JHEP **1004**, 019 (2010)

Chapter 30
Black Holes in Non-relativistic Holography

Stefan Janiszewski

Abstract Many non-relativistic quantum field theories with conserved particle number share a common set of symmetries: time dependent spatial diffeomorphisms acting on the background metric and U(1) invariance acting on the background field which couples to particle number. These symmetries are used to deduce a holographic gravity dual for any such theory in terms of Hořava gravity, a non-relativistic theory of gravity. The behavior of black holes in Hořava gravity are studied due to their importance in holography. The existence of causal horizons in this non-relativistic theory is shown, as well as examples of their thermodynamic properties.

30.1 Non-relativistic Holography from Hořava Gravity

The holographic principle is a powerful duality in theoretical physics. It has allowed the exploration of various strongly coupled systems, and has provided understanding of the information paradox in black hole physics. In its traditional form holography is inherently relativistic. Below we formulate a proposal to extend this duality to non-relativistic (NR) physics.

30.1.1 Holography

In its most general form, the holographic principle is the statement that any quantum theory of gravity in d dimensions is equivalent to a non-gravitational theory in $d-1$ dimensions. This means that any question asked in one theory can, in principle, be translated to the other theory, answered there, and then translated back. As with most dualities, this approach to problem solving is especially powerful when one of the theories in the correspondence is strongly coupled but the other is in a perturbative regime.

S. Janiszewski (✉)
University of Washington, Seattle, WA 98195-1560, USA
e-mail: stefanjj@uw.edu

P. Nicolini et al. (eds.), *1st Karl Schwarzschild Meeting on Gravitational Physics*, Springer Proceedings in Physics 170, DOI 10.1007/978-3-319-20046-0_30

The most understood holographic duality is that between general relativity (GR) on the Anti-de Sitter (AdS) spacetime and the conformal field theory (CFT) known as $\mathcal{N} = 4$ super [1]. This calculational tool has been used to great effect, pertinently in studying analogues of the strongly coupled plasma of QCD [2]. On the other hand, many interesting strongly coupled systems explorable in the lab are in a NR regime. Motivated by this, we will propose a form of NR holography below.

For the purposes in that direction, AdS/CFT provides us with two important lessons. The first is that the symmetry structure between the two dual theories needs to match. For example, relativistic quantum field theories (QFTs) have the global symmetry of coordinate changes. We can always work in what ever coordinates make a problem simpler, as long as we remember to correctly change things like the volume measure in integrals and any other coupling to the metric. In the dual gravitational theory, coordinate changes are of fundamental importance: reparametrization invariance is the defining gauge symmetry of the dynamic metric of general relativity. This matching of symmetries is necessary in the duality, relativistic or not.

The second important lesson to be gained from AdS/CFT is the importance of black holes to holography. Black holes have long been known to have a temperature and other thermodynamic properties. In holography these quantities match those of the dual field theory state, so studying finite temperature QFTs requires knowledge of black hole solutions. The nature of causality and horizons in NR theories such as Hořava gravity is distinct from the everyday black holes of GR, and are thoroughly explored in Sect. 30.2.

30.1.2 Non-relativistic Gravity

Hořava gravity [3], like GR, is a metric theory of a manifold $\mathcal{M}$. Unlike GR, the manifold comes fundamentally equipped with a co-dimension one foliation $\mathcal{F}$. The leaves of this foliation are events that take place simultaneously in a global time. Consequently, the theory is not Lorentz invariant as it has this preferred frame.

Coordinate changes that respect this foliation structure are gauge symmetries of Hořava gravity. These foliation preserving diffeomorphisms are arbitrary time dependent changes of the spatial coordinates, $x_I \to \tilde{x}_I(t, x_J)$, as well as reparametrizations of the global time, $t \to \tilde{t}(t)$. Spatially dependent temporal diffeomorphisms, $t \to \tilde{t}(t, x_J)$, are disallowed as they would mix space into time and violate the simultaneity set by the global time foliation.

There are some interesting properties of Hořava gravity that make it a promising candidate for a NR holographic duality. Foremost, the so-called Lifshitz spacetimes,

$$ds^2 = -\frac{dt^2}{r^{2z}} + \frac{dr^2}{r^2} + \frac{d\mathbf{x}^2}{r^2}, \tag{30.1}$$

which have an anisotropic scaling symmetry between time and space, are vacuum solutions to Hořava gravity, while GR requires somewhat unnatural additional matter fields to support these backgrounds. This scaling symmetry applies to many NR field theories, and therefore is something that would need to be captured in the dual gravitational theory.

Another interesting property of Hořava gravity, especially from a holographic viewpoint, concerns its high energy behavior. Unlike GR, which requires something like string theory as a UV completion, Hořava gravity is expected to be a UV complete theory, and therefore is a quantum theory of gravity all by itself. This, in principle, allows exploration of the full regime of the gravitational theory, whereas traditional holography has been limited to exploring the low energy perturbative sector given by GR. Such explorations are promising future work.

The final ingredient that leads to our proposal for a NR holography is an understanding of NR field theories. In [4] it was first discovered that foliation preserving diffeomorphisms play the role of a global symmetry of many NR QFTs. Guided by this, and in analogy to AdS/CFT, we propose that any holographic dual of such a NR QFT must have foliation preserving diffeomorphisms as its gauge symmetry. As discussed above, Hořava gravity is such a theory, and gives us a concrete framework for NR holography [5, 6].

30.2 Black Holes in Hořava Gravity

Due to its lack of Lorentz invariance, the notions of causality and black holes need to be reexamined in the context of Hořava gravity. Despite no light cone structure, causality is maintained by the preferred global time, while horizons that trap signals of any speed are seen to exist. Therefore, black holes can rightly be said to be part of Hořava gravity, and there properties can be explored.

30.2.1 Causality and Universal Horizons

The foliation of Hořava gravity breaks the Lorentz invariance that the manifold itself enjoyed. This means that there is no speed limit set by light, and signals can travel arbitrarily fast. One familiar with GR might think this also throws out any notion of causality, which is normally enforced by the time-like or space-like invariant separation of events, as determined by light cones. Luckily, the offending foliation itself comes into play to enforce causal behavior. Since the leaves of the foliation are events that happen at a constant global time, signals propagating arbitrarily fast will still propagate causally as long as they only move *forward* in global time. In this way, with respect to a given event, the spacetime divides itself into "past" and "future" along a leaf of the foliation, whereas GR causally divides spacetime into these categories inside the light cone, but has the additional "elsewhere" region outside of it.

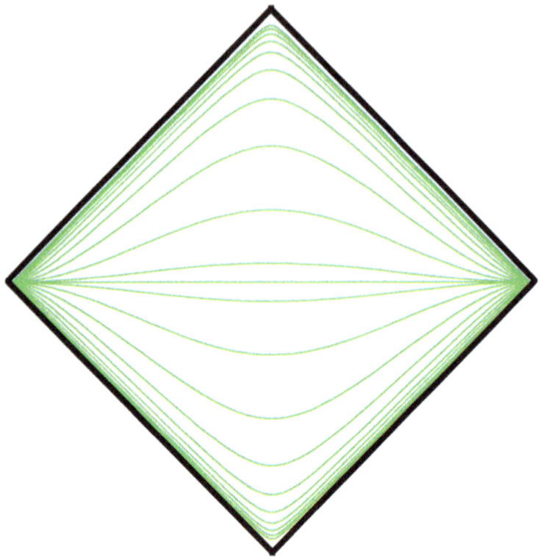

Fig. 30.1 A Penrose diagram of Minkowski space. The *thick diamond* is past and future infinity. The *thin lines* are the foliation by Minkowski time

This leads into the discussion of causal horizons in Hořava gravity. In GR a black hole is understood as a light-like surface from which behind no signal can propagate to future infinity. This is plausible from the above context: when inside a black hole future infinity is in your "elsewhere" region and therefore there is no causal way you can reach it. Such an object is well defined in GR because light cones are invariant. On the other hand, the causal division of spacetime into only "past" and "future" in Hořava gravity seems to leave no possibility for horizons to hide behind. To clear up this issue it pays to examine the foliations of some spacetimes, as this is what encodes the causal structure.

Figure 30.1 shows the foliation of Minkowski space by the canonical time coordinate. This is a solution of Hořava gravity, meaning not only is the manifold an acceptable spacetime, but the foliation by Minkowski time is a global time coordinate. The lack of causal horizons is apparent: from any event on the spacetime future infinity can be reached by only moving forward in global time, that is from one leaf to the next.

Figure 30.2 shows the foliation of the Schwarzschild black hole by the canonical time. Here we see that the foliation that covers the asymptotic region does not cover the entire manifold, the leaves all bunch up at the Killing horizon. In GR this is irrelevant, our choice of time coordinate is not invariant and a Lorentzian diffeomorphism can change the foliation to cover the entire manifold. Instead, the Killing horizon is an invariant causal horizon in GR because it is light-like. In Hořava gravity the situation is different. First off, as presented, Fig. 30.2 is *not* a solution. The problem can be seen to be that the foliation is not a global time: in Hořava gravity the foliation is dynamically determined and Schwarzschild time does not solve the equations of motion. The bunching of the foliation is irrelevant for this trivial reason.

Figure 30.3 shows the foliation of the AdS-Schwarzschild black hole. Here the foliation has been dynamically determined and therefore this *is* a solution of Hořava

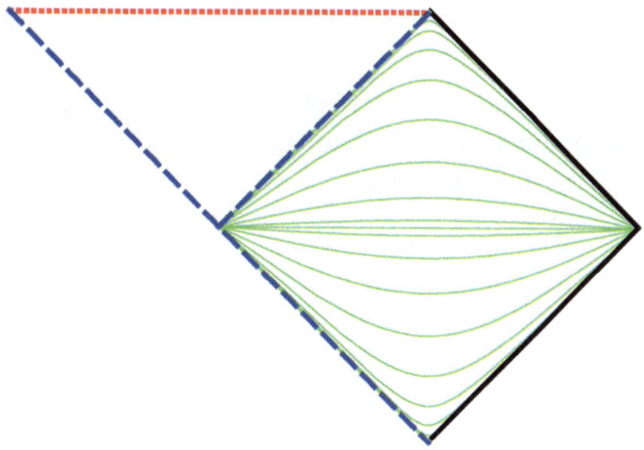

Fig. 30.2 A Penrose diagram of Schwarzschild space. The *solid thick line* is past and future infinity. The *thick dashed lines* are the Killing horizons, while the *thick dash-dot line* is the singularity. The *thin lines* are the foliation by Schwarzschild time

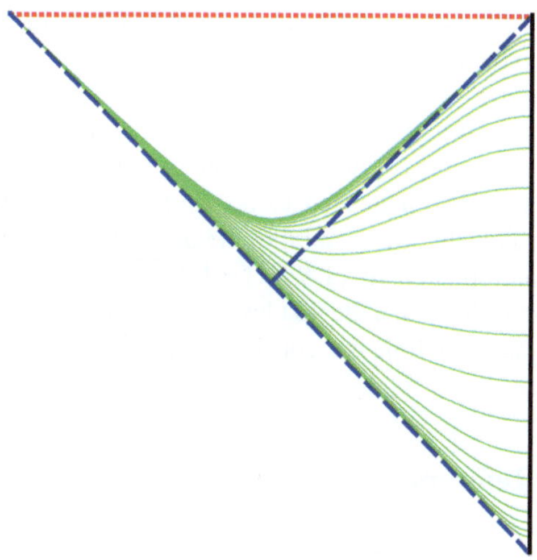

Fig. 30.3 A Penrose diagram of the AdS-Schwarzschild black hole. The *solid thick line* is asymptotic infinity. The *thick dashed lines* are the Killing horizons, while the *thick dotted line* is the singularity. The *thin lines* are the foliation by the dynamically determined global time

gravity [7]. The fact that the asymptotic foliation does not cover the entire manifold is an invariant statement: the allowed time reparametrizations cannot change this property. We now see how causal horizons can arise in Hořava gravity. Even though the foliation penetrates the Killing horizon, allowing fast signals to propagate past that surface, it does not penetrate the full interior. Events inside where the leaves all bunch up are to the future of any event on the foliation shown in Fig. 30.3, including the entire asymptotic boundary! This means that causal signals, traveling only forward in global time, can never escape this region to reach asymptotic infinity, even traveling arbitrarily fast. At last we have seen a black hole in Hořava gravity. Generically, they can be identified by foliations that cover the entire asymptotic region of a manifold, but do not penetrate the entire interior. The boundary of this foliation, where the leaves all bunch up, is called the universal horizon, and is the causal boundary defining the black hole.

30.2.2 Thermodynamics of an Analytic Black Hole

In [7] an analytic Hořava black hole is presented. This allows an exploration of the thermodynamics of these objects, crucial knowledge in the holographic context. The metric can be written as

$$ds^2 = -\frac{(r^3 - r_h^3)^2}{r^2 r_h^6} dt^2 + \frac{r_h^6}{r^2(r^3 - r_h^3)^2} \left(dr + \frac{r^3(r^3 - r_h^3)}{\sqrt{1 - c_3 r_h^6}} dt \right)^2 + \frac{d\mathbf{x}^2}{r^2}, \quad (30.2)$$

where c_3 is a coupling constant of Hořava gravity and r_h is the radial location of the universal horizon. The asymptotic region is as $r \to 0$, where it is AdS space. Of importance is that the time coordinate t is a global time, that is, the leaves given by its level sets is a preferred foliation of Hořava gravity.

By examining the Hamiltonian of this black hole its mass density can be calculated to be $M = 1/(4\pi(1 - c_3) r_h^3 G_H)$, where G_H is the gravitational constant of Hořava gravity. By examining the near-horizon geometry of the black hole its temperature can be argued to be $T = 3/(2\pi r_h \sqrt{1 - c_3})$. This can also be calculated by the method of [8], where the rate of particles quantum mechanically tunneling through the horizon is seen to be thermal. Lastly, the entropy of this black hole can be calculated by examining the on-shell action, which is the free energy. This gives an entropy density of $S = 1/(4\sqrt{1 - c_3} r_h^2 G_H)$. Bringing these properties together we can check that the first law of thermodynamics $dM = TdS$ is satisfied, and therefore the black hole is a consistent thermal object.

References

1. J.M. Maldacena, The Large N limit of superconformal field theories and supergravity. Int. J. Theor. Phys. **38**, 1113 (1999)
2. G. Policastro, D.T. Son, A.O. Starinets, The shear viscosity of strongly coupled N=4 supersymmetric Yang-Mills plasma. Phys. Rev. Lett. **87**, 081601 (2001)
3. P. Horava, Quantum gravity at a lifshitz point. Phys. Rev. D **79**, 084008 (2009)
4. D.T. Son, M. Wingate, General coordinate invariance and conformal invariance in nonrelativistic physics: Unitary Fermi gas. Ann. Phys. **321**, 197 (2006)
5. S. Janiszewski, A. Karch, String theory embeddings of nonrelativistic field theories and their holographic hoava gravity duals. Phys. Rev. Lett. **110**(8), 081601 (2013)
6. S. Janiszewski, A. Karch, Non-relativistic holography from Horava gravity. JHEP **1302**, 123 (2013)
7. S. Janiszewski, Asymptotically hyperbolic black holes in Horava gravity, arXiv:1401.1463 [hep-th]
8. M.K. Parikh, F. Wilczek, Hawking radiation as tunneling. Phys. Rev. Lett. **85**, 5042 (2000)

Chapter 31
Black Holes and Running Couplings: A Comparison of Two Complementary Approaches

Benjamin Koch, Carlos Contreras, Paola Rioseco and Frank Saueressig

Abstract Black holes appear as vacuum solutions of classical general relativity which depend on Newton's constant and possibly the cosmological constant. At the level of a quantum field theory, these coupling constants typically acquire a scale-dependence. A quantum treatment of a black hole should thus take this effect into account. In these proceedings we briefly summarize two complementary approaches to this problem: the renormalization group improvement of the classical solution based on the scale-dependent gravitational couplings obtained within the gravitational Asymptotic Safety program and the exact solution of the improved equations of motion including an arbitrary scale dependence of the gravitational couplings. Remarkably the picture of the "quantum black holes" obtained from these very different improvement strategies is strikingly similar.

31.1 Introduction

The emergence of scale-dependent couplings is one of the central phenomena encountered in quantum field theory. While the quest for a consistent and predictive quantum formulation for gravity is still ongoing, it is natural to expect that this feature will emerge in this case as well. This expectation is supported by perturbative computations in the framework of higher-derivative gravity [1–10] as well as the non-perturbative computations carried out within the gravitational Asymptotic Safety program [11–14].

B. Koch (✉) · P. Rioseco
Instituto de Fisica, Pontificia Universidad Católica de Chile,
Av. Vicuña Mackenna 4860, Santiago, Chile
e-mail: bkoch@fis.puc.cl

C. Contreras
Departamento de Física, Universidad Técnica Federico Santa María,
Casilla 110-V, Valparaíso, Chile

F. Saueressig
Institute for Mathematics, Astrophysics and Particle Physics,
Radboud University, Heyendaalseweg 135, 6525 AJ Nijmegen, The Netherlands

P. Nicolini et al. (eds.), *1st Karl Schwarzschild Meeting on Gravitational Physics*,
Springer Proceedings in Physics 170, DOI 10.1007/978-3-319-20046-0_31

An important testing ground for ideas related to modified theories of gravity or quantum gravity is given by the black hole solutions obtained from classical general relativity. Striving for a quantum description of these classical black holes, it is natural to study the effect of scale-dependent coupling constants on the physics of the black holes. In this proceedings paper we will focus on two complementary strategies for capturing these effects:

- The first approach discussed in Sect. 31.2 was pioneered in [15, 16] and performs a renormalization group (RG) improvement of the classical black hole solution. Here the classical coupling constants are promoted to scale-dependent couplings whose flow is governed by beta functions computed within Asymptotic Safety [17–26].
- The second approach covered in Sect. 31.3 has been advocated in [27] and looks for consistent solutions of the improved equations of motion. Those equations can be solved without making further assumptions on the actual scale dependence of the couplings, leading two a new, spherically symmetric metric. This metric can be seen as a promising candidate for a physical black hole metric that incorporates general effects of scale dependent couplings.

In Sect. 31.4 we will compare those results and conclude.

31.2 Improved Solutions from Asymptotic Safety

This section basically follows [28]. Thus, we will summarize concepts and results, rather than detailed calculations. For more details the reader is referred to [28] and references therein.

The key ingredient for investigating Weinberg's Asymptotic Safety conjecture [29] is the gravitational effective average action Γ_k [30], a Wilson-type effective action that provides an effective description of physics at the momentum scale k. As its main virtue, the scale-dependence of Γ_k is governed by an exact functional renormalization group equation [30]

$$\partial_k \Gamma_k = \frac{1}{2} \mathrm{Tr} \left[\left(\Gamma_k^{(2)} + \mathscr{R}_k \right)^{-1} \partial_k \mathscr{R}_k \right] . \tag{31.1}$$

Here $\mathscr{R}_k$ is an IR-regulator that renders the trace finite and peaked on fluctuations with momenta $p^2 \approx k^2$ (Fig. 31.1).

For the study of (A)dS black holes conducted in this work, we restrict the gravitational part of Γ_k to the (euclidean) Einstein-Hilbert action

$$\Gamma_k^{\mathrm{grav}}[g] = \frac{1}{16\pi G_k} \int d^4x \sqrt{g} \left[-R + 2\Lambda_k \right] \tag{31.2}$$

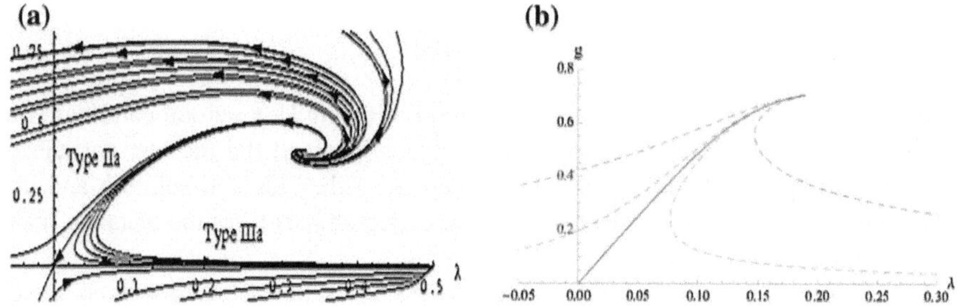

Fig. 31.1 **a** RG flow originating from the Einstein-Hilbert truncation (31.2) obtained with a sharp cut-off. The *arrows* point in the direction of increasing coarse-graining, i.e., of decreasing k. From [31]. **b** Schematic flow of the scale dependent couplings $\lambda_U(r)$ and $g_U(r)$ for $g_U^* = 0.707$, $\lambda_U^* = 0.193$, $g_I = 2.5$, and $G_0 = \Sigma = 1$. The different *curves* correspond to $l_I = \{-0.05, -0.005, 0, 0.005, 0.05\}$

and include two scale-dependent couplings, Newton's constant G_k and the cosmological constant Λ_k. The beta functions resulting from this truncation have first been derived in [30] and are most conveniently expressed in terms of the dimensionless coupling constants

$$g_k = G_k k^2, \quad \lambda_k = \Lambda_k k^{-2}. \tag{31.3}$$

The phase diagram resulting from the flow has been constructed in [31] and is shown below.

The flow is governed by the interplay of a Gaussian fixed point (GFP) located at the origin, $g_* = 0$, $\lambda_* = 0$ and an NGFP which, for an optimal cutoff, is located at

$$\lambda_* = 0.193 , \qquad g_* = 0.707 , \qquad g_* \lambda_* = 0.137 \ . \tag{31.4}$$

One way to investigate the implications of the scaling gravitational couplings on black hole physics is the improvement of the classical black hole solution pioneered by Bonanno and Reuter [15, 16] and subsequently refined by several groups [17, 19, 20, 23–26]. This procedure starts from the classical (Schwarzschild-de Sitter or anti-de Sitter) line-element

$$ds^2 = -f(r)\,dt^2 + f(r)^{-1}\,dr^2 + r^2 d\Omega_2^2 \tag{31.5}$$

with

$$f(r) = 1 - \frac{2GM}{r} - \frac{1}{3}\Lambda r^2 , \tag{31.6}$$

and replaces Newton's constant and the cosmological constant by their scale dependent counterparts, $G \to G_k$, $\Lambda \to \Lambda_k$. The crucial step following this improvement is the scale setting procedure, which relates the momentum scale k to the radial scale r

$$k(P(r)) = \frac{\xi}{d(P(r))} \quad , \tag{31.7}$$

where ξ is an a priory undetermined constant. Following [], a natural candidate for $d(P)$ is the radial proper distance between the origin and the the test particle is located at the point P. Based on the symmetries of the classical solution and the coordinate invariance of the identification we expect that $d(P)$ is the distance scale which is the relevant cutoff k.

Performing this analysis for improved (A)dS black holes led to various novel conclusions, which are largely independent of the details underlying the scale setting procedure:

(a) Including the effect of a scale-dependent cosmological constant in the RG-improvement process drastically affects the structure of the quantum-improved black holes *at short distances*. Thus a consistent RG-improvement procedure requires working in the class of Schwarzschild-(A)dS solutions of Einstein's equations.
(b) The short-distance structure of all quantum-improved black holes is governed by the non-Gaussian RG fixed point. This entails that the structure of light black holes is universal. In particular it is independently of the IR-value of Newton's constant and the cosmological constant and therefore identical for classical Schwarzschild, Schwarzschild-dS and Schwarzschild-AdS black holes.
(c) In the presence of the cosmological constant, the curvature singularity at $r = 0$ is not resolved.

31.3 Solving Improved Equations of Motion

An alternative strategy for modeling the quantum properties of the classical black holes, based on "improving the equations of motion", has been developed in [32]. In this case, the scale-setting procedure is carried out at the level of the (wick-rotated) Einstein-Hilbert action (31.2) where the k-dependence of the couplings is replaced by a generic r-dependence. The resulting equations of motion are [33, 34]

$$G_{\mu\nu} = -g_{\mu\nu}\Lambda(r) + 8\pi G(r)T_{\mu\nu} - \Delta t_{\mu\nu} \quad , \tag{31.8}$$

with

$$\Delta t_{\mu\nu} = G(r)\left(g_{\mu\nu}\Box - \nabla_\mu\nabla_\nu\right)\frac{1}{G(r)} \quad . \tag{31.9}$$

With the metric ansatz

$$ds^2 = -F(r)dt^2 + 1/F(r)dr^2 + r^2 d\theta^2 + r^2\sin(\theta)d\phi^2 \quad , \tag{31.10}$$

the equations of motion can be solved exactly, for the functions $F(r)$, $\Lambda(r)$, and $G(r)$. This solution is non-trivial, leading to four constants of integration c_1, c_2, c_3, c_4. These constants can be related to familiar properties of the classical solution such as M_0, G_0, Λ_0 together with a possible correction. Alternatively, they can be traded for the adimensional parameters g_I, g_U, λ_I, and λ_U which naturally appear in the induced coupling flow

$$g_U(r) = G(r)\Sigma^2 , \qquad \lambda_U(r) = -\Lambda(r)\frac{r}{\Sigma} , \tag{31.11}$$

where Σ is an arbitrary matching constant which has mass dimension one. The values of the UV fixed points of this "flow" are

$$g_U(r \to 0) = g_U^* , \qquad \lambda_U(r \to 0) = \lambda_U^* . \tag{31.12}$$

Plotting the "flow" for those two couplings induced be the solution of the equations of motion one finds a surprisingly similar shape to the flow shown in the attached figure. Some of the main conclusions of this research can be summarized as

(a) There exists a non-trivial solution of the improved equations of motion (31.8) which would not be possible to find without the cosmological term. Thus, including a scale dependent cosmological term is crucial for this approach.
(b) In the UV limit, the new solution is dominated by the fixed points g_U^* and λ_U^*.
(c) The new solution $F(r)$ exhibits a singularity at the origin, which is of the same grade as the singularity of the Schwarzschild solution. In this limit the solution is dominated by the non-trivial fixed point of the induced "flow".
(d) Interpreting the solution for the dimensionless couplings (31.11) as "flow" one finds interesting similarities with the RG flow derived from the effective average action Γ_k.

31.4 Comparison of the Two Approaches

We have summarized two strategies for modeling quantum corrections to classical black hole solutions based on implementing scale-dependent couplings. The first approach is based on improving the classical solutions and uses the beta functions obtained from Asymptotic Safety to fix the scale-dependence of the gravitational coupling constants (see Sect. 31.2). The second approach is based on simply solving the equations of motion that have to be fulfilled in the presence of scale dependent couplings $\Lambda(r)$ and $G(r)$ in a generally covariant theory (31.8). The most interesting overlapping results are

(a) Λ **matters**:
In both approaches the cosmological constant has a significant effect. In Sect. 31.2 this term strongly dominated the UV behavior of the improved solution, while

in Sect. 31.3 this term was actually crucial for obtaining a non-trivial solution at all.

(b) **Fixed points control UV**:
In both approaches the short distance behavior is dominated by the non-trivial fixed point of the true flow in Sect. 31.2 or of the induced "flow" in Sect. 31.3.

(c) **Singularity persists**:
Rather surprisingly, both approaches exhibit the same type of black hole singularity located at the origin. Since it is a general expectation that quantum gravity should be capable of resolving this singularity, it would be very interesting to understand this result in more detail.

To conclude we want to emphasize that we found several common main features of black holes in the context of scale dependent couplings, by using two very different approaches. These coincidences consolidates the findings of both approaches.

Acknowledgments The work of B.K. was supported proj. Fondecyt 1120360 and anillo Atlas Andino 10201; the research of F.S. is supported by the Deutsche Forschungsgemeinschaft (DFG) within the Emmy-Noether program (Grant SA/1975 1-1). The work of C.C. was supported proj. Fondecyt 1120360 and DGIP grant 11.11.05.

References

1. K.S. Stelle, Phys. Rev. D **16**, 953 (1977)
2. J. Julve, M. Tonin, Nuovo Cim. B **46**, 137 (1978)
3. E.S. Fradkin, A.A. Tseytlin, Phys. Lett. B **104**, 377 (1981)
4. E.S. Fradkin, A.A. Tseytlin, Nucl. Phys. B **201**, 469 (1982)
5. I.G. Avramidi, A.O. Barvinsky, Phys. Lett. B **159**, 269 (1985)
6. N.H. Barth, S.M. Christensen, Phys. Rev. D **28**, 1876 (1983)
7. G. de Berredo-Peixoto, I.L. Shapiro, Phys. Rev. D **71**, 064005 (2005). arXiv:hep-th/0412249
8. A. Codello, R. Percacci, Phys. Rev. Lett. **97**, 221301 (2006). arXiv:hep-th/0607128
9. M. Niedermaier, Phys. Rev. Lett. **103**, 101303 (2009)
10. K. Groh, S. Rechenberger, F. Saueressig, O. Zanusso, PoS EPS **-HEP2011**, 124 (2011). arXiv:1111.1743
11. M. Niedermaier, M. Reuter, Living Rev. Rel. **9**, 5 (2006)
12. R. Percacci, in *Approaches to Quantum Gravity: Towards a New Understanding of Space, Time and Matter*, ed. by D. Oriti (Cambridge University Press, Cambridge, 2009). arXiv:0709.3851
13. A. Codello, R. Percacci, C. Rahmede, Int. J. Mod. Phys. A **23**, 143 (2008). arXiv:0705.1769
14. M. Reuter, F. Saueressig, New J. Phys. **14**, 055022 (2012). arXiv:1202.2274
15. A. Bonanno, M. Reuter, Phys. Rev. D **60**, 084011 (1999). arXiv:gr-qc/9811026
16. A. Bonanno, M. Reuter, Phys. Rev. D **62**, 043008 (2000). arXiv:hep-th/0002196
17. H. Emoto. arXiv:hep-th/0511075
18. A. Bonanno, M. Reuter, Phys. Rev. D **73**, 083005 (2006). arXiv:hep-th/0602159
19. B. Koch, Phys. Lett. B **663**, 334 (2008). arXiv:0707.4644
20. T. Burschil, B. Koch, Zh Eksp, Teor. Fiz. **92**, 219 (2010). arXiv:0912.4517
21. M. Reuter, E. Tuiran. arXiv:hep-th/0612037
22. M. Reuter, E. Tuiran, Phys. Rev. D **83**, 044041 (2011). arXiv:1009.3528
23. K. Falls, D.F. Litim, A. Raghuraman, Int. J. Mod. Phys. A **27**, 1250019 (2012). arXiv:1002.0260
24. K. Falls, D.F. Litim. arXiv:1212.1821
25. D.F. Litim, K. Nikolakopoulos. arXiv:1308.5630

26. R. Casadio, S.D.H. Hsu, B. Mirza, Phys. Lett. B **695**, 317 (2011). arXiv:1008.2768
27. M. Reuter, H. Weyer, Phys. Rev. D **69**, 104022 (2004). arXiv:hep-th/0311196
28. B. Koch, F. Saueressig, Class. Quant. Grav. (to Appear). arXiv:1306.1546
29. S. Weinberg, in *General Relativity, an Einstein Centenary Survey*, eds. by S.W. Hawking, W. Israel (Cambridge University Press, Cambridge, 1979)
30. M. Reuter, Phys. Rev. D **57**, 971 (1998). arXiv:hep-th/9605030
31. M. Reuter, F. Saueressig, Phys. Rev. D **65**, 065016 (2002). arXiv:hep-th/0110054
32. C. Contreras, B. Koch, P. Rioseco, Class. Quant. Grav. **30**, 175009 (2013). arXiv:1303.3892
33. M. Reuter, H. Weyer, JCAP **0412**, 001 (2004). arXiv:hep-th/0410119
34. S. Domazet, H. Stefancic, Class. Quant. Grav. **29**, 235005 (2012). arXiv:1204.1483

Chapter 32
Quantum Harmonic Black Holes

Alessio Orlandi and Roberto Casadio

Abstract Inspired by the recent conjecture that black holes are condensates of gravitons, we investigate a simple model for the black hole degrees of freedom that is consistent both from the point of view of Quantum mechanics and of General Relativity. Since the two perspectives should "converge" into a unified picture for small, Planck size, objects, we expect our construction to be a useful step for understanding the physics of microscopic, quantum black holes. In particular, we show that a harmonically trapped condensate gives rise to two horizons, whereas the extremal case (corresponding to a remnant with vanishing Hawking temperature) is not contained in the spectrum.

32.1 Introduction

One of the major mysteries in modern theoretical physics is to understand the internal degrees of freedom of black holes (BHs). Our best starting point is the classical description of BHs provided by General Relativity [6], along with well established semiclassical results, such as the predicted Hawking radiation [15, 16].

It was recently proposed by Dvali and Gomez that BHs are Bose-Einstein Condensates (BECs) of gravitons at a critical point, with Bogoliubov modes that become degenerate and nearly gapless representing the holographic quantum degrees of freedom responsible for the BH entropy and the information storage [10–13]. In order to support this view, they consider a collection of objects (gravitons) interacting via the Newtonian gravitational potential $V_{\rm N} \sim -G_{\rm N}\,\mu/r$ and whose effective mass μ is

A. Orlandi (✉) · R. Casadio
Dipartimento di Fisica e Astronomia, Università di Bologna,
via Irnerio 46, 40126 Bologna, Italy
e-mail: alessio.j.orlandi@gmail.com

R. Casadio
Istituto Nazionale di Fisica Nucleare, Sezione di Bologna,
Bologna, Italy
e-mail: casadio@bo.infn.it

P. Nicolini et al. (eds.), *1st Karl Schwarzschild Meeting on Gravitational Physics*,
Springer Proceedings in Physics 170, DOI 10.1007/978-3-319-20046-0_32

related to their characteristic quantum mechanical size via the Compton/de Broglie wavelength $\ell \simeq \hbar/\mu = \ell_p\, m_p/\mu$.

These bosons can superpose and form a "ball" of radius ℓ, and total energy $M = N\,\mu$, where N is the total number of constituents. Within the Newtonian approximation, there is then a value of N for which the whole system becomes a BH. In details, given the coupling constant $\alpha = \ell_p^2/\ell^2 = \mu^2/m_p^2$ there exists an integer N such that no constituent can escape the gravitational well it contributed to create, and which can be approximately described by the potential

$$U(r) \simeq \mu N V_N(\ell) \simeq -N\,\alpha\,\frac{\hbar}{\ell}\,\Theta(\ell - r), \tag{32.1}$$

where Θ is the Heaviside step function. This implies that components in the depleting region are "marginally bound" when the kinetic energy given by $E_K \simeq \mu$ equals the potential energy

$$E_K + U \simeq 0 \quad \Longleftrightarrow N\,\alpha = 1. \tag{32.2}$$

Consequently, the effective boson mass and total mass of the BH scale according to

$$\mu \simeq \frac{m_p}{\sqrt{N}} \quad \text{and} \quad M = N\,\mu \simeq \sqrt{N}\,m_p. \tag{32.3}$$

Note that one has here assumed that the ball is of size ℓ (since bosons superpose) and, therefore, the constituents will interact at a maximum distance of order $r \sim \ell$, with fixed ℓ. The Hawking radiation and the negative specific heat spontaneously result from quantum depletion of the condensate for the states satisfying (32.2). This description is partly Quantum Mechanics and partly classical Newtonian physics, but no General Relativity is involved, in that geometry does not appear in the argument. In this work, we will show how this picture, which draws from the conjectured UV-self-completeness of gravity [9], can be both improved within Quantum Mechanics and reconciled with the usual geometric description of space-time in General Relativity. Some considerations about the possible existence of remnants will also follow.

32.2 Quantum Mechanical Model

We can improve on the former model by employing the Quantum Mechanical theory of the harmonic oscillator as a (better) mean field approximation for the Newtonian gravitational interaction acting on each boson inside the BEC.[1] The potential U in (32.1) is therefore replaced by[2]

[1] We shall use units such that $c = 1$, $\hbar = \ell_p\, m_p$ and the Newton constant $G_N = \ell_p/m_p$.

[2] This is nothing but the Newton oscillator, which would correspond to a homogenous BEC distribution in the Newtonian approximation.

$$V = \frac{1}{2}\mu\omega^2(r^2 - d^2)\Theta(d - r) \equiv V_0(r)\Theta(d - r) \tag{32.4}$$

and we further set $V(0) = U(0)$, so that $\mu\,\omega^2\,d^2/2 = N\,\alpha\,\hbar/\ell$. We also assume that the effective mass, length and frequency of a single graviton mode are related by $\mu = \hbar\,\omega = \hbar/\ell$, which leads to $d = \sqrt{2\,N\,\alpha}\,\ell = \sqrt{2\,N}\,\ell_{\rm p}$.

If we neglect the finite size of the well, the Schrödinger equation in polar coordinates yields the well-known eigenfunctions

$$\psi_{nlm}(r, \theta, \phi) = \mathcal{N}\,r^l\,e^{-\frac{r^2}{2\ell^2}}\,{}_1F_1(-n, l + 3/2, r^2/\ell^2)\,Y_{lm}(\theta, \phi), \tag{32.5}$$

where $\mathcal{N}$ is a normalization constant, ${}_1F_1$ the Kummer confluent hypergeometric function of the first kind and $Y_{lm}(\theta, \phi)$ are the usual spherical harmonics. The corresponding energy eigenvalues are given by $E_{nl} = \hbar\,\omega\,[2n + l + 3/2 - V(0)] = \hbar\,\omega\left[2n + l + \left(3 - d^2/\ell^2\right)/2\right]$, where n is the radial quantum number and l the angular momentum (not to be confused with ℓ). Following the idea in [10–13], we view the above spectrum as representing the effective Quantum Mechanical dynamics of depleting modes, which can be described by the first (non-rotating) excited state[3] $\psi_{100}(r) = \sqrt{2}\,e^{-\frac{r^2}{2\ell^2}}\left(2\,r^2 - 3\,\ell^2\right)/\sqrt{3\,\ell^7\,\sqrt{\pi}}$. The marginally binding condition (32.2), that is $E_{10} \simeq 0$, then leads to the desired scaling laws $\ell = \sqrt{2\,N/7}\,\ell_{\rm p}$ and $\mu = \sqrt{7/(2\,N)}\,m_{\rm p}$. We can now estimate the effect of the finite width of the potential well (32.4) by simply applying first order perturbation theory and obtaining $\Delta E_{10} = -\int_d^\infty r^2\,{\rm d}r\,\psi_{100}^2(r)\,V_0(r) \simeq -0.1/\sqrt{N}\,m_{\rm p}$. This can now be compared, for example, with the ground state energy $E_{00} = -\sqrt{14/N}\,m_{\rm p} \simeq -3.7\,m_{\rm p}/\sqrt{N}$. Since $|\Delta E_{10}| \ll |E_{00}|$, our approximation appears reasonable. We however remark that the ground state energy in this model has no physical meaning. Indeed, the Schrödinger equation must be viewed as describing the effective dynamics of BH constituents, and the total energy of the "harmonic black hole" is still given by the sum of the individual boson effective masses,

$$M = N\,\mu \simeq \sqrt{\frac{7\,N}{2}}\,m_{\rm p}, \tag{32.6}$$

in agreement with the "maximal packing" of (32.3) and the expected mass spectrum of quantum BHs (see, for example, [2, 14]).

32.3 Regular Geometry

It is now reasonable to assume that the actual density profile of the BEC gravitational source is related to the ground state wave function in (32.5) according to

---

[3] Note we have already integrated out the angular coordinates.

$$\rho(r) \simeq M\,\psi_{000}^2 \simeq \frac{7^2\,m_{\rm p}\,{\rm e}^{-\frac{7r^2}{2N\ell_{\rm p}^2}}}{\sqrt{\pi}\,N\,\ell_{\rm p}^3}\,. \tag{32.7}$$

Similar Gaussians profiles have been extensively studied in [17–19], where it was proven that such densities satisfy the Einstein field equations with a "de Sitter vacuum" equation of state, $\rho = -p$, where p is the pressure. Curiously, BECs can display this particular equation of state [7, 8, 20]. This feature provides a connection between Quantum Mechanics and the geometrical description.

Let us indeed take the static and normalised, energy density profile of [18],[4]

$$\rho(r) = \frac{M\,{\rm e}^{-\frac{r^2}{4\theta}}}{\sqrt{4\pi}\,\theta^{3/2}}\,, \tag{32.8}$$

where $\sqrt{\theta}$ is viewed as a fundamental length related to space-time noncommutativity, and r is the radial coordinate such that the integral inside a sphere of area $4\pi\,r^2$ gives the total Arnowitt-Deser-Misner (ADM) mass M of the object for $r \to \infty$, i.e.: $M(r) = \int_0^r \rho(\bar{r})\,\bar{r}^2\,{\rm d}\bar{r} = M\,\gamma(3/2, r^2/4\theta)/\Gamma(3/2)$. Here, $\Gamma(3/2)$ and $\gamma(3/2, r^2/4\theta)$ are the complete and upper incomplete Euler Gamma functions, respectively. This energy distribution then satisfies the Einstein field equations together with the Schwarzschild-like metric ${\rm d}s^2 = -f(r)\,{\rm d}t^2 + f^{-1}(r)\,{\rm d}r^2 + r^2\,{\rm d}\Omega^2$, where $f(r) = 1 - 2\,G_{\rm N}\,M(r)/r$. According to [18], one has a BH only if the *mass-to-characteristic length* ratio is sufficiently large, namely for

$$M \gtrsim 1.9\,\frac{\sqrt{\theta}}{G_{\rm N}} = 1.9\,m_{\rm p}\,\frac{\sqrt{\theta}}{\ell_{\rm p}} \equiv M_*. \tag{32.9}$$

If the above inequality is satisfied, the metric function $f = f(r)$ has two zeros and there are two distinct horizons. For $M = M_*$, $f = f(r)$ has only one zero which corresponds to an "extremal" BH, with two coinciding horizons (and vanishing Hawking temperature). The latter represents the minimum mass BH, and a candidate BH remnant of the Hawking decay [3]. Further, the classical Schwarzschild case is precisely recovered in the limit $G_{\rm N}\,M/\sqrt{\theta} \to \infty$, so that departures from the standard geometry become quickly negligible for very massive BHs.

Going back to the BEC model, whose total ADM mass is given in (32.6), and comparing the Gaussian profile (32.7) with (32.8), that is setting $\theta = N\,\ell_{\rm p}^2/14$, one finds that the condition in (32.9) reads $1.8\,\sqrt{N} \gtrsim 0.5\,\sqrt{N}$, and is always satisfied (for $N \geq 1$). We can therefore conclude that harmonic black holes always have two horizons, and the degenerate case is not realised in their spectrum. Although this mismatch might appear as a shortcoming of our construction, it is actually consistent with the idea that the extremal case should have vanishing Hawking temperature and

[4]The squared length θ should not be confused with one of the angular coordinates of the previous expressions. Also, note ρ has already been integrated over the angles.

therefore no depleting modes. It also implies that the final evaporation phase, if it ends in the extremal case, must be realised by a transition that most likely drives the BEC out of the critical point. The precise nature of such a "quantum black hole" state remains, however, unclear.

32.4 Conclusions and Outlook

We have shown that the scenario of [10–13], in which BH inner degrees of freedom (as well as the Hawking radiation) correspond to depleting states in a BEC, can be understood and recovered in the context of General Relativity by viewing a BH as made of the superposition of N constituents, with a Gaussian density profile, whose characteristic length is given by the constituents' effective Compton wavelength. From the point of view of Quantum Mechanics, such states straightforwardly arise from a binding harmonic oscillator potential. Moreover, requiring the existence of (at least) a horizon showed that the extremal case, corresponding to a remnant with vanishing Hawking temperature, is not realised in the harmonic spectrum (32.6). Such states will therefore have to be described by a different model.

At the threshold of BH formation (see, for example, [5]), for a total ADM mass $M \simeq m_\mathrm{p}$ (thus $N \simeq 1$), the above description should allow us to describe Quantum Mechanical processes involving BH intermediate (or metastable) states. However, we can already anticipate that quantum BHs with spin should be relatively easy to accommodate in our description, by simply considering states in (32.5) with $l > 0$. This should allow us to consider more realistic quantum BH formation from particle collisions.

Many questions are still left open. First of all, the discretisation of the mass has an important consequence in the classical limit. For example, let us look again at (32.6), and consider two non-rotating BHs with mass $M_1 = \sqrt{(7/2)\,N_1}\,m_\mathrm{p}$ and $M_2 = \sqrt{(7/2)\,N_2}\,m_\mathrm{p}$, where N_1 and N_2 are positive integers, which slowly merge in a head-on collision (with zero impact parameter). The resulting BH should have a mass M which is also given by (32.6). However, there is in general no integer N_3 such that $\sqrt{N_3} = \sqrt{N_1} + \sqrt{N_2}$. It therefore appears that either the mass should not be conserved, $M \neq M_1 + M_2$, or the mass spectrum described by (32.6) is not complete. This problem, which is manifestly more significant for small BH masses (or, equivalently, integers N), is shared by all those models in which the BH mass does not scale exactly like an integer. If we wish to keep (32.6), or any equivalent mass spectrum, we might then argue that a suitable amount of energy (of order $M_1 + M_2 - M_3$) should be expelled during the merging, in order to accommodate the overall mass into an allowed part of the spectrum. In this case, one may also wonder if this emission can be thought of as some sort of Hawking radiation,[5] or if it is completely different in nature.

---

[5]Note that for vanishing impact parameter, one does not expect any emission of classical gravitational waves.

Another issue regards the assumption in (32.7), i.e. the idea that the classical density profile corresponds to the square modulus of the (normalised) wavefunction. At the semiclassical level, this seems reasonable and intuitive, but necessarily removes the concept of "point-like test particle" from General Relativity, thus forcing us to reconsider the idea of geodesics only in terms of propagation of extended wave packets, which might show unexpected features or remove others from the classical theory. Also, elementary particles would not differ from extended massive objects and therefore should have an equation of state (see, for instance, the old shell model in [1, 4]).

Last but not least, there is the question of describing the formation of a BEC during a stellar collapse. Condensation is usually achieved at extremely low temperature, when the thermal de Broglie wavelength becomes comparable to the inter-particle spacing. Whereas one has no doubt that particles inside a BH are extremely packed, it is not clear how such a dramatic drop of temperature could occur.

Acknowledgments This work is supported in part by the European Cooperation in Science and Technology (COST) action MP0905 "Black Holes in a Violent Universe".

References

1. R. Arnowitt, S. Deser, C.W. Misner, Phys. Rev. Lett. **4**, 375 (1960)
2. J.D. Bekenstein, Lett. Nuovo Cim. **11**, 467 (1974)
3. R. Casadio, P. Nicolini, JHEP **0811**, 072 (2008)
4. R. Casadio, R. Garattini, F. Scardigli, Phys. Lett. B **679**, 156 (2009)
5. R. Casadio, O. Micu, A. Orlandi, Eur. Phys. J. C **72**, 2146 (2012)
6. S. Chandrasekhar, *The mathematical theory of black holes* (Oxford University Press, Oxford, 1992)
7. P.-H. Chavanis, Phys. Rev. D **84**, 043531 (2011)
8. P.-H. Chavanis, Astrophys. J. **537**, A127 (2012)
9. G. Dvali, C. Gomez, *Self-Completeness of Einstein Gravity*. arXiv:1005.3497 [hep-th]
10. G. Dvali, C. Gomez, Phys. Lett. B **716**, 240 (2012)
11. G. Dvali, C. Gomez, *Black Holes as critical point of quantum phase transition*. arXiv:1207.4059 [hep-th]
12. G. Dvali, C. Gomez, Fortsch. Phys. **61**, 742 (2013)
13. G. Dvali, C. Gomez, Phys. Lett. B **719**, 419 (2013)
14. G. Dvali, C. Gomez, S. Mukhanov, *Black Hole masses are quantized*. arXiv:1106.5894 [hep-ph]
15. S.W. Hawking, Nature **248**, 30 (1974)
16. S.W. Hawking, Commun. Math. Phys. **43**, 199 (1975) [Erratum-ibid. **46**, 206 (1976)]
17. P. Nicolini, Int. J. Mod. Phys. A **24**, 1229 (2009)
18. P. Nicolini, A. Smailagic, E. Spallucci, Phys. Lett. B **632**, 547 (2006)
19. P. Nicolini, A. Orlandi, E. Spallucci, The final stage of gravitationally collapsed thick matter layers. arXiv:1110.5332 [gr-qc]
20. L.P. Pitaevskii, S. Stringari, *Bose-Einstein condensation* (Oxford University Press, Oxford, 2003), Chap. 11

Chapter 33
Holographic Entanglement Entropy of Semi-local Quantum Liquids

Da-Wei Pang, Johanna Erdmenger and Hansjörg Zeller

Abstract We consider the holographic entanglement entropy of $(d+2)$-dimensional semi-local quantum liquids, for which the dual gravity background in the deep interior is $AdS_2 \times \mathbf{R}^d$ multiplied by a warp factor which depends on the radial coordinate. The entropy density of this geometry goes to zero in the extremal limit. For the case of an asymptotically AdS UV completion of this geometry, we show that the entanglement entropy of a strip and an annulus exhibits a phase transition as a typical length of the different shapes is varied, while there is no sign of such a transition for the entanglement entropy of a sphere.

33.1 Introduction

Gauge/gravity duality has been proven to be a powerful tool for studying the dynamics of strongly coupled field theories, and has in particular been applied to understand the low-temperature physics of strongly-coupled electron systems (AdS/CMT).

One extensively studied example is $(d + 2)$-dimensional Einstein-Maxwell-Dilaton theory, which admits a metric with hyperscaling violation as an exact solution. For this case, the entropy density at finite temperature scales as $s \sim T^{(d-\theta)/z}$, where z denotes the dynamical exponent and θ is the hyperscaling violation parameter (i.e. the spatial volume scales with reduced dimension $d-\theta$). It has been observed in [1] that by taking the limit $z \to \infty$, $\theta \to -\infty$ while keeping $\eta = -\theta/z$ fixed, the (undesirable) ground state entropy density vanishes and low energy modes at all momenta may exist. This behavior resembles features found in fermionic systems. In the same limit the resulting metric is $AdS_2 \times \mathbf{R}^d$ multiplied by a warp factor,

D.-W. Pang (✉) · J. Erdmenger · H. Zeller
Max-Planck-Institut für Physik (Werner-Heisenberg-Institut), Föhringer Ring 6, 80805 Munich, Germany
e-mail: dwpang@mppmu.mpg.de

J. Erdmenger
e-mail: jke@mppmu.mpg.de

H. Zeller
e-mail: zeller@mppmu.mpg.de

P. Nicolini et al. (eds.), *1st Karl Schwarzschild Meeting on Gravitational Physics*, Springer Proceedings in Physics 170, DOI 10.1007/978-3-319-20046-0_33

which depends on the radial coordinate. In [2] the $AdS_2 \times \mathbf{R}^d$ near horizon geometry of the $(d+2)$-dimensional extremal Reissner-Nordström-AdS black hole is referred to as a *holographic semi-local quantum liquid*, characterized by a finite spatial correlation length, an infinite correlation time and a non-trivial scaling behavior in the time direction. The background discussed here may be seen as a generalization of the dual of holographic semi-local quantum liquids.

We study the holographic entanglement entropy of $(d+2)$-dimensional semi-local quantum liquids for entangling regions with different shapes (strip, disk/sphere and annulus). We start by constructing the full $(d+2)$-dimensional geometry which is asymptotically AdS near the boundary and possesses semi-locality in the interior. Moreover, we compute the holographic entanglement entropy for the different shapes in this geometry. For the strip, the entanglement entropy for small values of the strip width $l \leq l_{\text{crit}}$ corresponds to a connected hypersurface in the bulk. For values larger than l_{crit} we see a phase transition to two parallel slabs going from the boundary into the deep interior of the bulk (disconnected solution). For the disk/sphere case we do not find such a transition. We also calculate the entanglement entropy for an annulus entangling surface in order to interpolate between the sphere and the strip case. We find that there is a phase transition taking place between two concentric spheres (disconnected solution) and a deformed annulus (connected solution) at a critical value of the difference between the outer and inner radius. Depending on the parameters the transition may be either first or second order.

33.2 The Full Solution

We start from the action of Einstein-Maxwell-Dilaton theory,

$$S = \int d^{d+2}x\sqrt{-g}\left[R - \frac{1}{2}(\nabla\Phi)^2 - V(\Phi) - \frac{1}{4}Z(\Phi)F_{\mu\nu}F^{\mu\nu}\right] \tag{33.1}$$

and the ansatz for the metric and the gauge field

$$ds^2_{d+2} = \frac{L^2}{z^2}\left[-f(z)dt^2 + g(z)dz^2 + \sum_{i=1}^{d} dx_i^2\right], \quad A_t = A_t(z). \tag{33.2}$$

A physically sensible solution is

$$f(z) = kz^{-p}, \quad g(z) = \frac{z_F^2}{z^2}, \tag{33.3}$$

where $p \equiv 2d/\eta$ and k is a positive constant. This solution is conformal to $AdS_2 \times \mathbf{R}^d$, which can be seen by taking the coordinate transformation $z = \xi^{2/p}$. We call this solution the IR solution. The UV completion with an asymptotically AdS geometry reads

$$f(z) = \frac{k}{k + z^p}, \quad g(z) = \frac{z_F^2}{z^2 + z_F^2}. \tag{33.4}$$

We see that this solution has $f(0) = g(0) = 1$, while $f(z)$ and $g(z)$ reduce to the IR solution (33.3) in the $z \to \infty$ limit.

33.3 The Holographic Entanglement Entropy

33.3.1 The Strip

Let us consider the strip case,

$$x_1 \equiv x \in \left[-\frac{l}{2}, \frac{l}{2}\right], \quad x_i \in [0, L_x], \; i = 2, \ldots, d, \tag{33.5}$$

where $l \ll L_x$. The minimal surface area is given by

$$A(\gamma) = 2L^d L_x^{d-1} \int \frac{dz}{z^d} \sqrt{g(z) + x'^2}, \tag{33.6}$$

where we have parameterized $x = x(z)$. We evaluate the boundary separation length l for the IR solution and we obtain the constant value

$$l = l_{\text{crit}} = \frac{\pi z_F}{d}. \tag{33.7}$$

For the full background solution the behavior of l versus z_* is plotted on the left side of Fig. 33.1 for $d = 2$ with $z_F = 1$. It can be seen that l is a smooth function of z_*. When z_* is small, significant difference between l and l_{crit} can be observed. However, when z_* is sufficiently large, l approaches l_{crit}.

The holographic entanglement entropy is determined by [3, 4]

$$S = \frac{A(\gamma)}{4G_N^{(d+2)}}, \tag{33.8}$$

where $A(\gamma)$ denotes the minimal surface area. The behavior of $\Delta A = A_{\text{finite}} - A_{\text{dis,finite}}$ for $d = 2$ is plotted in the right part of Fig. 33.1, where $A_{\text{dis,finite}}$ corresponds to the disconnected case given by substituting $x' = 0$ into (33.6). Note that we have subtracted the divergent terms following [4]. It is apparent that when $l < l_{\text{crit}}$, the connected surface dominates, as $l \to l_{\text{crit}}$, the difference tends to zero, which signifies that the disconnected surface will dominate. This behavior agrees with the proposal of [5].

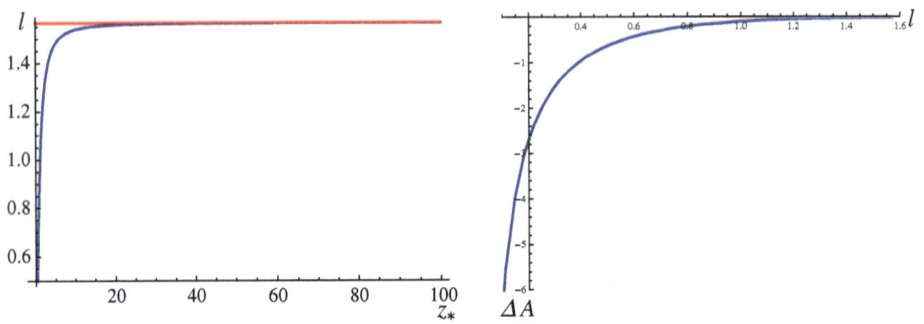

Fig. 33.1 *Left plot* The boundary separation length in the full solution (*blue curve*) and the IR solution (*red curve*). l and l_{crit} have significant differences when z_* is sufficiently small, which means that the minimal surface just probes the geometry near the UV. As z_* increases, the minimal surface goes deeper into the IR and l approaches l_{crit}. *Right plot* The difference between the HEE of the connected minimal surface and the disconnected one. When l is sufficiently small, the connected minimal surface dominates. As $l \to l_{\mathrm{crit}}$ the disconnected one dominates

33.3.2 The Sphere

The spherical entangling region is parameterized by $\sum_{i=1}^{d} x_i^2 = R^2$, which leads to the minimal surface area

$$A_{\mathrm{sphere}} = L^d \mathrm{Vol}(\Omega_{d-1}) \int \frac{dz}{z^d} \rho^{d-1} \sqrt{g(z) + \rho'^2}. \tag{33.9}$$

Note that in this case there is no trivial solution $\rho' = 0$. We are interested in the deviation of the finite part of HEE from the area law [5], which can be analyzed by performing a fit on the numerical data. The resulting behavior reads

$$A_{\mathrm{finite}} = -0.625328 - 0.796679R, \quad d = 2, \tag{33.10}$$

which may indicate that for large R the finite HEE is still governed by the area law, consistent with the conclusion in [5] (Fig. 33.2).

33.3.3 The Annulus

In the annulus case we expect to approximate a sphere in the limit of vanishing inner radius and the strip for both, the inner and outer radius, large in comparison to their difference.

We show generic results for the entanglement entropy for $d = 2$ in Fig. 33.3. There we plot A_{finite} versus the difference of the radii $\Delta\rho = \rho_2 - \rho_1$. We find two connected solutions (deformed annulus) for values of $\Delta\rho \leq (\Delta\rho)_{\mathrm{max}}$ and one disconnected solution (two concentric balls) for all values of $\Delta\rho$. Note that for each

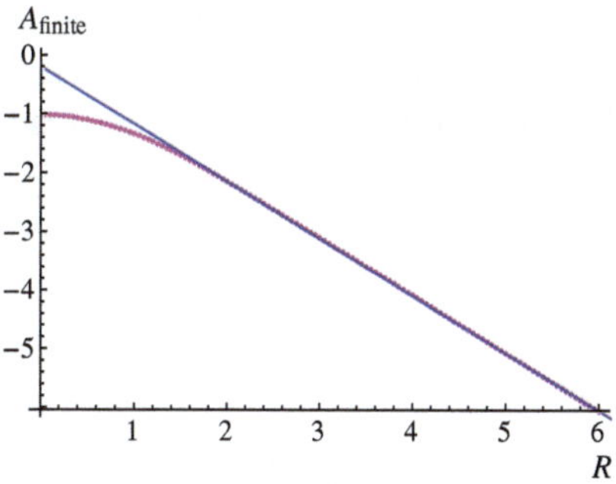

Fig. 33.2 The finite part with spherical entangling region. The *dots* are data from numerical evaluations and the *curves* denote the fits

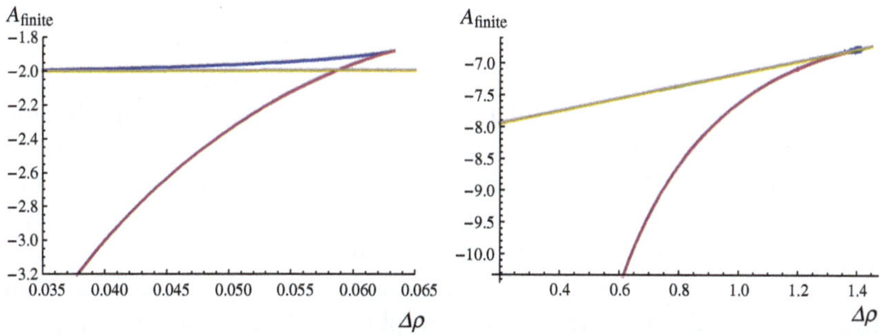

Fig. 33.3 The finite part of the annulus entangling region for $d = 2$ versus the difference of the radii $\Delta\rho = \rho_2 - \rho_1$ for the values $\rho_2 = 0.1z_F$ and $\rho_2 = 4z_F$. For small differences $\Delta\rho \leq (\Delta\rho)_{\text{max}}$ we see different solutions, two connected (deformed annulus) solutions with the *lower one* being preferred (*blue* and *red*) and the concentric balls solution (*yellow*). The transition between the connected and disconnected solutions at $(\Delta\rho)_{\text{crit}}$ is first order for small values of ρ_2 (swallow tail form of the *left plot*). For larger values of ρ_2 we have a second order transition (*right plot*). If $\Delta\rho > (\Delta\rho)_{\text{max}}$ the disconnected solution is the only solution

value of $\Delta\rho$ the preferred solution is the one with smaller value of A_{finite}. In the $d = 2$ case, at a value $(\Delta\rho)_{\text{crit}}$ we find a first order transition from the preferred connected to the disconnected solution for small values of the radii and a second order transition for larger ones. This behavior is very similar to the strip case discussed in Sect. 33.3.1, where there also only exists a connected solution for $l \leq l_{\text{crit}}$, however, there the transition is always second order. The analogy goes further: increasing the values of the radii ρ_1, ρ_2 leads to $(\Delta\rho)_{\text{crit}} \to \pi/d$, with $z_F = 1$. Looking closer at this limit in $d = 2$ we see the swallow tail becomes smaller turning into a second order transition (see right part of Fig. 33.3). From this behavior we deduce that the annulus tends towards the strip solution for large radii. The other limit, however, where we aim at approximating a sphere, does not work entirely as expected, since for each given pair of radii of the annulus solution, we always find a maximal difference $(\Delta\rho)_{\text{max}}$ between both which is smaller than the outer radius ρ_2. Therefore we can at most approximate two concentric spheres, but never one sphere alone. Finally, the similarity in most of the parameter space to the behavior seen in confining geometries is astonishing [6, 7]. It would be interesting to understand if there is a common origin to this resemblance.

33.4 Conclusions

We considered holographic entanglement entropy of semi-local quantum liquids, whose gravity dual in $(d+2)$-dimensions is described by a metric which is conformal to $AdS_2 \times \mathbf{R}^d$ in the IR. For the case of a strip entangling region we observe a second order phase transition: When the boundary separation length $l \leq l_{\text{crit}}$ the connected surface dominates, otherwise the disconnected one is preferred. Such a phase transition is not observed for the spherical entangling region. We also study the annulus case as an interpolating geometry between the sphere and the strip. For this shape we find two connected solutions, with one being preferred, i.e. it has lower entanglement entropy. For a certain value of $\Delta\rho$ (difference between the outer and inner radius of the annulus) we see a transition from the preferred connected solution to a disconnected solution (two concentric spheres). In the $d = 2$ case there always seems to be a transition, which is first order for small values of the outer radius and becomes a second order transition for increasing values of the radii. For a detailed analysis of the results presented here and the plots for the $d > 2$ case see [8].

Acknowledgments We thank A. Parnachev for discussions. DWP is supported by Alexander von Humboldt Foundation.

References

1. S.A. Hartnoll, E. Shaghoulian, JHEP **1207**, 078 (2012)
2. N. Iqbal, H. Liu, M. Mezei, JHEP **1204**, 086 (2012)
3. S. Ryu, T. Takayanagi, Phys. Rev. Lett. **96**, 181602 (2006)
4. S. Ryu, T. Takayanagi, JHEP **0608**, 045 (2006)
5. M. Kulaxizi, A. Parnachev, K. Schalm, JHEP **1210**, 098 (2012)
6. I.R. Klebanov, D. Kutasov, A. Murugan, Nucl. Phys. B **796**, 274 (2008)
7. A. Pakman, A. Parnachev, JHEP **0807**, 097 (2008)
8. J. Erdmenger, D.W. Pang, H. Zeller, arXiv:1311.1217

Chapter 34
Quadratic Palatini Gravity and Stable Black Hole Remnants

Diego Rubiera-Garcia, Francisco S.N. Lobo and Gonzalo J. Olmo

Abstract We present a four-dimensional Planck-scale corrected quadratic extension of General Relativity (GR) where no a priori relation between metric and connection is imposed (Palatini formalism). Static spherically symmetric electrovacuum solutions are obtained in exact analytical form. The macroscopic properties of these solutions are in excellent agreement with GR, though the region around the central singularity is modified. In fact, the singularity is generically replaced by a wormhole supported by the electric field, which provides a non-trivial topology to the spacetime. Moreover, for a certain charge-to-mass ratio the geometry is completely regular everywhere. For such regular solutions, the horizon disappears in the microscopic regime below a critical number of charges ($N < 17$), yielding a set of massive objects that could be naturally identified as black hole remnants.

34.1 Introduction

Almost one hundred years after its foundation, General Relativity (GR) has successfully passed a wide range of experimental tests [1]. However, there exist arguments suggesting that GR should be extended to encompass for new phenomena. One of them is represented by the existence of singularities both in the Big Bang and deep inside black holes, where the classical geometry breaks down. Such singularities are

D. Rubiera-Garcia (✉)
Departamento de Física, Universidade Federal da Paraíba, João Pessoa,
PB 58051-900, Brazil
e-mail: drubiera@fisica.ufpb.br

F.S.N. Lobo
Instituto de Astrofísica e Ciências do Espaço, Faculdade de Ciências da Universidade de Lisboa, Edifício C8, Campo Grande, P-1749-016 Lisbon, Portugal
e-mail: fslobo@fc.ul.pt

G.J. Olmo
Departamento de Física Teórica and Instituto de Física Corpuscular,
Centro Mixto Universidad de Valencia - Consejo Superior de Investigaciones Científicas.
Universidad de Valencia, 46100 Burjassot, Valencia, Spain
e-mail: gonzalo.olmo@csic.es

P. Nicolini et al. (eds.), *1st Karl Schwarzschild Meeting on Gravitational Physics*,
Springer Proceedings in Physics 170, DOI 10.1007/978-3-319-20046-0_34

believed to be an artifact of the classical description, to be resolved within a quantum theory of gravity. Even though the specific details of such a theory are not at hand, we could gain useful insights on its nature by studying classical gravity theories including higher-order curvature terms. Indeed such terms are needed for a high-energy completion of the Einstein-Hilbert Lagrangian [2], and moreover appear in some approaches to quantum gravity such as those based on string theory [3]. In this sense, if the underlying structure of spacetime is assumed to be a priori Riemmanian such terms usually give rise to fourth-order field equations and exhibit ghosts. However, the fact is that the metric and the connection are independent entities by their own, and thus, in the absence of further evidence for the physics beyond GR, one could apply the variational principle to both objects, instead of relying on a tradition or convention basis to keep the Riemannian assumption at all energies/scales. In this (Palatini) formulation of the theory, quadratic gravity Lagrangians satisfy second-order equations and contain Minkowski as a vacuum solution of the theory, which guarantees the absence of ghosts.

34.2 Theory and Solutions

We consider Palatini Lagrangian densities defined as

$$S = \frac{1}{2\kappa^2}\int d^4x\sqrt{-g}f(R,Q) + S_M[g_{\mu\nu},\Psi]\,, \tag{34.1}$$

with $R = g_{\mu\nu}R^{\mu\nu}$, $Q = R_{\mu\nu}R^{\mu\nu}$ and S_m the matter action. We apply the variation principle independently to the metric and connection, and impose vanishing torsion, $\Gamma^{\lambda}_{[\mu\nu]} = 0$, a posteriori, which leads to $R_{[\mu\nu]} = 0$ [4]. Under these conditions the field equations become

$$f_R R_{\mu\nu} - \frac{f}{2}g_{\mu\nu} + 2f_Q R_{\mu\alpha}R^{\alpha}{}_{\nu} = \kappa^2 T_{\mu\nu}\,, \tag{34.2}$$

$$\nabla_\beta\left[\sqrt{-g}\left(f_R g^{\mu\nu} + 2f_Q R^{\mu\nu}\right)\right] = 0\,. \tag{34.3}$$

To solve them, we note that the equation for the connection can be solved by defining a rank-two tensor $h_{\mu\nu}$ as $\nabla_\beta[\sqrt{-h}h^{\mu\nu}] = 0$, related to the metric $g_{\mu\nu}$ as

$$h^{\mu\nu} = \frac{g^{\mu\alpha}\Sigma_\alpha{}^{\nu}}{\sqrt{\det\Sigma}}\,, \quad h_{\mu\nu} = (\sqrt{\det\Sigma})[\Sigma^{-1}]_\mu{}^{\alpha}g_{\alpha\nu}\,, \quad \Sigma_\alpha^\nu = (f_R\delta_\alpha^\nu + 2f_Q P_\alpha{}^{\nu})\,, \tag{34.4}$$

and thus the independent connection $\Gamma^{\lambda}_{\mu\nu}$ is the Levi-Civita connection of $h_{\mu\nu}$. We have defined $P_\alpha{}^{\nu} \equiv R_{\alpha\mu}g^{\mu\nu}$ which, according to (34.2) is a function of $T_\mu{}^{\nu}$. In terms of $h_{\mu\nu}$ the equation for the metric (34.2) become [5]

$$R_\mu{}^{\nu}(h) = \frac{1}{\sqrt{\det\Sigma}}\left(\frac{f}{2}\delta_\mu^\nu + \kappa^2 T_\mu{}^{\nu}\right)\,. \tag{34.5}$$

This shows that the *auxiliary* metric $h_{\mu\nu}$ satisfies an Einstein-like set of partial differential equations. Moreover, since $g_{\mu\nu}$ is algebraically related with $h_{\mu\nu}$, the field equations for $g_{\mu\nu}$ are also second-order. Note also that in vacuum ${R_\mu}^\nu = \Lambda\delta_\mu^\nu$ and thus we recover GR with a cosmological constant term. This shows that these theories have the same number of propagating degrees of freedom as GR.

Next we consider the quadratic model $f(R,Q) = R + l_P^2(R + aQ)$, with $l_P^2 \equiv \hbar G/c^3$ representing the Planck length squared and a is a free parameter. To probe the theory, we study electrovacuum solutions, as given by the action

$$S_m[g,\psi_m] = -\frac{1}{16\pi}\int d^4x\sqrt{-g}F_{\alpha\beta}F^{\alpha\beta}, \tag{34.6}$$

where $F_{\mu\nu} = \partial_\mu A_\nu - \partial_\nu A_\mu$ is the field strength tensor. The energy-momentum tensor reads ${T_\mu}^\nu = -\frac{1}{4\pi}[{F_\mu}^\alpha {F_\alpha}^\nu - \frac{{F_\alpha}^\beta {F_\beta}^\alpha}{4}\delta_\mu^\nu]$, while the field equations $\nabla_\mu F^{\mu\nu} = 0$ in a spherically symmetric background $ds^2 = g_{tt}dt^2 + g_{rr}dr^2 + r^2 d\Omega^2$ satisfy $F^{tr} = \frac{q}{r^2}\frac{1}{\sqrt{-g_{tt}g_{rr}}}$. For this theory, the field equations (34.5) in the geometry $h_{\mu\nu}$ can be analytically solved, and putting that solution in terms of $g_{\mu\nu}$ using (34.4) leads to [5–7]

$$g_{tt} = -\frac{A(z)}{\sigma_+}, \quad g_{rr} = \frac{\sigma_+}{\sigma_- A(z)}, \quad A(z) = 1 - \frac{[1+\delta_1 G(z)]}{\delta_2 z \sigma_-^{1/2}}, \quad \frac{dG}{dz} = \frac{z^4+1}{z^4\sqrt{z^4-1}},$$

where $z = r/r_c$, $r_c \equiv \sqrt{r_q l_P}$, $r_q^2 \equiv \tilde{\kappa}^2 q^2 = \kappa^2 q^2/(4\pi)$, $\delta_1 = \frac{1}{2r_S}\sqrt{r_q^3/l_P}$, $\delta_2 = \sqrt{r_q l_P}/r_S$ and $\sigma_\pm = 1 \pm 1/z^4$. The function $G(z)$ behaves as follows: far from the center we have $G(z) \approx -\frac{1}{z} - \frac{3}{10z^5} + \cdots$, and thus the GR limit is quickly recovered for $z \gg 1$. However, at $z \to 1$ the metric undergoes important modifications, as here we obtain $G(z) \approx \beta + 2\sqrt{z-1} - \frac{11}{6}(z-1)^{3/2} + \cdots$, where $\beta \approx -1.74804$ is a constant needed to match the far and near expansions. At $z = 1$, in general, the curvature invariants become divergent, but for the particular choice $\delta_1 = \delta_1^* = -1/\beta$ they behave as

$$r_c^2 R(g) \approx \left(-4 + \frac{16\delta_c}{3\delta_2}\right) + \cdots \quad , \quad r_c^4 Q(g) \approx \left(10 + \frac{86\delta_c^2}{9\delta_2^2} - \frac{52\delta_c}{3\delta_2}\right) + \cdots$$

$$r_c^4 K(g) \approx \left(16 + \frac{88\delta_c^2}{9\delta_2^2} - \frac{64\delta_c}{3\delta_2}\right) + \cdots$$

and thus all of them are finite.

34.3 Wormhole Structure and Black Hole Remnants

Using Eddington-Finkelstein type coordinates, $ds^2 = g_{tt}dv^2 + 2dvdr^* + r^2(r^*)d\Omega^2$, the line element at the surface $z = 1$ becomes

$$g_{tt} = \frac{(1-\delta_1/\delta_c)}{4\delta_2\sqrt{z-1}} - \frac{1}{2}\left(1 - \frac{\delta_1}{\delta_2}\right) + O(\sqrt{z-1}), \tag{34.7}$$

with $z(z^*)$ defined by $(dz/dz^*)^2 = \sigma_-$, being $z^* = r^*/r_c$. For $\delta_1 = \delta_c$, we find that g_{tt} is smooth everywhere, and r^* can be extended to the negative axis by considering the two branches of the equation $(dz/dz^*) = \pm\sigma_-^{1/2}$. This equation indicates that the radial function has a minimum at $r = r_c$ (or $z = 1$), which can be integrated to yield $z^* = {}_2F_1[-\frac{1}{4}, \frac{1}{2}, \frac{3}{4}, \frac{1}{z^4}]z$ if $z^* \geq z_c^*$ and $z^* = 2z_c^* - {}_2F_1[-\frac{1}{4}, \frac{1}{2}, \frac{3}{4}, \frac{1}{z^4}]z$ if $z^* \leq z_c^*$, where $z_c^* \approx 0.59907$. The bounce in the radial function $z(z^*)$ is reminiscent of a wormhole geometry, with $z = 1$ representing the wormhole throat. This wormhole structure naturally leads to the concept of geon originally introduced by Wheeler in 1955 as "sourceless self-consistent gravitational electromagnetic entities" with a multiply-connected geometry [8]. This follows from the fact that the lines of force of the electric field enter through one of the mouths of the wormhole and exit through the other, creating the illusion of a negatively charged object on one side and a negatively charged on the other but without having *real* sources or charges generating the field. The locally measured electric charge arises from the computation of the flux $\Phi \equiv \int_S *F = 4\pi q$ through any 2-surface S enclosing the wormhole throat. This implies that the electric flux per surface unit at $z = 1$, which represents the density of lines of force crossing the wormhole throat, namely, $\Phi/(4\pi r_c^2) = q/r_c^2 = \sqrt{c^7/(2(\hbar G)^2)}$, turns out to be a *universal quantity*, independent of the specific amounts of charge and mass. Moreover, the dependence of $\frac{\Phi}{4\pi r_c^2}$ on fundamental constants (irrespective of M and q) indicates that the geon structure persists for those solutions showing curvature divergences ($\delta_1 \neq \delta_1^*$).

The evaluation of the action on the solutions yields

$$S_T = \frac{q^2}{r_c\delta_1^*}\int dt = 2\frac{\delta_1}{\delta_1^*}Mc^2\int dt, \tag{34.8}$$

which can be interpreted as the addition of the electromagnetic plus *gravitational binding energy*. This value, which is finite for arbitrary δ_1, coincides with the action of an isolated particle of mass M at rest, $S_{p.p.} = mc^2\int dt\sqrt{1-v^2/c^2}$, when $\delta_1 = \delta_c$.

From (34.5), one may easily find the conditions for the existence of an event horizon when $\delta_1 = \delta_c$. It turns out that the sign of $(1 - \delta_1/\delta_c)$ determines if there is an event horizon. Expressing $q = N_q e$, we find that $r_q = 2l_P N_q/N_q^c$, where $N_q^c \equiv \sqrt{2/\alpha_{em}} \approx 16.55$. Thus, if $\delta_c/\delta_2 = N_q/N_q^c > 1$ there is an event horizon, which is absent otherwise. Therefore, for $\delta_1 = \delta_c$ we find a family of horizonless objects not affected by Hawking's quantum instability and with a mass spectrum given by $M \approx 1.23605(N_q/N_q^c)^{3/2}m_P$. The topological nature of the charge of these solutions makes them stable against arbitrary classical perturbations that preserve the topology.

It is worth mentioning that Hawking's estimates on the generation of primordial black holes by large density perturbations in the early universe [9], which are based entirely on the process of classical collapse, predict that objects with $M \geq m_P$ and $N_q < 30$ could have been formed in the early universe. Such numbers are in excellent quantitative agreement with the characteristics of the horizonless $\delta_1 = \delta_c$ solutions found here. Thus, since large numbers of objects with $M \sim m_P$ and $N_q < 30$ could have been formed in the early universe, a fraction of them could reach the stability conditions found here. The evaporation of more massive objects could also yield this type of particles. The existence of stable states in the lowest part of the mass and charge spectrum which can be continuously connected with black hole states, supports the view that these objects can be naturally identified as black hole remnants. The existence of these objects suggests that a maximum temperature should be reached in the evaporation process, which is compatible with the lack of observations of black hole explosions.

We note that the robustness of our results can be tested by considering nonlinearities (coming e.g. from pair production) in the matter sector through nonlinear electrodynamics. Proceeding in this way, we find that exact solutions, provided by Born-Infeld electrodynamics, are in qualitative agreement with the Maxwell case [11]. Moreover, it turns out that the mass spectrum can be reduced by many orders of magnitude.

34.4 Dynamical Generation

The above wormhole solutions can be dynamically generated as follows. Consider a presure less flux of ingoing charged matter with stress-energy tensor $T_{\mu\nu}^{flux} = \rho_{in}k_\mu k_\nu$. This flux conveys a current $J^\nu \equiv \Omega(v)k^\nu$ (with $\Omega(v)$ a function to be determined) in the electromagnetic equations of motion as $\nabla_\mu F^{\mu\nu} = 4\pi J^\nu$. The field equations (34.5) admit a solution [12], which is formally similar to that of (34.5) with the replacements $t \to v$, $\delta_1 \to \delta_1(v)$, $\delta_2 \to \delta_2(v)$, and $g_{tt} \to g_{tt} + 2l_P^2\kappa^2\rho_{in}/(\sigma_-(1-2r_c^4/r^4))$, where v is the advanced time coordinate.

If we assume the initial state to be Minkowski space, a charged perturbation of compact support propagating within an interval $[v_i, v_f]$ makes the area of the 2-spheres of constant v to never become smaller than $r_c^2(v)$. Once the flux is switched off ($v > v_f$), the result is a static geometry identical to that given by (34.2). This change in the geometry can be interpreted as the formation of a wormhole whose throat has an area $A_{WH} = 4\pi r_c^2(v_f)$. The existence or not of curvature divergences at $r = r_c(v_f)$ depends on the (integrated) charge-to-mass ratio of the flux. This result seems to imply a change in the global properties of space-time and, in particular, in its topology. However, a different viewpoint emerges if one accepts that curvature divergences are not necessarily as pathologically as usually regarded. Indeed, the existence of a well-defined electric flux flowing across the wormhole throat supports the idea that the wormhole structure is present even in those cases with curvature divergences. Within this interpretation, the region $v < v_i$ would be made of two disconnected pieces of Minkowski space-time which are separated by a vanishing radius throat that, once electric charge comes into play, is opened up.

34.5 Ending Comments

An important lesson that follows from our analysis is that given the puzzles that Nature presents us in understanding the physics beyond GR, the consideration of foundational aspects of gravity, such as the validity of the Riemannian assumption on the space-time at all scales/energies, cannot be overlooked any longer. This important aspect has consequences on the formulation of classical gravity theories, as the metric and Palatini approach yield, in general, inequivalent results [13] (an important exception is precisely GR). Working in the Palatini approach we have found the existence of wormhole solutions with a geonic structure that can be interpreted as black hole remnants. The existence of a non-trivial topology conveyed by these solutions raises several questions on the true meaning of curvature divergences in this theory, and on the very nature of space-time that requires further investigation.

Work supported by FCT grants CERN/FP/123615/2011 and CERN/FP/123618/2011 (F.S.N.L.); grant FIS2011-29813-C02-02, the Consolider Programme CPAN (CSD2007-00042), and the JAE-doc program of CSIC (G.J.O.); CNPq grant 561069/2010-7 (D.R.-G.).

References

1. C.M. Will, The confrontation between general relativity and experiment. Living Rev. Rel. **9**, 3 (2006) [gr-qc/0510072]
2. N.D. Birrell, P.C.W. Davies, *Quantum Fields in Curved Space* (Cambridge University Press, Cambridge, 1982)
3. M. Green, J. Schwarz, E. Witten, *Superstring Theory* (Cambridge University Press, Cambridge, 1987)
4. G.J. Olmo, D. Rubiera-Garcia, Importance of torsion and invariant volumes in Palatini theories of gravity. Phys. Rev. D **88**, 084030 (2013). arXiv:1306.4210 [hep-th]
5. G.J. Olmo, D. Rubiera-Garcia, Reissner-Nordström black holes in extended Palatini theories. Phys. Rev. D **86**, 044014 (2012). arXiv:1207.6004 [gr-qc]
6. G.J. Olmo, D. Rubiera-Garcia, Nonsingular black holes in quadratic Palatini gravity. Eur. Phys. J. C **72**, 2098 (2012). arXiv:1112.0475 [gr-qc]
7. G.J. Olmo, D. Rubiera-Garcia, Nonsingular charged black holes à la Palatini. Int. J. Mod. Phys. D **21**, 1250067 (2012). arXiv:1207.4303 [gr-qc]
8. J.A. Wheeler, Geons. Phys. Rev. **97**, 511 (1955)
9. S. Hawking, Gravitationally collapsed objects of very low mass. Mon. Not. Roy. Astron. Soc. **152**, 75 (1971)

10. F.S.N. Lobo, G.J. Olmo, D. Rubiera-Garcia, Semiclassical geons as solitonic black hole remnants. JCAP **1307**, 011 (2013)
11. G.J. Olmo, D. Rubiera-Garcia, Semiclassical geons at particle accelerators, JCAP **1402**, 010 (2014), arXiv:1306.6537 [hep-th]
12. F.S.N. Lobo, J. Martinez-Asencio, G.J. Olmo, D. Rubiera-Garcia, Dynamical generation of wormholes with charged fluids in quadratic Palatini gravity, Phys. Rev. D **90**, 024033 (2014), arXiv:1403.0105 [hep-th]
13. M. Borunda, B. Janssen, M. Bastero-Gil, Palatini versus metric formulation in higher curvature gravity. JCAP **0811**, 008 (2008), arXiv:0804.4440 [hep-th]

Chapter 35
Kermions

Elizabeth Winstanley

Abstract In the framework of quantum field theory in curved space-time, we study the quantization of a massless fermion field on a non-extremal Kerr black hole. The key theme in this note is the fundamental difference between scalar and fermion fields for the process of defining quantum states. In particular, we define two new states for fermions on Kerr which cannot be defined for quantum scalar fields on Kerr. These two states are the analogues of the standard Boulware and Hartle-Hawking states on a Schwarzschild black hole.

35.1 Canonical Quantization

In the canonical quantization of a free field on a curved space-time, an object of fundamental importance is the vacuum state $|0\rangle$. On a general curved space-time, there is no unique vacuum state.

For a quantum scalar field, the process starts by expanding the classical field Φ in terms of an orthonormal basis of field modes, which are split into positive frequency modes ϕ_j^+ and negative frequency modes ϕ_j^-:

$$\Phi = \sum_j a_j \phi_j^+ + a_j^\dagger \phi_j^- . \tag{35.1}$$

The choice of positive/negative frequency modes is constrained by the fact that positive frequency modes must have positive Klein-Gordon norm and negative frequency modes have negative Klein-Gordon norm (by "norm", here we mean the inner product of a field mode with itself). With this restriction, quantization proceeds by promoting the expansion coefficients in (35.1) to operators satisfying the usual commutation relations. The $\hat{a}_j$ are interpreted as particle annihilation operators and the $\hat{a}_j^\dagger$ as par-

E. Winstanley (✉)
Consortium for Fundamental Physics, School of Mathematics and Statistics,
The University of Sheffield, Hicks Building, Hounsfield Road, Sheffield S3 7RH, UK
e-mail: E.Winstanley@sheffield.ac.uk

P. Nicolini et al. (eds.), *1st Karl Schwarzschild Meeting on Gravitational Physics*,
Springer Proceedings in Physics 170, DOI 10.1007/978-3-319-20046-0_35

ticle creation operators. The vacuum state $|0\rangle$ is then defined as that state annihilated by all the particle annihilation operators: $\hat{a}_j\,|0\rangle = 0$. The definition of a vacuum state is therefore dependent on how the field modes are split into positive and negative frequency modes, which is restricted for a quantum scalar field by the fact that positive frequency (particle) modes must have positive norm.

For a fermion field Ψ, we again start with an expansion in terms of an orthonormal basis of field modes analogous to (35.1):

$$\Psi = \sum_j b_j \psi_j^+ + c_j^\dagger \psi_j^- . \tag{35.2}$$

In this case, both positive and negative frequency fermion modes have positive Dirac norm, so the split of the field modes into positive and negative frequency is much less constrained for a fermion field compared with a scalar field. As in the scalar field case, the expansion coefficients are promoted to operators but now they satisfy anti-commutation relations. The vacuum state $|0\rangle$ is again defined as that state annihilated by all the particle annihilation operators: $\hat{b}_j\,|0\rangle = 0 = \hat{c}_j\,|0\rangle$. Compared with the scalar field case, there is much more freedom in how the vacuum state is defined for a fermion field, because there is much more freedom in how positive frequency modes can be chosen.

35.2 Quantum Field Theory on Schwarzschild Space-Time

Before studying the construction of quantum states on a Kerr black hole, we briefly review the Boulware [1] and Hartle-Hawking [3] states on a Schwarzschild black hole. We emphasize that the construction of these two states is the same for quantum fermion and scalar fields.

Modes for both a massless scalar field and massless fermion field are indexed by the quantum numbers ω, ℓ (a total angular momentum quantum number) and m (the azimuthal quantum number). The quantum number ω is the frequency of the modes as seen by a static observer either near the event horizon or at infinity. For scalar field modes, the Klein-Gordon norm is proportional to ω. We emphasize that *all* fermion modes have positive Dirac norm. A suitable basis of field modes consists of the "in" and "up" modes shown in Fig. 35.1a, b.

35.2.1 Boulware State $|B\rangle$ [1]

To define this state we choose positive frequency modes as seen by a static observer at infinity. The resulting vacuum state $|B\rangle$ contains no particles in the "in" and "up" modes with $\omega > 0$. This state is as empty as possible at infinity $I^{\pm}$ but diverges on

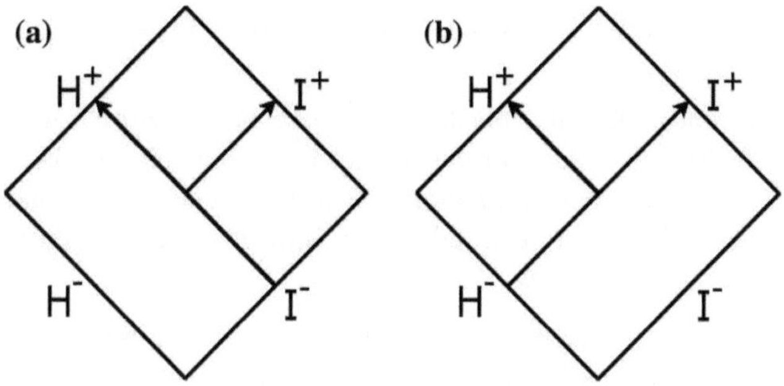

Fig. 35.1 **a** "In" modes: a wave is incident from infinity; part of the wave is reflected back to infinity and part goes down the event horizon of the black hole. **b** "Up" modes: a wave is outgoing from the event horizon; part of the wave is reflected back down the event horizon and part escapes to infinity

the event horizon $H^{\pm}$. The unrenormalized expectation value of the stress-energy tensor (UEVSET) for a fermion field in this state is

$$\langle B|\hat{\mathsf{T}}_{\mu\nu}|B\rangle = \frac{1}{2}\sum_{\ell=\frac{1}{2}}^{\infty}\sum_{m=-\ell}^{\ell}\int_{0}^{\infty} d\omega \left\{\mathsf{T}_{\mu\nu}\left[\psi^{\text{in}}_{\omega\ell m}\right] + \mathsf{T}_{\mu\nu}\left[\psi^{\text{up}}_{\omega\ell m}\right]\right\}. \tag{35.3}$$

Here, and in subsequent expressions, we write the UEVSET as a mode sum to show the differences between the various states considered. The expression $\mathsf{T}_{\mu\nu}[\psi^{\text{in/up}}_{\omega\ell m}]$ denotes a classical field mode contribution to the total UEVSET.

35.2.2 Hartle-Hawking State $|H\rangle$ [3]

In this case we choose positive frequency modes with respect to Kruskal time near the event horizon $H^{\pm}$. The "in" and "up" modes with $\omega > 0$ become thermally populated with energy ω. The resulting state $|H\rangle$ is regular at both the event horizon and infinity; however it is not empty at infinity. It represents a black hole in thermal equilibrium with a heat bath at the Hawking temperature T_H. The UEVSET for a fermion field in this state is

$$\langle H|\hat{\mathsf{T}}_{\mu\nu}|H\rangle = \frac{1}{2}\sum_{\ell=\frac{1}{2}}^{\infty}\sum_{m=-\ell}^{\ell}\int_{0}^{\infty} d\omega \tanh\left(\frac{\omega}{2T_H}\right)\left\{\mathsf{T}_{\mu\nu}\left[\psi^{\text{in}}_{\omega\ell m}\right] + \mathsf{T}_{\mu\nu}\left[\psi^{\text{up}}_{\omega\ell m}\right]\right\}. \tag{35.4}$$

For a quantum scalar field, the thermal "tanh" factor becomes a "coth".

35.3 Quantum Field Theory on Kerr Space-Time

The Kerr space-time represents a black hole whose event horizon rotates with angular velocity Ω_H. Some key features of the geometry are shown in Fig. 35.2a, b. In each figure, the central black region is the interior of the event horizon, and the axis of rotation runs vertically down the middle of each diagram. Two further surfaces are important in our later discussion:

- The *stationary limit surface* (the short-dashed surface in Fig. 35.2a, b) is the surface inside which it is not possible for an observer to remain at rest relative to infinity;
- The *speed of light surface* (the long-dashed surface in Fig. 35.2a, b) is the surface outside which it is not possible for an observer to co-rotate with the black hole event horizon.

A suitable basis of field modes is made up of "in" and "up" modes as in the Schwarzschild case, indexed by the quantum numbers ω, ℓ and m. A static observer near infinity measures ω to be the frequency of a particular field mode. Choosing modes with positive frequency at infinity therefore corresponds to choosing $\omega > 0$. Due to the rotation of the black hole, the corresponding frequency near the horizon is no longer ω but is shifted by the angular velocity of the black hole to be $\tilde{\omega} = \omega - m\Omega_H$. Choosing modes with positive frequency with respect to Kruskal time near the horizon therefore corresponds to thermally populating the "in" and "up" modes with energy $\tilde{\omega}$ rather than ω. A further complication is that, for scalar fields, the "in" modes have positive norm only if $\omega > 0$ whereas the "up" modes have positive norm only if $\tilde{\omega} > 0$. As previously, for a fermion field all modes have positive norm.

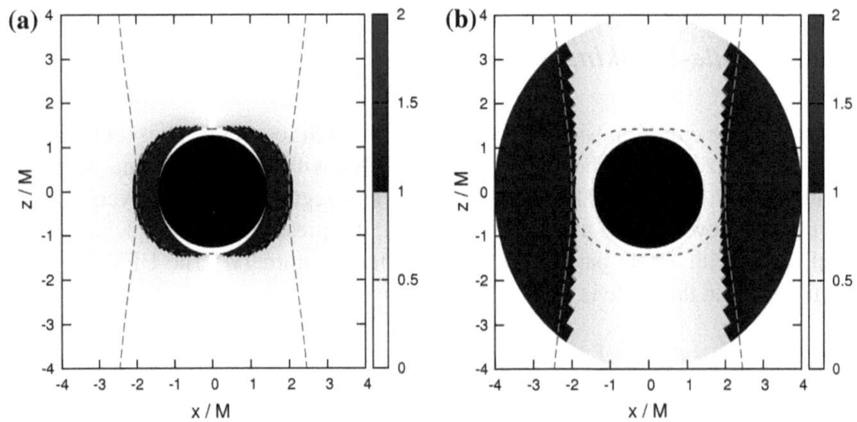

Fig. 35.2 **a** Regularity properties of $|B\rangle$. The UEVSET (35.5) is subtracted from (35.6). *Dark* regions indicate where this difference is divergent and *light* regions where it is finite. Taken from [2]. **b** Regularity properties of $|H\rangle$. The UEVSET (35.5) is subtracted from (35.7). *Dark* regions indicate where this difference is divergent and *light* regions where it is finite. Taken from [2]

We now describe three quantum states on Kerr space-time, including analogues of the Boulware and Hartle-Hawking states defined above for Schwarzschild black holes.

35.3.1 "Past" Boulware State $\left|B^{-}\right\rangle$ [5]

Based on the norms of the scalar field modes, the natural frequency for the "in" modes is ω, while for the "up" modes it is $\tilde{\omega}$. Choosing $\omega > 0$ for the "in" modes and $\tilde{\omega} > 0$ for the "up" modes as the definition of positive frequency leads to the "past" Boulware [4, 5] state, for which the UEVSET for a fermion field is [2]

$$\langle B^{-}|\hat{\mathsf{T}}_{\mu\nu}|B^{-}\rangle = \frac{1}{2}\sum_{\ell=\frac{1}{2}}^{\infty}\sum_{m=-\ell}^{\ell}\left\{\int_{0}^{\infty} d\omega\,\mathsf{T}_{\mu\nu}\left[\psi_{\omega\ell m}^{\mathrm{in}}\right] + \int_{0}^{\infty} d\tilde{\omega}\,\mathsf{T}_{\mu\nu}\left[\psi_{\omega\ell m}^{\mathrm{up}}\right]\right\}. \tag{35.5}$$

Like the Boulware state on a Schwarzschild black hole, the "past" Boulware state is divergent on the event horizon of a Kerr black hole, but it is regular everywhere outside the event horizon. Unlike the Boulware state on a Schwarzschild black hole, it is not empty at infinity; there is an outgoing flux of radiation at I^{+} [5]. The construction of the "past" Boulware state is identical for both scalar and fermion fields.

35.3.2 "Boulware" State $|B\rangle$ [2]

We next seek to construct an analogue of the Boulware state on Schwarzschild space-time, which is as empty as possible at both future and past null infinity, $I^{\pm}$. We would therefore like to choose $\omega > 0$ as positive frequency for both "in" and "up" modes. For scalar fields there is an immediate problem: the "up" modes have positive norm for $\tilde{\omega} > 0$, not $\omega > 0$. As a result for scalar fields no state empty at both future and past null infinity $I^{\pm}$ can be defined [4]. However, for fermion fields all modes have positive norm and in this case we can construct a state, the "Boulware" state $|B\rangle$, by taking $\omega > 0$ as the definition of positive frequency for all modes. The UEVSET for a fermion field in this state is [2]

$$\langle B|\hat{\mathsf{T}}_{\mu\nu}|B\rangle = \frac{1}{2}\sum_{\ell=\frac{1}{2}}^{\infty}\sum_{m=-\ell}^{\ell}\int_{0}^{\infty} d\omega\left\{\mathsf{T}_{\mu\nu}\left[\psi_{\omega\ell m}^{\mathrm{in}}\right] + \mathsf{T}_{\mu\nu}\left[\psi_{\omega\ell m}^{\mathrm{up}}\right]\right\}. \tag{35.6}$$

To determine where the state $|B\rangle$ is regular, we subtract from (35.6) the UEVSET (35.5), since the state $\left|B^{-}\right\rangle$ is regular everywhere outside the event horizon. It can be seen from Fig. 35.2a that this difference in expectation values is regular outside the stationary limit surface of the Kerr black hole, but diverges in the ergosphere. We therefore deduce that the state $|B\rangle$ is also regular outside the stationary limit surface but divergent in the ergosphere.

35.3.3 "Hartle-Hawking" State $|H\rangle$ [2]

To define an analogue of the Hartle-Hawking state on Schwarzschild space-time, we would like to construct a state representing a Kerr black hole in thermal equilibrium with a heat bath at the Hawking temperature T_H. Since the frequency of the modes near the horizon is $\tilde{\omega}$, this would correspond to thermally populating the "in" and "up" modes with energy $\tilde{\omega}$. For a quantum scalar field, this cannot be done because the "in" modes are defined in terms of frequency ω, not $\tilde{\omega}$.

On the other hand, for fermions on Kerr we are able to define a "Hartle-Hawking" state for which both the "in" and "up" modes are thermalized with energy $\tilde{\omega}$. In this case the UEVSET for the fermion field is [2]

$$\langle H|\hat{\mathsf{T}}_{\mu\nu}|H\rangle = \frac{1}{2}\sum_{\ell=\frac{1}{2}}^{\infty}\sum_{m=-\ell}^{\ell}\int_0^{\infty} d\tilde{\omega}\tanh\left(\frac{\tilde{\omega}}{2T_H}\right)\left\{\mathsf{T}_{\mu\nu}\left[\psi^{\text{in}}_{\omega\ell m}\right]+\mathsf{T}_{\mu\nu}\left[\psi^{\text{up}}_{\omega\ell m}\right]\right\}. \tag{35.7}$$

To determine where the state $|H\rangle$ is regular, we subtract from (35.7) the UEVSET (35.5). It can be seen from Fig. 35.2b that this difference in expectation values is regular between the event horizon and the speed-of-light surface, but diverges outside the speed-of-light surface. We therefore deduce that the state $|H\rangle$ is also regular between the event horizon and the speed-of-light surface but diverges outside the speed-of-light surface.

Acknowledgments This note discusses work completed in collaboration with Marc Casals, Sam Dolan, Brien Nolan and Adrian Ottewill. This work was supported by the Lancaster-Manchester-Sheffield Consortium for Fundamental Physics under STFC Grant No. ST/J000418/1, by an International Visitor Programme Grant from the Office of the Vice President for Research in Dublin City University and by EU COST Action MP0905 "Black Holes in a Violent Universe". We thank Victor Ambrus for helpful discussions.

References

1. D.G. Boulware, Spin 1/2 quantum field theory in Schwarzschild space. Phys. Rev. D **12**, 350–367 (1975)
2. M. Casals, S.R. Dolan, B.C. Nolan, A.C. Ottewill, E. Winstanley, Quantization of fermions on Kerr space-time. Phys. Rev. D **87**, 064027 (2013)
3. J.B. Hartle, S.W. Hawking, Path integral derivation of black hole radiance. Phys. Rev. D **13**, 2188–2203 (1976)
4. A.C. Ottewill, E. Winstanley, The renormalized stress tensor in Kerr space-time: general results. Phys. Rev. D **62**, 084018 (2000)
5. W.G. Unruh, Second quantization in the Kerr metric. Phys. Rev. D **10**, 3194–3205 (1974)

Part IV
Other Topics in Contemporary Gravitation

Chapter 36
Quantum Gravity and the Cosmological Constant Problem

John W. Moffat

Abstract A finite and unitary nonlocal formulation of quantum gravity is applied to the cosmological constant problem. The entire functions in momentum space at the graviton-standard model particle loop vertices generate an exponential suppression of the vacuum density and the cosmological constant to produce agreement with their observational bounds.

36.1 Introduction

A nonlocal quantum field theory and quantum gravity theory has been formulated that leads to a finite, unitary and locally gauge invariant theory [1–14]. For quantum gravity the finiteness of quantum loops avoids the problem of the non-renormalizabilty of local quantum gravity [15, 16].

The finiteness of the nonlocal quantum field theory draws from the fact that factors of $\exp[\mathscr{K}(p^2)/\Lambda^2]$ are attached to propagators which suppress any ultraviolet divergences in Euclidean momentum space, where Λ is an energy scale factor. An important feature of the field theory is *that only the quantum loop graphs have nonlocal properties*; the classical tree graph theory retains full causal and local behavior.

Consider first the 4-dimensional spacetime to be approximately flat Minkowski spacetime. Let us denote by f a generic local field and write the standard local Lagrangian as

$$\mathscr{L}[f] = \mathscr{L}_F[f] + \mathscr{L}_I[f], \tag{36.1}$$

where $\mathscr{L}_F$ and $\mathscr{L}_I$ denote the free part and the interaction part of the action, respectively, and

J.W. Moffat (✉)
Perimeter Institute for Theoretical Physics,
Waterloo, ON N2L 2Y5, Canada
e-mail: jmoffat@perimeterinstitute.ca

J.W. Moffat
Department of Physics and Astronomy, University of Waterloo,
Waterloo, ON N2L 2Y5, Canada

P. Nicolini et al. (eds.), *1st Karl Schwarzschild Meeting on Gravitational Physics*,
Springer Proceedings in Physics 170, DOI 10.1007/978-3-319-20046-0_36

$$\mathscr{L}_F[f] = \frac{1}{2} f_i \mathscr{K}_{ij} f_j. \tag{36.2}$$

In a gauge theory the action would be the Becchi, Rouet, Stora, Tyutin (BRST) gauge-fixed action including ghost fields in the invariant action required to fix the gauge [17, 18]. The kinetic operator $\mathscr{K}$ is fixed by defining a Lorentz-invariant distribution operator:

$$\mathscr{E} \equiv \exp\left(\frac{\mathscr{K}}{2\Lambda^2}\right) \tag{36.3}$$

and the operator:

$$\mathscr{O} = \frac{\mathscr{E}^2 - 1}{\mathscr{K}} = \int_0^1 \frac{d\tau}{\Lambda^2} \exp\left(\tau \frac{\mathscr{K}}{\Lambda^2}\right). \tag{36.4}$$

The regularized interaction Lagrangian takes the form

$$\hat{\mathscr{L}}_I = -\sum_n (-g)^n f \mathscr{I}[\mathscr{F}^n, \mathscr{O}^{(n-1)})] f, \tag{36.5}$$

where g is a coupling constant and $\mathscr{F}$ is a vertex function form factor. The decomposition of $\mathscr{I}$ in order $n = 2$ is such that the operator $\mathscr{O}$ splits into two parts $\mathscr{F}^2/\mathscr{K}$ and $-1/\mathscr{K}$. For Compton amplitudes the first such term cancels the contribution from the corresponding lower order channel, while the second term is just the usual local field theory result for that channel. The action is then invariant under an extended nonlocal gauge transformation. The precise results for QED were described in [2].

The regularized action is found by expanding $\hat{\mathscr{L}}_I$ in an infinite series of interaction terms. Since $\mathscr{F}$ and $\mathscr{O}$ are entire function of $\mathscr{K}$ the higher interactions are also entire functions of $\mathscr{K}$. This is important for preserving the Cutkosky rules and unitarity, for an entire function does not possess any singularities in the finite complex momentum plane.

The Feynman rules are obtained as follows: Every leg of a diagram is connected to a local propagator,

$$D(p^2) = \frac{i}{\mathscr{K}(p^2) + i\varepsilon} \tag{36.6}$$

and every vertex has a form factor $\mathscr{F}^k(p^2)$, where p is the momentum attached to the propagator $D(p^2)$, which has the form

$$\mathscr{F}^k(p^2) \equiv \mathscr{E}^k(p^2) = \exp\left(\frac{\mathscr{K}}{2\Lambda_k}\right), \tag{36.7}$$

where k denotes the particle nature of the external leg in the vertex. The formalism is set up in Minkowski spacetime and loop integrals are formally defined in Euclidean space by performing a Wick rotation. This facilitates the analytic continuation; the whole formalism could from the outset be developed in Euclidean space.

We will demonstrate how the nonlocal transcendental entire function in momentum space that generates the finite and unitary standard model (SM) and quantum gravity (QG) loops to all orders of perturbation theory, produces an exponential suppression of the estimated very large vacuum density and cosmological constant in local quantum field theory. This can solve the severe fine-tuning cosmological constant problem, avoiding a naturalness problem and the need for an anthropic and multiverse solution.

36.2 Nonlocal Quantum Gravity

We expand the metric around a smooth fixed background spacetime:

$$g_{\mu\nu} = \bar{g}_{\mu\nu} + h_{\mu\nu}. \tag{36.8}$$

By restricting ourselves to perturbation theory and a fixed geometrical background, we lose general covariance (diffeomorphism invariance). However, we still maintain gauge invariance of the gravitational calculations under the gauge group of the fixed background metric, e.g., for a fixed Minkowski metric background the action is invariant under local Poincaré transformations, while for a de Sitter background metric the action will be invariant under the group of de Sitter transformations. Although we lose general covariance in our perturbation calculations of gravitational scattering amplitudes, the basic physical properties such as finiteness of loop amplitudes, gauge invariance and unitarity will be expected to lead to correct and reliable physical conclusions. For the sake of simplicity, we shall only consider expansions about Minkowski spacetime.

Let us define $\mathbf{g}^{\mu\nu} = \sqrt{-g}g^{\mu\nu}$, where $\mathbf{g} = \det(\mathbf{g}^{\mu\nu})$ and $\partial_\rho \mathbf{g} = \mathbf{g}_{\alpha\beta}\partial_\rho \mathbf{g}^{\alpha\beta}\mathbf{g}$. We can then write the local gravitational action S_{grav} in the form [19]:

$$\begin{aligned} S_{\text{grav}} = \int d^4x \mathcal{L}_{\text{grav}} = \frac{1}{2\kappa^2}\int d^4x[(\mathbf{g}^{\rho\sigma}\mathbf{g}_{\lambda\mu}\mathbf{g}_{\kappa\nu} \\ -\frac{1}{2}\mathbf{g}^{\rho\sigma}\mathbf{g}_{\mu\kappa}\mathbf{g}_{\lambda\nu} - 2\delta^\sigma_\kappa\delta^\rho_\lambda\mathbf{g}_{\mu\nu})\partial_\rho\mathbf{g}^{\mu\kappa}\partial_\sigma\mathbf{g}^{\lambda\nu} \\ -\frac{2}{\alpha}\partial_\mu\mathbf{g}^{\mu\nu}\partial_\kappa\mathbf{g}^{\kappa\lambda}\eta_{\nu\lambda} + \bar{C}^\nu\partial^\mu X_{\mu\nu\lambda}C^\lambda], \end{aligned} \tag{36.9}$$

where $\kappa^2 = 32\pi G$ and we have added a gauge fixing term with the parameter α, C^μ is the Fadeev-Popov ghost field and $X_{\mu\nu\lambda}$ is a differential operator:

$$X_{\mu\nu\lambda} = \kappa(-\partial_\lambda\gamma_{\mu\nu} + 2\eta_{(\mu\lambda}\gamma_{\kappa\nu)}\partial^\kappa) + (\eta_{(\mu\lambda}\partial_{\nu)} - \eta_{\mu\nu}\partial_\lambda). \tag{36.10}$$

We expand the local interpolating graviton field $\mathbf{g}^{\mu\nu}$ as

$$\mathbf{g}^{\mu\nu} = \eta^{\mu\nu} + \kappa\gamma^{\mu\nu} + O(\kappa^2). \tag{36.11}$$

Then,

$$\mathbf{g}_{\mu\nu} = \eta_{\mu\nu} - \kappa\gamma_{\mu\nu} + \kappa^2\gamma_\mu{}^\alpha\gamma_{\alpha\nu} + O(\kappa^3). \tag{36.12}$$

The gravitational Lagrangian density is expanded as

$$\mathscr{L}_{\rm grav} = \mathscr{L}^{(0)} + \kappa\mathscr{L}^{(1)} + \kappa^2\mathscr{L}^{(2)} + \cdots \tag{36.13}$$

In the limit $\alpha \to \infty$, the Lagrangian density $\mathscr{L}_{\rm grav}$ is invariant under the gauge transformation

$$\delta\gamma_{\mu\nu} = X_{\mu\nu\lambda}\xi^\lambda, \tag{36.14}$$

where ξ^λ is an infinitesimal vector quantity.

To implement nonlocal quantum gravity, we introduce the "stripping" graviton propagator in the gauge $\alpha = -1$:

$$\tilde{D}_{\alpha\beta\mu\nu}(p) = \frac{1}{2}(\eta_{\alpha\mu}\eta_{\beta\nu} + \eta_{\alpha\nu}\eta_{\beta\mu} - \eta_{\alpha\beta}\eta_{\mu\nu})\mathscr{O}_0(p), \tag{36.15}$$

while the ghost stripping propagator is given by

$$\tilde{D}^{\rm ghost}_{\mu\nu}(p) = \eta_{\mu\nu}\mathscr{O}_0(p), \tag{36.16}$$

where

$$\mathscr{O}_0(p) = \frac{\mathscr{E}_0^2 - 1}{p^2}. \tag{36.17}$$

We choose $\mathscr{E}_0^2 = \exp(-p^2/2\Lambda_G^2)$ and we see that the local propagator can be obtained from the nonlocal propagator minus the stripping propagator

$$\frac{1}{p^2} = \frac{\exp(-p^2/2\Lambda_G^2)}{p^2} - \mathscr{O}_0(p). \tag{36.18}$$

The stripping propagators are used to guarantee that the tree-level graviton-graviton scattering amplitudes are identical to the local, point-like tree-level amplitudes, which couple only to physical gravitons.

The graviton propagator in the fixed de Donder gauge $\alpha = -1$ [20, 21] in momentum space is given by

$$D_{\mu\nu\rho\sigma}(p) = \frac{\eta_{\mu\rho}\eta_{\nu\sigma} + \eta_{\mu\sigma}\eta_{\nu\rho} - \eta_{\mu\nu}\eta_{\rho\sigma}}{p^2 + i\varepsilon}, \tag{36.19}$$

while the graviton ghost propagator in momentum space is

$$D^{\rm ghost}_{\mu\nu}(p) = \frac{\eta_{\mu\nu}}{p^2 + i\varepsilon}. \tag{36.20}$$

The on-shell vertex functions are unaltered from their local antecedents, while virtual particles are attached to nonlocal vertex function form factors. This destroys the gauge invariance of e.g. graviton-graviton scattering and requires an iteratively defined series of "stripping" vertices to ensure the decoupling of all unphysical modes. Moreover, the local gauge transformations have to be extended to nonlinear, nonlocal gauge transformations to guarantee the over-all invariance of the regularized amplitudes. The quantum gravity perturbation theory is invariant under generalized, nonlinear field representation dependent transformations, and it is finite to all orders. At the tree graph level all unphysical polarization states are decoupled and nonlocal effects will only occur in graviton and graviton-matter loop graphs. Because the gravitational tree graphs are purely local there is a well-defined classical GR limit. The finite quantum gravity theory is well-defined in four real spacetime dimensions or in any higher D-dimensional spacetime.

We quantize by means of the path integral operation

$$\langle 0|T^*(O[\mathbf{g}])|0\rangle_{\mathscr{E}} = \int [D\mathbf{g}]\mu[\mathbf{g}](\text{gauge fixing})O[\mathbf{g}]\exp(i\hat{S}_{\text{grav}}[\mathbf{g}]). \qquad (36.21)$$

The quantization is carried out in the functional formalism by finding a measure factor $\mu[\mathbf{g}]$ to make $[D\mathbf{g}]$ invariant under the classical symmetry. Because we have extended the gauge symmetry to nonlinear, nonlocal transformations, we must also supplement the quantization procedure with an invariant measure

$$\mathscr{M} = \Delta(\mathbf{g}, \bar{C}, C)D[\mathbf{g}_{\mu\nu}]D[\bar{C}_\lambda]D[C_\sigma] \qquad (36.22)$$

such that $\delta\mathscr{M} = 0$.

36.3 The Cosmological Constant Problem

The cosmological constant problem is considered to be the most severe hierarchy problem in modern physics [22–25].

We can define an effective cosmological constant

$$\lambda_{\text{eff}} = \lambda_0 + \lambda_{\text{vac}}, \qquad (36.23)$$

where λ_0 is the 'bare' cosmological constant in Einstein's classical field equations, and λ_{vac} is the contribution that arises from the vacuum density $\lambda_{\text{vac}} = 8\pi G\rho_{\text{vac}}$. The observational bound on ρ_{vac} is

$$\rho_{\text{vac}} \leq 10^{-47}\,(\text{GeV})^4, \qquad (36.24)$$

corresponding to the bound on $\lambda_{\rm vac}$:

$$\lambda_{\rm vac} \leq 10^{-84}\,{\rm GeV}^2. \tag{36.25}$$

Zeldovich [26] showed that the zero-point vacuum fluctuations must have a Lorentz invariant form

$$T_{{\rm vac}\,\mu\nu} = \lambda_{\rm vac} g_{\mu\nu}, \tag{36.26}$$

consistent with the equation of state $\rho_{\rm vac} = -p_{\rm vac}$. Thus, the vacuum within the framework of particle quantum physics has properties identical to the cosmological constant. In quantum theory, the second quantization of a classical field of mass m, treated as an ensemble of oscillators each with a frequency $\omega(k)$, leads to a zero-point energy $E_0 = \sum_k \frac{1}{2}\hbar\omega(k)$. An evaluation of the vacuum density obtained from a summation of the zero-point energy modes gives

$$\rho_{\rm vac} = \frac{1}{(2\pi)^2}\int_0^{M_c} dk k^2(k^2+m^2)^{1/2} \sim \frac{M_c^4}{16\pi^2}, \tag{36.27}$$

where M_c is the cutoff. Taking $M_c \sim M_{\rm Planck} \sim 10^{19}$ GeV, we get $\rho_{\rm vac} \sim 122$ orders of magnitude greater than the observed value. Already at the level of the standard model, we get $\rho_{\rm vac} \sim (10^2\,{\rm GeV})^4$ which is 55 orders of magnitude larger than the bound (36.24). To agree with the experimental bound (36.24), we would have to invoke a very finely tuned cancellation of $\lambda_{\rm vac}$ with the 'bare' cosmological constant λ_0, which is generally conceded to be theoretically unacceptable.

We adopt a model consisting of a photon field A_μ coupled to gravity. We have for the effective field Lagrangian density:

$$\mathscr{L}_A = -\frac{1}{4}(-\mathbf{g})^{-1/2}\mathbf{g}^{\mu\nu}\mathbf{g}^{\alpha\beta}F_{\mu\alpha}F_{\nu\beta}, \tag{36.28}$$

where

$$F_{\mu\nu} = \partial_\nu A_\mu - \partial_\mu A_\nu. \tag{36.29}$$

We have

$$\mathscr{L}_A^{(0)} = -\frac{1}{4}\eta^{\mu\nu}\eta^{\alpha\beta}F_{\mu\alpha}F_{\nu\beta}, \tag{36.30}$$

and

$$\mathscr{L}_A^{(1)} = -\frac{1}{4}\left(\eta^{\mu\nu}\gamma^{\alpha\beta} + \eta^{\alpha\beta}\gamma^{\mu\nu} - \frac{1}{2}\eta^{\mu\nu}\eta^{\alpha\beta}\gamma\right)F_{\mu\alpha}F_{\nu\beta}. \tag{36.31}$$

We include in the Lagrangian density $\mathscr{L}_A^{(0)}$ an additional gauge-fixing piece $-\frac{1}{2}(\partial^\mu A_\mu)^2$. For a particular gauge no Faddeev-Popov ghost particles and diagrams contribute to the lowest order photon-graviton self-energy calculation. The local photon propagator has the form

$$D^{\rm A}_{\mu\nu}(p) = \frac{\eta_{\mu\nu}}{p^2 + i\varepsilon}. \tag{36.32}$$

The graviton-A-A vertex in momentum space is given by

$$\begin{aligned}\mathscr{V}_{\alpha\beta\lambda\sigma}(q_1, q_2) &= \eta_{\lambda\sigma} q_{1(\alpha} q_{2\beta)} - \eta_{\sigma(\beta} q_{1\alpha)} q_{2\lambda} - \eta_{\lambda(\alpha} q_{1_\sigma} q_{2\beta)} \\ &+ \eta_{\sigma(\beta}\eta_{\alpha)\lambda} q_1{\cdot}q_2 - \frac{1}{2}\eta_{\alpha\beta}(\eta_{\lambda\sigma} q_1{\cdot}q_2 - q_{1\sigma} q_{2\lambda}),\end{aligned} \tag{36.33}$$

where q_1, q_2 denote the momenta of the two Vs connected to the graviton with momentum p.

The lowest order correction to the graviton vacuum loop will have the form

$$\begin{aligned}\Pi^{\rm GA}_{\mu\nu\rho\sigma}(p) &= -\kappa^2 \exp(-p^2/\Lambda_G^2) \int d^4q \mathscr{V}_{\mu\nu\lambda\alpha}(p, q) \mathscr{F}(q^2) D^{A\,\lambda\delta}(q) \\ &\times \mathscr{V}_{\rho\sigma\kappa\delta}(p, q-p) \mathscr{F}((q-p)^2) D^{A\,\alpha\kappa}(q-p).\end{aligned} \tag{36.34}$$

Let us adopt the entire functions $\mathscr{F}(p^2)_{\rm SM} = \exp(-p^2/2\Lambda_{\rm SM}^2)$ and $\mathscr{F}(p^2) = \exp(-p^2/2\Lambda_G^2)$ in Euclidean momentum space, scaled by the SM energy scale $\Lambda_{\rm SM}$ and the gravitational energy scale Λ_G, respectively. We obtain

$$\begin{aligned}\Pi^{\rm GV}_{\mu\nu\rho\sigma}(p) &= -\kappa^2 \exp(-p^2/\Lambda_G^2) \int \frac{d^4q \eta^{\lambda\delta}\eta^{\alpha\kappa}}{q^2(q-p)^2} \mathscr{V}_{\mu\nu\lambda\alpha}(p, q) \\ &\times \mathscr{V}_{\rho\sigma\kappa\delta}(p, q-p) \exp\left(-q^2/2\Lambda_{\rm SM}^2\right) \exp\left(-(q-p)^2/2\Lambda_{\rm SM}^2\right).\end{aligned} \tag{36.35}$$

As usual, we must add to (36.35) the contributions from the tadpole vector-graviton diagrams and the invariant measure diagram.

We observe that from power counting of the momenta in the integral (36.35), we obtain

$$\Pi^{\rm GA}_{\mu\nu\rho\sigma}(p) \sim \kappa^2 \exp(-p^2/\Lambda_G^2) N_{\mu\nu\rho\sigma}(\Lambda_{\rm SM}, p^2), \tag{36.36}$$

where $N_{\mu\nu\rho\sigma}(\Lambda_{\rm SM}, p^2)$ is a finite contribution to $\Pi^{\rm GA}_{\mu\nu\rho\sigma}(p)$. $\Pi^{\rm GA}{}_\mu{}^{\mu\sigma}{}_\sigma(p)$ vanishes at $p^2 = 0$, as it should because of gauge invariance to this order and the massless graviton.

The vector field vertex form factor, *when coupled to SM gauge bosons*, will have the form

$$\mathscr{E}^{\rm SM}(p^2) = \exp\left(-p^2/2\Lambda_{SM}^2\right). \tag{36.37}$$

If we choose $\Lambda_{SM} \gtrsim 1$ TeV, then we will reproduce the low energy SM experimental results and $\mathscr{F}^{\rm SM}(p^2)$ becomes $\mathscr{F}^{\rm SM}(0) = 1$ on the mass shell $p^2 = 0$ [1, 2].

36.4 Cosmological Constant Problem and Quantum Gravity

The cosmological constant problem is considered to be the most severe hierarchy problem in modern physics [22–25]. Can our quantum gravity theory solve the cosmological constant problem? The cosmological constant is a non-derivative coupling in the Lagrangian density $\mathscr{L}_{\rm grav}$:

$$\mathscr{L}_\lambda = -\frac{4}{\kappa^2}\lambda\sqrt{-g}. \tag{36.38}$$

In diagrammatic terms, it is a sum of zero momentum and zero temperature vacuum fluctuation loops coupled to external gravitons. The problem is to explain why the magnitude of λ is suppressed to be zero or a very small value when compared to observation.

Let us initially consider the basic lowest order vacuum fluctuation diagram computed from the matrix element in flat Minkowski spacetime:

$$\begin{aligned}\rho_{\rm vac} &\sim \rho^{(2)}_{\rm vac} \sim g^2\int d^4pd^4p'd^4k\delta(k+p-p')\delta(k+p-p')\\ &\times\frac{1}{k^2+m^2}\mathrm{Tr}\left(\frac{i\gamma^\sigma p_\sigma - m_f}{p^2+m_f^2}\gamma^\mu\frac{i\gamma^\sigma p'_\sigma - m_f}{p'^2+m_f^2}\gamma_\mu\right)\\ &\exp\left[-\left(\frac{p^2+m_f^2}{2\Lambda_{SM}^2}\right)-\left(\frac{p'^2+m_f^2}{2\Lambda_{SM}^2}\right)-\left(\frac{k^2+m^2}{2\Lambda_{SM}^2}\right)\right],\end{aligned} \tag{36.39}$$

where g is a coupling constant associated with the standard model. We have considered a closed loop made of a SM fermion of mass m_f, an anti-fermion of the same mass and an internal SM boson propagator of mass m; the scale $\Lambda_{\rm SM}\sim 1$ TeV. This leads to the result

$$\begin{aligned}\rho_{\rm vac} &\sim \rho^{(2)}_{\rm vac} \sim 16\pi^4g^2\delta^4(a)\int_0^\infty dpp^3\\ &\int_0^\infty dp'p'^3\left[\frac{-P^2+p^2+p'^2+4m_f^2}{(P+a)(P-a)}\right]\frac{1}{(p^2+m_f^2)(p'^2+m_f^2)}\\ &\times\exp\left[-\frac{(p^2+p'^2+2m_f^2)}{2\Lambda_{SM}^2}-\frac{P^2+m^2}{2\Lambda_{SM}^2}\right],\end{aligned} \tag{36.40}$$

where $P = p - p'$ and a is an infinitesimal constant which formally regularizes the infinite volume factor $\delta^4(0)$. We see that $\rho_{\rm vac}\sim\Lambda^4_{\rm SM}$. By choosing our nonlocal energy scale for the standard model, $\Lambda_{\rm NL}\sim\Lambda_{\rm SM}\sim 1\,\mathrm{TeV} = 10^3$ GeV, we have reduced the magnitude of the vacuum density by 64 orders of magnitude compared to having $\Lambda_{\rm SM}\sim\Lambda_{\rm Planck}\sim 10^{19}$ GeV.

In Minkowski spacetime, the sum of all *disconnected* vacuum diagrams is a constant factor C in the scattering S-matrix $S' = SC$. Since the S-matrix is unitary $|S'|^2 = 1$, then we must conclude that $|C|^2 = 1$, and all the disconnected vacuum graphs can be ignored. This result is also known to follow from the Wick ordering of the field operators. However, due to the equivalence principle *gravity couples to all forms of energy*, including the vacuum energy density $\rho_{\rm vac}$, so we can no longer ignore these virtual quantum fluctuations in the presence of a non-zero gravitational field.

We can view the cosmological constant as a non-derivative coupling of the form $\lambda_0\sqrt{-g}$ in the Einstein-Hilbert action (See Fig. 36.1). Quantum corrections to λ_0 come from loops formed from massive SM states, coupled to external graviton lines at essentially zero momentum. The massive SM states are far off-shell. Experimental tests of the standard model involving gravitational couplings to the SM states are very close to being on-shell. Important quantum corrections to λ_0 are generated by a huge extrapolation to a region in which gravitons couple to SM particles which are far off-shell.

To reduce the size of the vacuum density to agree with the observational bound, we must discover how gravity can couple to the vacuum energy density and generate an exponential damping of the very large $\rho_{\rm vacSM}$. This exponential suppression of $\rho_{\rm vacSM}$ can be produced by nonlocal QG. There will be virtual graviton legs connected to the quantum gravity-standard model loops by a nonlocal vertex entire function, $\exp(-p_G^2/2\Lambda_G^2)$. We see from (36.36) that the standard model vacuum polarization and vacuum density are reduced by the nonlocal graviton vertex interaction:

$$\rho_{\rm vac} \sim \exp(-\bar{p}_G^2/2\Lambda_G^2)\rho_{\rm vacSM}, \tag{36.41}$$

where $\bar{p}_G$ is an average mean of the virtual graviton momentum p_G. If we choose $\bar{p}_G = 16.49\Lambda_G$, then we have

$$\rho_{\rm vac} \sim \exp(-\bar{p}_G^2/2\Lambda_G^2)\rho_{\rm vacSM} \sim 10^{-47}\,{\rm GeV}^4, \tag{36.42}$$

and we have reduced the cosmological constant contribution, $\lambda_{\rm vac} = 8\pi G\rho_{\rm vac}$, to the observed bound $\lambda_{\rm vacObs} \leq 10^{-84}\,{\rm GeV}^2$, where we have used the nonlocal energy scale $\Lambda_{SM} \sim 1$ TeV in the coupling to standard model particles. The size of Λ_G

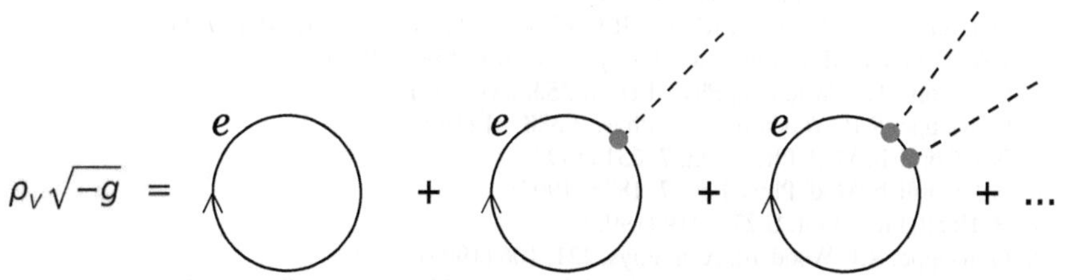

Fig. 36.1 Electron vacuum fluctuation loops coupled to gravitons generating a vacuum density

should be small enough to allow for soft graviton momenta. This can be achieved by choosing $\Lambda_G < 1$ MeV, so that the mean virtual graviton momentum $\bar{p}_G = 16.5\Lambda_g < 17$ MeV. The size of the exponential suppression of the vacuum energy in (36.42) can be related to a violation of the weak equivalence principle through the electrostatic energy associated with the vacuum polarization of atomic hydrogen coupled to external gravitons [23, 24], so the choice of Λ_G can play an important role. However, the violation of the equivalence principle can be affected by the material environment, namely, the difference between the atomic matter environment versus the vacuum energy density in empty space at extra-galactic distance scales.

36.5 Conclusions

The nonlocal formulation of quantum gravity provides a finite, unitary and locally gauge invariant perturbation theory. The vertex functions associated with point-like interactions in local quantum field theory are replaced by smeared out nonlocal vertex functions controlled by transcendental entire functions. The choice of entire function in momentum space $\exp(-p^2/2\Lambda^2)$, where $\Lambda = \Lambda_{\rm SM} \sim 1$ TeV and $\Lambda = \Lambda_G$ for the standard model and quantum gravity, respectively, guarantees the finiteness of all quantum loops. We have demonstrated how the vacuum fluctuations involving SM loops can be exponentially dampened by the entire functions for the graviton-standard model particle vertex functions. For a mean value of the virtual graviton momenta the exponential suppression can reduce the vacuum density fluctuations and the cosmological constant to agree with the cosmological observational bounds.

Acknowledgments The John Templeton Foundation is thanked for its generous support of this research. The research was also supported by the Perimeter Institute for Theoretical Physics. Research at the Perimeter Institute for Theoretical Physics is supported by the Government of Canada through industry Canada and by the Province of Ontario through the Ministry of Research and Innovation (MRI). I thank Martin Green and Viktor Toth for helpful discussions.

References

1. J.W. Moffat, Phys. Rev. D **41**, 1177 (1990)
2. D. Evens, J.W. Moffat, G. Kleppe, R.P. Woodard, Phys. Rev. D **43**, 49 (1991)
3. J.W. Moffat, S.M. Robbins, Mod. Phys. Lett. A **6**, 1581 (1991)
4. G. Kleppe, R.P. Woodard, Phys. Lett. B **253**, 331 (1991)
5. G. Kleppe, R.P. Woodard, Nucl. Phys. B **388**, 81 (1992)
6. N.J. Cornish, Mod. Phys. Lett. **7**, 631 (1992)
7. N.J. Cornish, Mod. Phys. Lett. **7**, 1895 (1992)
8. B. Hand, Phys. Lett. B **275**, 419 (1992)
9. G. Kleppe, R.P. Woodard, Ann. Phys. **221**, 106 (1993)
10. M.A. Clayton, L. Demopolous, J.W. Moffat, Int. J. Mod. Phys. A **9**, 4549 (1994)
11. J. Paris, Nucl. Phys. B **450**, 357 (1995)
12. J. Paris, W. Troost, Nucl. Phys. B **482**, 373 (1996)

13. G. Saini, S.D. Joglekar, Z. Phys. C **76**, 343 (1997)
14. J.W. Moffat, Eur. Phys. J. Plus **126**, 43 (2011). arXiv:1008.2482 [gr-qc]
15. G. 't Hooft, M. Veltman, Ann. Inst. Henri Poincaré **30**, 69 (1974)
16. M. Goroff, A. Sagnotti, Nucl. Phys. B **266**, 709 (1986)
17. C. Becchi, A. Rouet, R. Stora, Comm. Math. Phys. **42**, 127 (1975)
18. I.V. Tyutin, Lebedev Institute preprint N39 (1975)
19. J.N. Goldberg, Phys. Rev. **111**, 315 (1958)
20. T. de Donder, *La Grafique Einsteinienne* (Gauthier-Villars, Paris, 1921)
21. V.A. Fock, *Theory of Space, Time and Gravitation* (Pergamon, New York, 1959)
22. S. Weinberg, Rev. Mod. Phys. **61**, 1 (1989)
23. J. Polchinski, arXiv:0603249 [hep-th]
24. J. Martin, arXiv:1205.3365 [gr-qc]
25. C.P. Burgess, arXiv:1309.4133 [hep-th]
26. Ya.B. Zeldovich, Pis'ma Zh. Eksp. Teor. Fiz. **6**, 883 [JETP Lett. **6**, 316 (1967)]

Chapter 37
Emergent Gravity and the Cosmological Constant

Tanu Padmanabhan

Abstract Several recent results suggest that gravity is an emergent phenomenon with its field equations having the same status as the equations of fluid dynamics or elasticity. Interestingly, these investigations have provided fresh insights into several aspects of classical gravitational dynamics itself, including the problem of cosmological constant.

37.1 A Route to Gravitational Dynamics

Recent research [1] motivated by the emergent paradigm of gravity shows that the description of classical gravity simplifies significantly if we use the variables $f^{ab} \equiv \sqrt{-g}g^{ab}$ and the corresponding canonical momentum N^c_{ab}. These 'old variables' were studied in the early days of general relativity (and were used sporadically later on in the literature) but have not acquired the popularity they deserve. My first task is to advertise the virtues of this description.[1]

The Hamilton's equations in classical mechanics can be obtained from an action principle based on *either of the two* Lagrangians: $L_q(p, q, \dot{q}) = p\dot{q} - H(p, q)$ or $L_p(q, p, \dot{p}) = -q\dot{p} - H(p, q)$ which differ by a time derivative, i.e., $L_p = L_q - (d(pq)/dt)$. In either action principle, we vary *both* q and p as independent variables and let the action principle itself tell us what should be held fixed at the end points. We will find that we get the standard equations if we keep q fixed with L_q and p fixed with L_p (which explains the subscripts).

What is not stressed in textbooks is that we can do the same thing in field theory. The field equations, $\Box\phi = -V'(\phi)$ for a scalar field ϕ, say, can be obtained using

[1] I will concentrate and elaborate here on some results of interest to relativists, which I couldonly briefly mention in my lecture at KSM-I, rather than review older results in emergent gravity paradigm which exist in published literature [2, 3]. I have omitted detailed references and proofs of the assertions due to lack of space [4].

T. Padmanabhan (✉)
Inter-University Centre for Astronomy and Astrophysics,
Pune University Campus, Ganeshkhind, Pune 7, India
e-mail: nabhan@iucaa.ernet.in

P. Nicolini et al. (eds.), *1st Karl Schwarzschild Meeting on Gravitational Physics*,
Springer Proceedings in Physics 170, DOI 10.1007/978-3-319-20046-0_37

either of the two Lagrangians: $L_\phi = p^a \partial_a \phi - H$ or $L_p = -\phi \partial_a p^a - H = L_\phi - \partial_a(\phi p^a)$ with $H = (1/2) p^a p_a + V(\phi)$. These Lagrangians, again, differ by a total divergence. In contrast to text book description, we are here treating H as a *Lorentz scalar* rather than as time component of a four-vector; further, full Lorentz invariance is maintained without any (1+3) split. Note that, while $p^a \partial_a \phi$ term does not involve a metric, $p^a p_a = \eta_{ab} p^a p^b$ needs a background metric to 'lower the index' on p^a. It is straightforward to generalize the description to a curved spacetime by the usual prescriptions. We can also work out the dynamics of a vector field and, in particular, $U(1)$ gauge field in an analogous manner.

It becomes really interesting when we consider the theory of a symmetric second rank tensor field described by a matrix f with elements denoted by f^{ab}. The "momenta" corresponding to f^{ab} will be denoted by N^c_{ab} which, in turn, can be thought of as elements of four matrices N^c. As in the previous cases, we can consider two possible Lagrangians (which differ from each other by a four-divergence) to describe the theory:

$$L_f = N^c_{ab} \partial_c f^{ab} - \mathcal{H}_g(f^{ij}, N^k_{lm}); \quad L_N = -f^{ab} \partial_c N^c_{ab} - \mathcal{H}_g = L_f - \partial_c(f^{ab} N^c_{ab}) \tag{37.1}$$

with a specific choice:

$$\mathcal{H}_g = f^{ab}\left(N^c_{ad} N^d_{bc} - \frac{1}{3} N^c_{ac} N^d_{bd}\right) \tag{37.2}$$

These Lagrangians can be defined without the use of any metric (in contrast to the scalar/vector field theories) and the contractions in (37.1) and (37.2) are purely combinatorics operations involving only δ^i_j. In fact, instead of thinking of f^{ab} as a field in spacetime etc., we can think of these Lagrangians as describing the (abstract) dynamics of five 4×4 matrices (f, N^c), the elements of each of which depends on four parameters q^i. The matrix action is determined by integrating the L_f or L_N over d^4q. The variational principle using L_f (with f fixed at the boundary) or with L_N (with N^c fixed at the boundary) will lead to the dynamical equations for the matrices.

Incredibly enough, the resulting equations are identical to those of Einstein's gravity (without sources; it can be easily extended to take care of matter sources [1]) if we identify the arbitrary curved spacetime coordinates x^i with the parameters q^i and set $f^{ab} \equiv \sqrt{-g} g^{ab}$. The equations then imply the further identification:

$$N^a_{bc} = -\Gamma^a_{bc} + \frac{1}{2}(\Gamma^d_{bd} \delta^a_c + \Gamma^d_{cd} \delta^a_b), \tag{37.3}$$

(For a detailed proof, see [1].) Thus, gravitational dynamics in an *arbitrary* coordinate system with labels x^i and metric functions $g_{ab}(x^i)$ can be obtained from the dynamics of five matrices (f, N^c). I will now comment on several aspects of this description, which are novel from a conceptual point of view.

(1) We achieve "general covariance without general covariance" by not using the metric in the action. If we choose another set of coordinates in the spacetime, the metric functions will change. *But we do not have to change anything as regards the abstract matrix description; in particular we do not have to change d^4q by a Jacobian.* We simply identify the new coordinates with q^i and new $\sqrt{-g}g^{ab}$ with f^{ab} and quite trivially everything will work out. This separation of kinematics (coordinates, explicit form of metric functions, their change under coordinate transformation etc.) from the dynamics (described once and for all by the equations satisfied by the *matrices* (f, N^a)) is an attractive conceptual feature of this approach.

(2) One can easily introduce more 'pre-geometric' variables λ_A, say, into the matrix dynamics and treat $f^{ab}(q^i)$ as some kind of coarse-grained object after averaging over λ_A, like $f^{ab}(q^i) = \langle f^{ab}(q^i;\lambda_A)\rangle$ where the averaging is over the pre-geometric variables λ_A with some measure. This opens up interesting avenues for further work [4].

(3) Another surprise in the case of gravity is that L_N is numerically just $\sqrt{-g}R$ where R is the curvature scalar! That is, treating N^i_{jk} as a function of f^{ab} given by (37.3), one can show [1] that $L_N = L_f - \partial_c(f^{ab}N^c_{ab}) = -f^{ab}\partial_c N^c_{ab} - \mathscr{H}_g = \sqrt{-g}R$. The reason Einstein-Hilbert action leads to second order equations is now clear: It is a momentum space Lagrangian of a special structure.

(4) The structure of gravitational dynamics simplifies tremendously when we use the variables (f^{ab}, N^c_{ab}). For example, it is known that $\mathscr{L}_{EH} = \sqrt{-g}\,Q^{ab}_{cd}R^{cd}_{ab}$ (where $Q^{cd}_{ab} = (1/2)(\delta^c_a\delta^d_b - \delta^d_a\delta^c_b)$ is the determinant tensor) can be decomposed into a quadratic Lagrangian (called Γ^2 Lagrangian) and a divergence term as: $\mathscr{L}_{\rm EH} = \mathscr{L}_{\rm quad} + \mathscr{L}_{\rm sur}$, with $\mathscr{L}_{\rm quad} \equiv 2\sqrt{-g}\,Q_a^{\ bcd}\Gamma^a_{dk}\Gamma^k_{bc}$ and $\mathscr{L}_{\rm sur} \equiv 2\partial_c\left[\sqrt{-g}\,Q_a^{\ bcd}\Gamma^a_{bd}\right]$. These expressions take simpler forms when we use the variables f^{ab} and N^a_{bc} treating N^a_{bc} as a specified function of f^{ab} given by (37.3). We again have

$$\mathscr{L}_{EH} = \mathscr{L}_{quad} + \mathscr{L}_{sur}; \quad \mathscr{L}_{quad} = \frac{1}{2}N^a_{bc}\partial_a f^{bc}; \quad \mathscr{L}_{sur} = -\partial_c(f^{ab}N^c_{ab}) \tag{37.4}$$

(5) Though N^a_{bc} is not a tensor, its variation δN^a_{bc} is a tensor which can be related to the variation of Ricci tensor δR_{ab} in a remarkably simple form: $\delta R_{ab} = -\nabla_c(\delta N^c_{ab})$. This, in turn, allows us to formulate gravitational field equations in a novel manner. Suppose we have a source $\Im_{ab}(s) \equiv (T_{ab} - (1/2)g_{ab}T)$ where $T_{ab}(s)$ is the stress-tensor which depends on some parameter s (mass, charge, cosmological constant, anything....). For every value of s we can, in principle, solve the field equations and obtain $g_{ab}(s), f_{ab}(s), N^c_{ab}(s), \ldots$ etc. Let an overdot denote derivative of any of these wrt s like $\dot{\Im}_{ab} = d\Im_{ab}/ds$ etc. Then the gravitational field equations can be written simply as:

$$\nabla_c(\dot{N}^c_{ab}) = -8\pi\dot{\Im}_{ab} \tag{37.5}$$

which tells us how the geometry *changes* when we *change* the source. Given $\dot{\Im}_{ab}$, we can solve (37.5) for $\dot{N}^a_{bc}(s)$ and find $N^a_{bc}(s)$ by integrating $\dot{N}^a_{bc}(s)$ over s with flat spacetime as initial condition, say. We again see, from (37.5), the importance of N^a_{bc}; the field equations directly determine how it changes when the source changes.

(6) In the variation of the Einstein-Hilbert action we have:

$$\delta(\sqrt{-g}R) = \delta(R_{ab}f^{ab}) = R_{ab}\delta f^{ab} + f^{ab}\delta R_{ab} = R_{ab}\delta f^{ab} - \partial_c[f^{ik}\delta N^c_{ik}], \tag{37.6}$$

So that the boundary term which arises in the variation of action has the form:

$$(16\pi)\delta A_{sur} = -\int_{\mathscr{V}} d^4x\partial_c[f^{ab}\delta N^c_{ab}] = -\int_{\partial\mathscr{V}} d^3x\sqrt{h}n_c g^{ab}\delta N^c_{ab} \tag{37.7}$$

It turns out that this variation can be given a simple thermodynamic interpretation. Consider a null surface with temperature $T = \kappa/2\pi$ and entropy density $s = \sqrt{\sigma}/4$ attributed to it by local Rindler observers who perceive it as a horizon with κ being the surface gravity defined using a suitable null congruence and $\sqrt{\sigma}$ being the area element of the local horizon surface: Then, (see [1] for details) the following results hold:

- The boundary term in the action, evaluated over a null surface, can be interpreted as its heat content Ts (if we ignore unimportant contributions at the end points $\lambda = \lambda_1, \lambda_2$); that is:

$$\frac{1}{16\pi}\int d^3\Sigma_c(N^c_{ab}f^{ab}) = \int d\lambda d^2x \left(\frac{\kappa}{2\pi}\right)\left(\frac{\sqrt{\sigma}}{4}\right) = \int d\lambda d^2x\; Ts \tag{37.8}$$

- More remarkably, the variations $f\delta N$ and $N\delta f$ have corresponding thermodynamic interpretations for a class of variations which preserve the null surface:

$$\frac{1}{16\pi}\int d^3\Sigma_c(N^c_{ab}\delta f^{ab}) = \int d\lambda d^2x \left(\frac{\kappa}{2\pi}\right)\delta\left(\frac{\sqrt{\sigma}}{4}\right) = \int d\lambda d^2x\; T\delta s; \tag{37.9}$$

$$\frac{1}{16\pi}\int d^3\Sigma_c(f^{ab}\delta N^c_{ab}) = \int d\lambda d^2x \left(\frac{\sqrt{\sigma}}{4}\right)\delta\left(\frac{\kappa}{2\pi}\right) = \int d\lambda d^2x\; s\delta T \tag{37.10}$$

We see that (f, N^c) are not only *dynamically conjugate* variables but their variations $(\delta f, \delta N^c)$ exhibit *thermodynamic conjugacy* in terms of corresponding variations in $(\delta T, \delta S)$. For example, if we consider the gravitational action principle in a region of spacetime ('causal diamond') bound by null surfaces, then the boundary condition on the surface corresponds to isothermality ($\delta T = 0$) with respect to the local Rindler observers who perceive the null surfaces as a local Rindler horizons. It is obvious how thermodynamic considerations enhance our understanding of standard gravitational dynamics.

37.2 Conserved Currents from Geometry

I will next show how the variables (f, N^a) are closely related to conserved (Noether) currents for vector fields purely because of some—rather trivial—identities in differential geometry. This delinks the existence of conserved (Noether) currents from the invariance properties of any action principle or such dynamical considerations.

We begin by noting that the derivative $\nabla_k v_j$ of *any* vector field v^k can be decomposed into the anti-symmetric and symmetric parts by $\nabla^j v^k + \nabla^k v^j \equiv S^{jk}$ and $\nabla^j v^k - \nabla^k v^j \equiv J^{jk}$. The antisymmetric part J^{lm} immediately leads to a conserved current $J^i \equiv \nabla_k J^{ik}$; *in other words, from every vector field v^k in the spacetime we can trivially obtain a conserved current!* This has obviously nothing to do with diffeomorphism invariance or action principles.

If we substitute the decomposition, $\nabla^j v^k = (1/2)\left(J^{jk} + S^{jk}\right)$, in the standard identity $\nabla_k\left(\nabla^j v^k\right) - \nabla^j\left(\nabla_k v^k\right) = R^j_l v^l$, we get $J^j + \nabla_k(S^{jk} - g^{jk}S) = 2R^j_l v^l$. The first term is the conserved (Noether) current, while the second term can be related to the Lie derivative of N^c_{ab} (which is a tensor) by: $g^{ab}\pounds_v N^c_{ab} = -\nabla_b(S^{bc} - g^{cb}S) \equiv B^c[v]$. We thus get an explicit expression for the conserved (Noether) current in terms of the Lie derivative of N^c_{ab} as:

$$J^a[v] = \nabla_b J^{ab}[v] = 2R^a_b v^b + g^{ij}\pounds_v N^a_{ij} = 2R^a_b v^b + B^a[v] \tag{37.11}$$

An alternate way of obtaining the same result is as follows: From the Lie derivative of the connection $\pounds_v \Gamma^a_{bc} = \nabla_b\nabla_c v^a + R^a{}_{cmb}v^m$, one can obtain, on using (37.3) the relation

$$2Q^{adc}{}_e \pounds_v \Gamma^e_{cd} = g^{bc}\pounds_v N^a_{bc} = \nabla_b J^{ab} - 2R^a_b v^b = J^a - 2R^a_b v^b \tag{37.12}$$

which is the same as the one (37.11).

So there are two ways of obtaining the standard Noether current (either (i) from the antisymmetric part of $\nabla_i v_j$ using identity for $R^i_j v^j$ or (ii) by contracting the Lie derivative of Γ suitably) associated with a vector field v^i *without ever mentioning the action principle for gravity or its diffeomorphism invariance!*[2] This emphasizes the fact that (37.12), for example, is just a differential geometric identity and *has nothing to do with the gravitational dynamics*. Delinking the *form* of Noether current from any action principle allows us to reinterpret gravitational dynamics later on using Noether currents without introducing any circuitous reasoning.

[2]The connection with the usual approach can be made using the fact that the symmetric part S^{jk} is the change in the metric tensor g^{jk} under the diffeomorphism $x^a \to x^a + v^a$; i.e., $S^{ab} = \delta g^{ab} = -\pounds_v g^{ab}$; this identifies $B^c[v]$ with the boundary term in the variation of the Hilbert action under a diffeomorphism.

37.3 Noether Charge as Surface Heat Content

Consider a spacetime foliated by a series of space-like hypersurfaces defined by constant values for the scalar field $t(x)$. We define the unit normal to t = constant surfaces by $u_a = -N\nabla_a t$; if we choose our time coordinate to coincide with the hypersurface label, then $u_a = -N\delta^0_a$. we next define a 'time-development' vector ζ^a by the invariant condition $\zeta^a\nabla_a t = 1$. (In the preferred coordinate system, we can choose $\zeta^a = \delta^a_0$). In general, ζ^a and u^a will not be in the same direction and we have $\zeta^a = -(\zeta^b u_b)u^a + \beta^a$ where $u_a\beta^a = 0$ and $\beta^a = h^a_b\zeta^b$. In component form $\zeta^a = Nu^a + \beta^a$ with $\beta^a = (0, N^\alpha)$. This decomposition also introduces the vector $\xi^a \equiv -(\zeta^b u_b)u^a = Nu^a$ which will turn out to be of considerable importance in what follows. When we impose the coordinate condition such that $g_{0\alpha} = 0$ in a local region, we will have $\zeta^a = \xi^a$ and if the spacetime is static we can identify ξ^a with the time-like Killing vector.

Thus, in *any* spacetime, there exists a *natural* diffeomorphism, with vector field $\xi^a = Nu^a = -N^2\nabla_a t$. The Noether current for ξ^a leads to interesting consequences. To calculate this current, it is convenient to use an identity connecting Noether currents for two vector fields q^a and $v^a \equiv f(x)q^a$. It can be shown that [4]

$$q_a J^a[fq] - fq_a J^a[q] = \nabla_b\left[\left(q^a q^b - q^2 g^{ab}\right)\nabla_a f\right] \tag{37.13}$$

which is particularly useful if $q_a = \nabla_a\phi$ so that $J^{ab}[q] = 0$. If we use (37.13) with $q_a = -\xi_a/N^2,\ f = -N^2$ one can obtain a nice result for the Noether charge density:

$$u_a J^a(\xi) = 2D_\alpha(Na^\alpha) \tag{37.14}$$

where $a^i \equiv u^j\nabla_j u^i$ is the acceleration and $D_i a^i = D_\alpha a^\alpha = \nabla_i a^i - a^2$ where D_i is the covariant derivative on the t = constant surface. Integrating (37.14) over $\sqrt{h}d^3x$ to obtain the total Noether charge, we find that the flux of the acceleration is essentially the total Noether charge contained inside a volume. Noting that we have set $16\pi G = 1$ and adding the correct proportionality constant (with $G = 1$!), we get:

$$\int_{\mathscr{V}} \sqrt{h}d^3x u_a J^a[\xi] = \int_{\mathscr{V}} d\Sigma_a J^a[\xi] = \int_{\partial\mathscr{V}} \frac{\sqrt{\sigma}d^2x}{8\pi}(Nr_\alpha a^\alpha) \tag{37.15}$$

This result is valid for any region $\mathscr{V}$ in any spacetime. If we choose $\partial\mathscr{V}$ to be an N = constant 2-surface (within a t = constant 3-surface), then $r_i \propto D_i N \propto h^j_i\nabla_j N$ is in the direction of a_i and $r_\alpha a^\alpha = a$. So we can interpret $T = Nr_\alpha a^\alpha/2\pi = Na/2\pi$ as the (Tolman redshifted) Unruh temperature and $s = \sqrt{\sigma}/4$ as the entropy density. Then we get a remarkably simple interpretation of the Noether charge as the heat (enthalpy) content (TS) of the surface:

$$\int_{\mathscr{V}} \sqrt{h} d^3x u_a J^a[\xi] = \int_{\partial \mathscr{V}} d^2x \, Ts \tag{37.16}$$

So far we have only used results from quantum field theory in curved space and have not used gravitational field equations. I will now show how gravitational field equations acquire an interesting interpretation in this language.

37.4 Gravitational Dynamics and Holographic Equipartition

Since the Noether charge is related to TS while δN is related to δT, it is obvious that one can interpret the gravitational dynamics in terms of thermodynamic variables using Noether currents. To do this, we rewrite (37.11) using (37.14), integrate the result over a 3-dimensional region $\mathscr{R}$ with the measure $\sqrt{h}\, d^3x$, and introduce gravitational dynamics by substituting $R_{ab} = 8\pi \bar{T}_{ab}$ to get

$$\int_{\mathscr{R}} \frac{d^3x}{8\pi} \sqrt{h}\, u_a g^{ij} \pounds_\xi N^a_{ij} = \int_{\partial \mathscr{R}} d^2x \sqrt{\sigma} \left(\frac{N a_\alpha r^\alpha}{4\pi} \right) - \int_{\mathscr{R}} d^3x \, N \sqrt{h}\, (2u^a u^b \bar{T}_{ab}) \tag{37.17}$$

where r_α is the normal to the boundary of the 3-dimensional region. Again, taking $\partial\mathscr{R}$ to be a $N =$ constant surface, the first term on the right hand side can be interpreted as an integral over $dA(1/2)k_B T_{loc}$ where $T_{loc} \equiv (N a_\alpha r^\alpha/2\pi)$ is the local Unruh temperature. In the second term, we identify $2N \bar{T}_{ab} u^a u^b = (\rho + 3p)N$ as the Komar energy density. Thus the above result can be summarized in the form

$$\frac{1}{8\pi} \int d^3x \sqrt{h}\, u_a g^{ij} \pounds_\xi N^a_{ij} = \int_{\partial \mathscr{R}} d^2x \sqrt{\sigma} \left(\frac{1}{2} k_B T_{\text{loc}} \right) - \int_{\mathscr{R}} d^3x \sqrt{h}\, \rho_{\text{Komar}} \tag{37.18}$$

This result, again, has a remarkable physical meaning. If the spacetime is static and we choose the foliation such that ξ^a is a Killing vector, then $\pounds_\xi N^a_{ij} = 0$ and the left hand side vanishes. The equality of two terms on the right hand side can be thought of as representing the *holographic equipartition* [5–7] if we define the bulk and surface degrees of freedom along the following lines: We take the number of surface degrees of freedom to be:

$$N_{\text{sur}} \equiv \frac{A}{L_P^2} = \int_{\partial \mathscr{V}} \frac{\sqrt{\sigma}\, d^2x}{L_P^2} \tag{37.19}$$

We next define an *average* temperature T_{avg} to the surface $\partial\mathscr{V}$ by:

$$T_{\text{avg}} \equiv \frac{1}{A} \int_{\partial \mathscr{V}} \sqrt{\sigma}\, d^2x \, T_{\text{loc}} \tag{37.20}$$

and the number of bulk degrees of freedom by

$$N_{\rm bulk} \equiv \frac{|E|}{(1/2)k_B T_{\rm avg}} = \pm\frac{1}{(1/2)k_B T_{\rm avg}} \int_{\mathscr{V}} \sqrt{h} d^3x \; \rho_{Komar} \tag{37.21}$$

where E is the total Komar energy in the bulk region $\mathscr{V}$ contributing to gravity. *If* the energy E in the bulk region has reached equipartition with the average surface temperature, *then*, this is indeed the correct number for the bulk degrees of freedom. Our result in (37.18) then says that comoving observers in any static spacetimes will find:

$$N_{\rm sur} = N_{\rm bulk} \quad \text{(Holographic equipartion)} \tag{37.22}$$

What is more (37.18) suggests that *the discrepancy from holographic equipartition—resulting in a non-zero value for the right hand side—is what drives the dynamical evolution of the spacetime.* (Even in a static spacetime non-static observers e.g., freely falling observers will perceive a departure from holographic equipartition because (37.18) is generally covariant but foliation dependent through u_i.) A straightforward computation also shows that

$$\sqrt{h} u_a g^{ij} \pounds_\xi N^a_{ij} = -h_{ab} \pounds_\xi p^{ab}; \quad p^{ab} \equiv \sqrt{h}(K h^{ab} - K^{ab}) \tag{37.23}$$

So we can rewrite (37.18) as

$$-\int \frac{d^3x}{8\pi} h_{ab} \pounds_\xi p^{ab} = \frac{1}{2} k_B T_{\rm avg}(N_{\rm sur} - N_{\rm bulk}) \tag{37.24}$$

which clearly shows how the departure from holographic equipartition drives the evolution of geometry [4, 8], providing an interesting alternative description of spacetime dynamics.

37.5 Where Did We Go Wrong with Gravity?

The results described above, as well as several other pieces of work [2, 3] in this area, suggest an intriguing connection between (i) the thermodynamics attributed to null surfaces by local Rindler observers and (ii) dynamics of gravity. But in the description given above, I have only *re-interpreted* the standard gravitational dynamics—which, I think, is actually flawed—in the thermodynamic language. I will now show how it is possible to provide a completely independent, stand-alone, derivation of the *correct* gravitational field equations from a thermodynamic perspective.

To motivate this, I begin by stressing the *single most important fact* about gravitational dynamics which—purely by historical accident—is usually completely ignored: *Gravity does not couple to bulk energy density arising from the addition of a constant to matter Lagrangian.* Any attempt to describe gravity without incorporating this feature is bound to be wrong.

This fact, in turn, requires that gravitational field equations *must be* invariant under the symmetry transformation $L_{\rm matter} \rightarrow L_{\rm matter} + \text{constant}$, resulting in $T^a_b \rightarrow T^a_b + (\text{constant})\ \delta^a_b$. (The electroweak symmetry breaking, for example, is equivalent to shifting of the standard model Lagrangian by a large constant and we know that the evolution of the universe was unaffected by this transition.) It can be proved that this invariance cannot be maintained in any action principle satisfying the following two criteria: (a) The action is local and the Lagrangian is generally covariant. (b) The metric is varied in the action in an unrestricted form. Since we do not want to sacrifice condition (a), we must modify condition (b) to achieve our result.

A simple way to incorporate this symmetry is to demand that $[R_{ab} - 8\pi T_{ab}]\ell^a\ell^b = 0$ for *all* null vectors ℓ^a. This equation is clearly invariant under $T^a_b \rightarrow T^a_b + (\text{constant})\ \delta^a_b$ and has the first integral $R^a_b - 8\pi T^a_b = f(x)\delta^a_b$ which, on using $\nabla_a R^a_b = 0 = \nabla_a T^a_b$ leads to Einstein's equations *with an undetermined cosmological constant term arising as an integration constant.* It is the freedom to choose this constant that allows us to incorporate the shifting of the matter Lagrangian by a constant.

It is possible to derive the demand $[R_{ab} - 8\pi T_{ab}]\ell^a\ell^b = 0$ as a thermodynamic variational principle [9]. A useful version of such a variational principle can be obtained by noting that for a null congruence ℓ^a on a null surface we have: $R_{ab}\ell^a\ell^b = -\nabla_i(\theta\ell^i) - \mathcal{S}$ where $\nabla_i\ell^i = \theta + \kappa$ and we have defined

$$\mathcal{S} = [\nabla_i\ell^j\nabla_j\ell^i - (\nabla_i\ell^i)^2] \tag{37.25}$$

which we shall later identify with the heat (enthalpy) density associated with the null surface. While integrating over a null surface with the measure $d\lambda d^2x\sqrt{\sigma}$, we can ignore terms of the kind $\nabla_i(\phi\ell^i)$ (for any scalar ϕ) since they produce only boundary contributions. (This follows from the fact that, for any scalar ϕ, we have $\sqrt{\sigma}\nabla_i(\phi\ell^i) = d/d\lambda(\sqrt{\sigma}\,\phi)$). Ignoring the boundary contributions, we then have

$$\int_{\lambda_1}^{\lambda_2} \frac{d\lambda d^2x}{16\pi}\sqrt{\sigma}\ [-2R_{ab} + 16\pi T_{ab}]\,\ell^a\ell^b = \int_{\lambda_1}^{\lambda_2} \frac{d\lambda d^2x}{16\pi}\,[2\mathcal{S} + 16\pi T_{ab}\ell^a\ell^b] \tag{37.26}$$

where we have reintroduced $16\pi G$ with $G = 1$. It is easy to show that extremising this functional with respect to all ℓ^a, subject to the constraint $\ell^2 = 0$, will lead to $R^a_b - 8\pi T^a_b = f(x)\delta^a_b$ which, on using $\nabla_a R^a_b = 0 = \nabla_a T^a_b$ leads to Einstein's equations with an undetermined cosmological constant term arising as an integration constant. It follows that we now have an alternative variational principle (in which we vary ℓ^a) based on the expression:

$$Q \equiv \int_{\lambda_1}^{\lambda_2} \frac{d\lambda d^2x}{16\pi}\sqrt{\sigma}\,[2\mathcal{S} + 16\pi T_{ab}\ell^a\ell^b] \tag{37.27}$$

Since $T_{ab}\ell^a\ell^b$ can be thought of as the heat (enthalpy) density $TS/V = Ts$ of matter, we can again think of $(\mathscr{S}/8\pi)$ as essentially the heat density of the null surface.

It is possible to re-express this result in terms $\pounds_\ell N^c_{ab}$. Let ℓ_a be a null congruence defining a null surface which may not be affinely parametrized. If we take $\ell_a = A(x)\nabla_a B(x)$, then it is easy to prove that $\ell^i\nabla_i\ell_j = \kappa\ell_j$ where $\kappa = \nabla_i A\nabla^i B = \ell^a\nabla_a \ln A$. We can compute the Noether current for ℓ_a using our (37.13) and noting that the Noether current for $q_a = \ell_a/A$ is zero. Straightforward calculation gives $\ell_a J^a(\ell) = \nabla_b(\kappa\ell^b) - \kappa^2$. One can further show that $\nabla_a(\kappa\ell^a) - \kappa^2 = \mathscr{D}_a(\kappa\ell^a) + d\kappa/d\lambda$ where $\mathscr{D}_a$ is the covariant derivative operator on the 2 dimensional cross-section of the null surface. We thus find the Noether charge corresponding to the null congruence to be:

$$\ell_a J^a(\ell) = 2R_{ab}\ell^a\ell^b + \ell_a g^{ij}\pounds_\ell N^a_{ij} = \mathscr{D}_a(\kappa\ell^a) + \frac{d\kappa}{d\lambda} = \nabla_a(\kappa\ell^a) - \kappa^2 \quad (37.28)$$

If we integrate (37.28) over the null surface with the measure $d\lambda d^2x\sqrt{\sigma}$ and ignore the pure boundary contribution, we get:

$$\int d\lambda d^2x\sqrt{\sigma}\,\ell_a J^a(\ell) = \int d\lambda d^2x\sqrt{\sigma}\,\frac{d\kappa}{d\lambda} \quad (37.29)$$

Using these results we find that we can obtain the field equations by varying ℓ_a in the functional

$$\begin{aligned} Q &\equiv \int_{\lambda_1}^{\lambda_2} d\lambda d^2x\,\sqrt{\sigma}\left[\frac{1}{16\pi}\left(g^{ij}\ell_a\pounds_\ell N^a_{ij} - \pounds_\ell\kappa\right) + T_{ab}\ell^a\ell^b\right] \\ &= \int_{\lambda_1}^{\lambda_2} d\lambda d^2x\,\sqrt{\sigma}\left[\frac{1}{16\pi}\left(g^{ij}\ell_a\pounds_\ell N^a_{ij} - \frac{d\kappa}{d\lambda}\right) + T_{ab}\ell^a\ell^b\right] \end{aligned} \quad (37.30)$$

Since $T_{ab}\ell^a\ell^b$ can be thought of as the heat (enthalpy) density $TS/V = Ts$ of matter, we can think of the rest as essentially the heat density of the null surface. When ℓ_a is affinely parametrized with $\kappa = 0$ (which is a choice we can always make) then the variational principle can be based on the integral

$$Q_1 \equiv \int_{\lambda_1}^{\lambda_2} d\lambda d^2x\,\sqrt{\sigma}\left[\frac{1}{16\pi}g^{ij}\ell_a\pounds_\ell N^a_{ij} + T_{ab}\ell^a\ell^b\right] \quad (37.31)$$

This shows that the quantity $g^{ij}\ell_a\pounds_\ell N^a_{ij}$ plays a vital role even in the derivation of field equations from an alternative extremum principle. We see from (37.10) that the integral of this term on a null surface has a very simple physical meaning in terms of the heat content of the null surface. This term, therefore, can be interpreted in thermodynamic language. (The fact that $\kappa^2 \propto T^2_{loc}$ in (37.30) is proportional to the energy density of 2D thermal gas has interesting implications [4] which I will not discuss here).

37.6 The Value of the Cosmological Constant

Once we accept that gravitational field equations are invariant under $T^a_b \to T^a_b +$ (constant) δ^a_b, the solution will have an undetermined cosmological constant arising as integration constant. We then need a *new* physical principle to determine its value which I will now describe [10, 11].

Observations indicate that our universe can be characterized by (i) an early inflationary phase with approximately constant density ρ_{inf}, (ii) a phase dominated by radiation and matter, with $\rho = \rho_{eq}[x^{-4} + x^{-3}]$ where $x(t) \equiv a(t)/a_{\rm eq}$ and ρ_{eq} is another constant and (iii) an accelerated phase expansion at late time dominated by the energy of the cosmological constant ρ_Λ. Thus, there are three undetermined densities $[\rho_{inf}, \rho_{eq}, \rho_\Lambda]$ which describe the dynamics of our universe. It is generally believed that high energy physics will eventually determine ρ_{inf} and ρ_{eq} but we need a new principle to fix the value of ρ_Λ.

It turns out that, a universe with these three phases has a *conserved* quantity, viz. the number N of length scales which cross the Hubble radius during each of these phases. It can be shown that $N(a_2, a_1) = (2/3\pi)\ln(H_2a_2/H_1a_1)$ during $a_1 < a < a_2$. Any physical principle which fixes the value of N during the radiation-matter dominated phase, say, will relate ρ_Λ to $[\rho_{inf}, \rho_{eq}]$. We have given arguments elsewhere [10, 11] as to why we expect $N = 4\pi$ which leads to remarkable relation connecting the three densities:

$$\rho_\Lambda \approx \frac{4}{27}\frac{\rho_{\rm inf}^{3/2}}{\rho_{\rm eq}^{1/2}}\exp(-36\pi^2) \tag{37.32}$$

For the observed range of ρ_{eq}, and the range of inflationary energy scale $\rho_{\rm inf}^{1/4} =$ (1.084 – 1.241) $\times 10^{15}$ GeV, we get $\rho_\Lambda L_P^4 = (1.204 - 1.500) \times 10^{-123}$, which is consistent with observational results! I will conclude with two brief comments; more details can be found in [10, 11].

(a) This is a very novel approach to solving the cosmological constant problem based on a unified view of cosmic evolution, connecting all the three phases through (37.32). This is in contrast to standard cosmology where the three phases are put together in an unrelated, ad hoc, manner.

(b) It is difficult to incorporate $N = 4\pi$ into the standard cosmological paradigm. But it fits naturally into the concept of holographic equipartition [8]. I described earlier and writing the Friedmann equation in the form $dV/dt = L_P^2(N_{\rm sur} - N_{\rm bulk})$; with $V = (4\pi/3H^3)$, $N_{\rm sur} = (4\pi/L_P^2H^2)$, $T = H/2\pi$, $N_{\rm bulk} = -\varepsilon(2(\rho_{Komar}V/k_BT)$ and $\varepsilon = +1$ if $(\rho + 3p) < 0$ and $\varepsilon = -1$ if $(\rho + 3p) > 0$. In this approach, the assumption $N = 4\pi$ arises very naturally.

Acknowledgments I thank the organizers of KSM-1 for their excellent hospitality. My work is partially supported by J.C. Bose fellowship of DST, India.

References

1. K. Parattu, B. Ranjan Majhi, T. Padmanabhan. Phys. Rev. D **87**, 124011 (2013). arXiv:1303.1535
2. T. Padmanabhan, Rep. Prog. Phys. **73**, 046901 (2010). arXiv:0911.5004
3. T. Padmanabhan, J. Phys. Conf. Ser. **306**, 012001 (2011). arXiv:1012.4476 and references therein
4. T. Padmanabhan (2014) papers in preparation
5. T. Padmanabhan, Class. Quan. Grav. **21**, 4485 (2004). arXiv:gr-qc/0308070
6. T. Padmanabhan, Mod. Phys. Letts. A **25**, 1129–1136 (2010). arXiv:0912.3165
7. T. Padmanabhan, Phys. Rev. D **81**, 124040 (2010). arXiv:1003.5665
8. T. Padmanabhan, Res. Astro. Astrophys. **12**, 891 (2012). arXiv:1207.0505
9. T. Padmanabhan, A. Paranjape, Phys. Rev. D **75**, 064004, (2007). arXiv:gr-qc/0701003
10. H. Padmanabhan, T. Padmanabhan, Int. J. Mod. Phys. D **22**, 1342001 (2013). arXiv:1302.3226
11. T. Padmanabhan, arXiv:1210.4174

Chapter 38
Tunnelling Methods and Unruh-DeWitt Detectors in Curved Spacetimes

Giovanni Acquaviva

Abstract In this contribution we describe some interesting interplay between quantum theory, general relativity and thermodynamics. In order to highlight the connection between these theories, we describe two approaches that allow to calculate thermal features as perceived by different observers in curved spacetimes. the tunnelling method and the Unruh-DeWitt detector. In this context, the semi-classical tunnelling approach is applied to the issue of Hawking radiation and allows the calculation of the horizon temperature in a wide variety of scenarios. The Unruh-DeWitt model is instead a quantum field-theoretical approach that should give a more exact answer in terms of transition rates between energy levels of an idealized detector.

38.1 Introduction

Since the theoretical discovery of thermal radiance from black holes made by Hawking [3], the connections between gravitational systems (GR), quantum theory (QT) and thermodynamics (TD) have become undeniable matter of interest, mainly because of the longstanding purpose of syncretizing GR and QT in a coherent way. The problems that arise when trying to directly (and naively) quantize GR are just one indication that a new theory is desirable which includes both original theories in its very foundations. Leaving aside the task of formulating such a theory (a task that the present work does not deal with), one is lead to consider instead scenarios in which semi-classical or even quantized fields are coupled to a classical geometric background. This kind of approach could be regarded only as an *effective* version of the full (yet hypotetical) quantum-gravitational theory, while at the same time it could be useful in order to highlight a first degree of interaction between GR and QT.

Many approaches have been brought forward in order to reproduce and extend Hawking's result, which essentially states that

G. Acquaviva (✉)
Dipartimento di Fisica, Università degli Studi di Trento, Trento, Italy
e-mail: gioacqua@gmail.com

P. Nicolini et al. (eds.), *1st Karl Schwarzschild Meeting on Gravitational Physics*, Springer Proceedings in Physics 170, DOI 10.1007/978-3-319-20046-0_38

$$\frac{\Gamma_{em}}{\Gamma_{abs}} = \mathrm{e}^{-\omega/T_H}\,, \tag{38.1}$$

where the probability for a quantum of energy ω to be emitted from a gravitational horizon is found to have a Boltzmannian form: hence the outgoing radiation can be expressed as a thermal state and the associated temperature is given by the surface gravity evaluated on the horizon, i.e., $T_H = \kappa_H/2\pi$. One of the limitations of this result is the stationarity of the geometric background needed for its derivation.

In this contribution we present two approaches that are able to confirm this result as well as to extend its range of validity to more general scenarios. In Sect. 38.2 we sketch the so-called tunnelling method: this is essentially a variant of the original method used by Hawking and—equipped with suitable ingredients—can be applied to dynamical (both black hole's and cosmological) horizons. In Sect. 38.3 instead a model of Unruh-DeWitt detector is presented: this approach makes use of quantum field-theoretical tools in order to build a *quantum thermometer* endowed with a trajectory in a curved spacetime. In Sect. 38.4 we draw some conclusions.

38.2 Tunnelling Methods

It is well known that in a WKB approximation, a tunnelling probability rate is given by

$$\Gamma \propto \mathrm{e}^{-2\,Im(S)} \tag{38.2}$$

where S is the classical action along the trajectory. The presence of a non-vanishing imaginary contribution in S is thus linked to a non-zero probability associated to the tunnelling trajectory.

Historically one can actually identify two different (but equivalent in the stationary regime) approaches that fall under the denomination of tunnelling method:

- the *null-geodesic method*, introduced by Kraus, Parikh and Wilczek [6, 7] and
- the *Hamilton-Jacobi method*, formulated by Padmanabhan and collaborators [8]

Here we will focus on the latter.[1] The procedure to be followed is easily spelled in the following steps:

1. the action S of the massive tunnelling particle is assumed to satisfy the relativistic Hamilton-Jacobi equation

$$g^{\mu\nu}\,\partial_\mu S\,\partial_\nu S + m^2 = 0 \tag{38.3}$$

   where $g^{\mu\nu}$ is the inverse metric of the spacetime considered;

[1] See [10] for details of both methods and a discussion on why the H-J method is preferable.

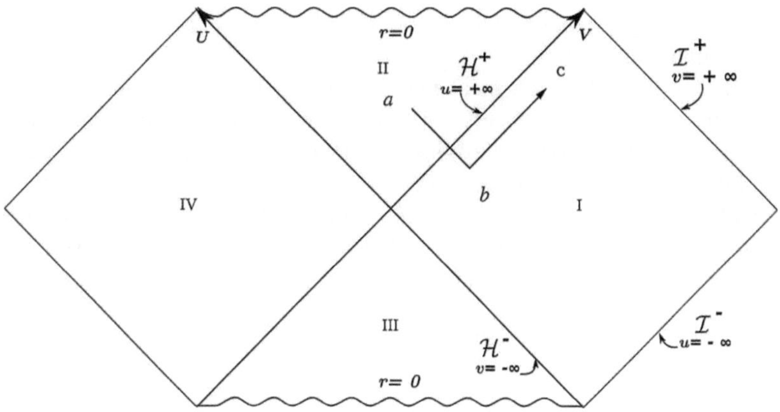

Fig. 38.1 Eternal Schwarzschild black hole. The oriented null path *abc* crosses the horizon $\mathscr{H}^+$ once, before escaping towards null infinity

2. then one makes use of the identity

$$S = \int_\gamma \partial_\mu S \, \mathrm{d}x^\mu \tag{38.4}$$

giving an ansatz for the form of the action based on the symmetries of the metric; the integration is carried along an oriented, piecewise null curve γ crossing the horizon at least in one point, as shown in Fig. 38.1 for an eternal black hole spacetime;
3. eventually one expresses the integrand of (38.4) through (38.3) and performs a near-horizon approximation, treating the divergence through Feynman's $-\Im\varepsilon$ prescription.[2]

The solution of the integral acquires a non-vanishing imaginary part,

$$Im(S) = \frac{\pi \, \omega}{\kappa} \tag{38.5}$$

where ω is the energy of the tunnelling particle and κ is the surface gravity of the horizon. Inserting the expression for the imaginary contribution in (38.2), and by comparison with the Boltzman distribution (38.1), one can identify the temperature associated to the emitted radiation:

$$T_H = \frac{\kappa}{2\pi} \tag{38.6}$$

[2]The choice of the sign in the prescription is related to the choice of positive-energy particles propagating towards infinity.

The result is easily extended to the whole Kerr-Newman class, for both scalar and fermionic tunnelling particles.

In order to treat the dynamical metric regime, the introduction of Kodama-Hayward theoretical results (see [4, 5]) has been shown to be determinant in the spherically symmetric case. Identifying the areal radius of the metric spheres with R, the definition of a Kodama vector is fairly simple: $K^{\mu} = \varepsilon^{\mu\nu}\partial_{\nu}R$. This vector field can be regarded as a sort of generalization of what the Killing vector field in stationary cases is: in fact it allows to define a particle's conserved energy through its flow and a surface gravity associated to the *trapping horizon*, respectively

$$\omega_H = -K^i\,\partial_i S \tag{38.7}$$

$$\kappa_H = \frac{1}{2\sqrt{-\gamma}}\partial_i\left(\sqrt{-\gamma}\,\gamma^{ij}\partial_j\,R\right)|_H \tag{38.8}$$

where γ_{ij} is the $(1+1)$ metric normal to the spheres of symmetry. Applying the same procedure as before with these quantities in mind, the invariant imaginary contribution $Im(S) = \pi\omega_H/\kappa_H$ is found and one can thus identify the temperature $T_H = \kappa_H/2\pi$ associated to the dynamical trapping horizon.

The main features regarding the tunnelling picture include:

- the possibility of proving the covariance of the method as well as the equivalence of its two aforementioned versions;
- the consistency of the result in a wide variety of situations: higher-dimensional solutions, Taub and Taub-NUT solutions, decay of unstable particles, emission from cosmological horizons and naked singularities.

38.3 Unruh-DeWitt Detectors

As it is often reasonably argued, in the context of diffeomorphism invariant theories like GR one can not ignore the fact that *particle* is an observer-dependent concept. As a consequence, the description of phenomena like the Hawking effect should be better treated from the point of view of specific observers.

The so-called Unruh-DeWitt detector [2, 9] can in fact be used in order to distinguish whether an observer is in a vacuum or not by means of three ingredients: (i) a real scalar field $\phi(x^0, \mathbf{x})$ coupled to (ii) a curved metric $g_{\mu\nu}$ and (iii) a localized two-level quantum system with energy eigenstates $\{|E_0\rangle, |E_1\rangle\}$ and endowed with a trajectory $\left(x^0(\tau), \mathbf{x}(\tau)\right)$. The idea is to evaluate the probability for the transition

$$|0\rangle|E_0\rangle(\tau_0) \quad\rightarrow\quad |\phi\rangle|E_1\rangle(\tau_1) \tag{38.9}$$

irrespective of the final state of the field $|\phi\rangle$, i.e., to witness whether or not the two-level system gets excited under the action of field modes while moving along a given trajectory in the curved metric. The interpretation of such excitations in terms of particles is not straightforward: nevertheless, in certain cases, one can associate a thermodynamical meaning to the result.

Leaving aside the details of the derivation (see [1]), the relevant quantity that allows such a calculations is the *transition rate* given in general form by

$$\dot{F}(E) = 2\int_0^{\Delta\tau} Re\left[e^{-\Im E s}\, W\left(x(\tau), x(\tau - s)\right)\right] ds \tag{38.10}$$

where E is the energy gap of the detector, $\Delta\tau$ is the duration of the detection (the temporal window during which the coupling is switched on, from τ to $\tau' = \tau - s$) and W is the Wightman function evaluated on the trajectory. The particular form of the Wightman function clearly depends on the metric of the spacetime and on the trajectory of the detector. Considering a conformally coupled scalar field in a 4-dimensional, conformally flat spacetime with conformal factor $a(x)$, it acquires the form

$$4\pi^2\, W(\tau, s) = \frac{1}{\sigma^2(\tau, s)} = \frac{1}{a(\tau)a(\tau - s) \times \sigma_M^2(\tau, s)} \tag{38.11}$$

where σ^2 (resp. σ_M^2) is the separation between $x^\mu(\tau)$ and $x^\mu(\tau')$ in the $g_{\mu\nu}$ metric (resp. in Minkowski spacetime). The problem of the divergence of W for $s \to 0$ can be dealt with in a comfortable way through a pole-subtraction scheme and the resulting expression of the transition rate for this case is

$$\begin{aligned}\dot{F} = &-\frac{E}{2\pi}\theta(-E) + \frac{1}{2\pi^2}\int_0^\infty \cos(E\, s)\left(\frac{1}{\sigma^2(\tau, s)} + \frac{1}{s^2}\right) ds\; + \\ &-\frac{1}{2\pi^2}\int_{\Delta\tau}^\infty \frac{\cos(E\, s)}{\sigma^2(\tau, s)}\, ds\end{aligned} \tag{38.12}$$

In this expression it is possible to identify three contributions: the first term regards the process of spontaneous emission; the second term is an asymptotic contribution, which evaluates the response of the detector during an infinite time of detection; the last term modifies the second one by taking into account the effects of a finite-time window of detection.

We now specialize to the case of two stationary spacetimes, de Sitter and Schwarzschild, considering a particular observer which is sitting at constant distance from the horizon (a so-called Kodama observer, in the terminology of [1]). In both cases the Wightman function has the form

$$W(s) = \frac{1}{4\pi^2}\frac{\kappa^2}{4V\,\sinh^2\left(\frac{\kappa}{2\sqrt{V}}\, s\right)} \tag{38.13}$$

where κ is the horizon's surface gravity and V is the Tolman redshift factor. Equation (38.13) is (i) stationary and (ii) periodic in imaginary time, two conditions that qualify the Wightman function as *thermal*. This denomination is justified: the contribution of the asymptotic term—calculated by summing the residues of the infinite poles in the lower half complex s-plane—is given by

$$\dot{F}_{\infty}(E) = \frac{1}{2\pi} \frac{E}{\exp\left(\frac{2\pi\sqrt{V}}{\kappa} E\right) - 1} \tag{38.14}$$

which is a Planckian distribution. Hence the detector registers a thermal radiation at temperature $T = \frac{\kappa}{2\pi\sqrt{V}}$. Some comments regarding this result are in order:

- the redshift factor that appears in the expression of the temperature is consistent with Tolman's theorem on thermodynamical equilibrium in a gravitational field: the product $T\sqrt{-g_{00}}$ is constant—in this case proportional to the horizon's surface gravity;
- thanks to the redshift factor, the expression for the temperature can be separated in two contributions $T^2 = T_A^2 + T_H^2$: the first one is proportional to the acceleration of the detector and hence related to its proper motion (Unruh effect), while the second one is purely related to the presence of the horizon.

Eventually, one can also evaluate the contribution to the transition rate coming from the finite-time term in (38.12). The result is an oscillating behaviour exponentially damped in time. No thermal contribution comes from this term, representing only a transient towards the equilibrium.

38.4 Conclusions

Thet wo methods presented in this contribution have been widely adopted throughout the literature in order to shed some light on possible thermodynamical interpretations of the gravitational field (or specific observables thereof). The results obtained in both frameworks are clearly consistent if one limits the analysis to stationary space-times. Moreover, the main feature arising from the Unruh-DeWitt method is the explicit observer-dependence of the result (through the Tolman factor), to be compared with the observer-independence of the tunnelling's outcome.

It is in the non-stationary cases that the two methods show less consistency (see [1]). The roots of this deviation in more general cases could be due (i) to an actually different interpretation of the results or perhaps (ii) to the role of the observer, which has a crucial weight in the Unruh-DeWitt picture. In the latter case, it should be possible to identify a particular observer whose measurement gives the tunnelling result as an outcome.

Acknowledgments GA would like to thank Luciano Vanzo, Sergio Zerbini and Roberto Di Criscienzo for valuable discussions and groupwork that lead to the results presented in this contribution.

References

1. G. Acquaviva, R. Di Criscienzo, M. Tolotti, L. Vanzo, S. Zerbini, Unruh-DeWitt detectors in spherically symmetric dynamical space-times. Int. J. Theor. Phys. **51**, 1555–1571 (2011)
2. B.S. DeWitt, Gravitational radiation, in *General Relativity: An Einstein Centenary Survey*, eds. by S.W. Hawking, W. Israel (Cambridge University Press, Cambridge, 1979), pp. 680–745
3. S.W. Hawking, Black hole explosions? Nature **248**, 30–31 (1974)
4. S.A. Hayward, Unified first law of black-hole dynamics and relativistic thermodynamics. Class. Quant. Grav. **15**, 3147–3162 (1998)
5. H. Kodama, Conserved energy flux for the spherically symmetric system and the backreaction problem in the black hole evaporation. Prog. Theor. Phys. **63**, 1217–1228 (1980)
6. P. Kraus, F. Wilczek, Self-interaction correction to black hole radiance. Nucl. Phys. B **13**, 403–420 (1995)
7. M.K. Parikh, F. Wilczek, Hawking radiation as tunneling. Phys. Rev. Lett. **85**, 5042–5045 (2000)
8. K. Srinivasan, T. Padmanabhan, Particle production and complex path analysis. Phys. Rev. D **60**, 24007 (1999)
9. W.G. Unruh, Notes on black-hole evaporation. Phys. Rev. D **14**, 870 (1976)
10. L. Vanzo, G. Acquaviva, R. Di Criscienzo, Tunnelling methods and Hawking's radiation: achievements and prospects. Class. Quant. Grav. **28**, 183001 (2011)

Chapter 39
Fermions on AdS

Victor E. Ambruş and Elizabeth Winstanley

Abstract We construct the Feynman propagator for Dirac fermions on anti-de Sitter space-time and present an analytic expression for the bi-spinor of parallel transport. We then renormalise the vacuum expectation value of the stress-energy tensor and end by analysing its renormalised expectation value at finite temperatures.

39.1 Introduction

Quantum field theory (QFT) on curved spaces (CS) is a semi-classical theory for the investigation of quantum effects in gravity. Due to its simplicity, the scalar field has been the main focus of QFT on CS. However, due to the fundamental difference between the quantum behaviour of fermions and bosons, it is important to also study fermionic fields. In this paper, we consider the propagation of Dirac fermions on the anti de Sitter (adS) background space-time, where the maximal symmetry can be used to obtain analytic results.

We start this paper by presenting in Sect. 39.2 an expression for the spinor parallel propagator [7]. Using results from geodesic theory [1, 7], an exact expression for the Feynman propagator is obtained in Sect. 39.3. Section 39.4 is devoted to Hadamard's regularisation method [8], while, in Sect. 39.5, the result for the renormalised vacuum expectation value (v.e.v.) of the stress-energy tensor (SET) is presented using two methods: the Schwinger-de Witt method [4] and the Hadamard method [6]. The exact form of the bi-spinor of parallel transport is then used in Sect. 39.6 to calculate the thermal expectation value (t.e.v.) of the SET for massless spinors. More details on the current work, as well as an extension to massive spinors, can be found in [2].

V.E. Ambruş (✉) · E. Winstanley
Consortium for Fundamental Physics, School of Mathematics and Statistics,
The University of Sheffield, Hicks Building, Hounsfield Road, Sheffield S3 7RH, UK
e-mail: app10vea@sheffield.ac.uk

E. Winstanley
e-mail: E.Winstanley@sheffield.ac.uk

P. Nicolini et al. (eds.), *1st Karl Schwarzschild Meeting on Gravitational Physics*,
Springer Proceedings in Physics 170, DOI 10.1007/978-3-319-20046-0_39

39.2 Geometric Structure of adS

Anti-de Sitter space-time (adS) is a vacuum solution of the Einstein equation with a negative cosmological constant, having the following line element:

$$ds^2 = \frac{1}{\cos^2 \omega r}\left[-dt^2 + dr^2 + \frac{\sin^2 \omega r}{\omega^2}\left(d\theta^2 + \sin^2\theta d\varphi^2\right)\right]. \tag{39.1}$$

The time coordinate t runs from $-\infty$ to ∞, thereby giving the covering space of adS. The radial coordinate r runs from 0 to the space-like boundary at $\pi/2\omega$, while θ and φ are the usual elevation and azimuthal angular coordinates. In the Cartesian gauge, the line element (39.1) admits the following natural frame [5]:

$$\omega^{\hat{t}} = \frac{\mathrm{d}t}{\cos\omega r}, \qquad \omega^{\hat{i}} = \frac{\mathrm{d}x^j}{\cos\omega r}\left[\frac{\sin\omega r}{\omega r}\left(\delta_{ij} - \frac{x^i x^j}{r^2}\right) + \frac{x^i x^j}{r^2}\right], \tag{39.2}$$

such that $\eta_{\hat{\alpha}\hat{\beta}}\omega^{\hat{\alpha}}_{\mu}\omega^{\hat{\beta}}_{\nu} = g_{\mu\nu}$, where $\eta_{\hat{\alpha}\hat{\beta}} = \mathrm{diag}(-1, 1, 1, 1)$ is the Minkowski metric.

A key role in the construction of the propagator of the Dirac field is played by the bi-spinor of parallel transport $\Lambda(x, x')$, which satisfies the parallel transport equation $n^\mu D_\mu \Lambda(x, x') = 0$ [7]. On adS, the explicit form of $\Lambda(x, x')$ is [2]:

$$\begin{aligned}\Lambda(x, x') = {} & \frac{\cos(\omega\Delta t/2)}{\cos(\omega s/2)\sqrt{\cos\omega r\cos\omega r'}}\left\{\cos\frac{\omega r}{2}\cos\frac{\omega r'}{2} + \frac{\mathbf{x}\cdot\hat{\gamma}}{r}\frac{\mathbf{x}'\cdot\hat{\gamma}}{r'}\sin\frac{\omega r}{2}\sin\frac{\omega r'}{2}\right. \\ & \left. - \gamma^{\hat{t}}\tan\frac{\omega\Delta t}{2}\left(\frac{\mathbf{x}\cdot\hat{\gamma}}{r}\sin\frac{\omega r}{2}\cos\frac{\omega r'}{2} - \frac{\mathbf{x}'\cdot\hat{\gamma}}{r'}\cos\frac{\omega r}{2}\sin\frac{\omega r'}{2}\right)\right\},\end{aligned} \tag{39.3}$$

where $\gamma^{\hat{\alpha}} = (\gamma^{\hat{t}}, \hat{\gamma})$ are the gamma matrices in the Dirac representation and s is the geodesic distance between x and x'.

39.3 Feynman Propagator on adS

The Feynman propagator $S_F(x, x')$ for a Dirac field of mass m can be defined as the solution of the inhomogeneous Dirac equation, with appropriate boundary conditions:

$$(i\not{D} - m)S_F(x, x') = (-g)^{-1/2}\delta^4(x - x'), \tag{39.4}$$

where D_μ denotes the spinor covariant derivative and g is the determinant of the background space-time metric. Due to the maximal symmetry of adS, the Feynman propagator can be written in the following form [7]:

$$S_F(x, x') = \left[\alpha_F(s) + \not{n}\,\beta_F(s)\right]\Lambda(x, x'). \tag{39.5}$$

The functions α_F and β_F can be determined using (39.4):

$$\alpha_F = \frac{\omega^3 k}{16\pi^2}\cos\frac{\omega s}{2}\left\{-\frac{1}{\sin^2\frac{\omega s}{2}} + 2(k^2-1)\ln\left|\sin\frac{\omega s}{2}\right| {}_2F_1\left(2+k, 2-k; 2; \sin^2\frac{\omega s}{2}\right)\right.$$
$$\left. + (k^2-1)\sum_{n=0}^{\infty}\frac{(2+k)_n(2-k)_n}{(2)_n n!}\left(\sin^2\frac{\omega s}{2}\right)^n \Psi_n\right\}, \tag{39.6}$$

$$\beta_F = \frac{i\omega^3}{16\pi^2}\sin\frac{\omega s}{2}\left\{\right.$$
$$\frac{1+k^2\sin^2(\omega s/2)}{[\sin(\omega s/2)]^4} - k^2(k^2-1)\ln\left|\sin\frac{\omega s}{2}\right| {}_2F_1\left(2+k, 2-k; 3; \sin^2\frac{\omega s}{2}\right)$$
$$\left. - \frac{k^2(k^2-1)}{2}\sum_{n=0}^{\infty}\frac{(2+k)_n(2-k)_n}{(3)_n n!}\left(\sin^2\frac{\omega s}{2}\right)^n\left(\Psi_n - \frac{1}{2+n}\right)\right\}, \tag{39.7}$$

where $a_n = \Gamma(a+n)/\Gamma(a)$ is the Pochhammer symbol, $\Gamma(z) = \int_0^\infty x^{z-1}e^{-x}dx$ is the gamma function, $k = m/\omega$,

$$\Psi_n = \psi(k+n+2) + \psi(k-n-1) - \psi(n+2) - \psi(n+1) \tag{39.8}$$

and $\psi(z) = \mathrm{d}\ln\Gamma(z)/\mathrm{d}z$ is the digamma function.

39.4 Hadamard Renormalisation

To regularise S_F, it is convenient to use the auxiliary propagator $\mathscr{G}_F$, defined by analogy to flat space-time [8]:

$$S_F(x, x') = (i\not{D} + m)\mathscr{G}_F. \tag{39.9}$$

On adS, $\mathscr{G}_F$ can be written using the bi-spinor of parallel transport:

$$\mathscr{G}_F(x, x') = \frac{\alpha_F}{m}\Lambda(x, x'), \tag{39.10}$$

where α_F is given in (39.6).

According to Hadamard's theorem, the divergent part $\mathscr{G}_H$ of $\mathscr{G}_F$ is state-independent, having the form [8]:

$$\mathscr{G}_H(x, x') = \frac{1}{8\pi^2}\left[\frac{u(x, x')}{\sigma} + v(x, x')\ln\mu^2\sigma\right], \tag{39.11}$$

where $u(x, x')$ and $v(x, x')$ are finite when x' approaches x, $\sigma = -s^2/2$ is Synge's world function and μ is an arbitrary mass scale. The functions u and v can be found by solving the inhomogeneous Dirac equation (39.4), requiring that the regularised auxiliary propagator $\mathscr{G}_F^{\text{reg}} \equiv \mathscr{G}_F - \mathscr{G}_H$ is finite in the coincidence limit:

$$u(x, x') = \sqrt{\Delta(x, x')}\Lambda(x, x'), \tag{39.12}$$

$$v(x, x') = \frac{\omega^2}{2}(k^2 - 1)\cos\frac{\omega s}{2}\,{}_2F_1\left(2 - k, 2 + k; 2; \sin^2\frac{\omega s}{2}\right)\Lambda(x, x'), \tag{39.13}$$

where the Van Vleck-Morette determinant $\Delta(x, x') = (\omega s/\sin \omega s)^3$ on adS.

39.5 Renormalised Vacuum Stress-Energy Tensor

To remove the traditional divergences of quantum field theory, we employ two regularisation methods: the Schwinger–de Witt method in Sect. 39.5.1 and the Hadamard method in Sect. 39.5.2. Due to the symmetries of adS, the regularised v.e.v. of the SET takes the form $\langle \mathsf{T}_{\mu\nu}\rangle_{\text{vac}}^{\text{reg}} = \frac{1}{4}\mathsf{T} g_{\mu\nu}$, where $\mathsf{T} = \mathsf{T}^{\mu}{}_{\mu}$ is its trace. The renormalisation process has the profound consequence of shifting T for the massless (hence, conformal) Dirac field to a finite value, referred to as the conformal anomaly.

39.5.1 Schwinger–de Witt Regularisation

By using the Schwinger–de Witt approach to investigate the singularity structure of the propagator of the Dirac field in the coincidence limit, Christensen [4] calculates a set of subtraction terms which only depend on the geometry of the background space-time, using the following formula:

$$\langle \mathsf{T}_{\mu\nu}\rangle = \lim_{x'\to x} \operatorname{tr}\left\{\frac{i}{2}\left[\gamma_{(\mu}D_{\nu)} - \gamma_{(\mu'}D_{\nu')}\right]S_F(x, x')\right\}\Lambda(x', x). \tag{39.14}$$

After subtracting Christensen's terms, we exactly recover the result obtained by Camporesi and Higuchi [3] using the Pauli-Villars regularisation method:

$$\langle \mathsf{T}\rangle_{\text{vac}}^{\text{SdW}} = -\frac{\omega^4}{4\pi^2}\left\{\frac{11}{60} + k - \frac{k^2}{6} - k^3 + 2k^2(k^2 - 1)\left[\ln\frac{\mu}{\omega} - \psi(k)\right]\right\}, \tag{39.15}$$

where μ is an arbitrary mass scale.

39.5.2 Hadamard Regularisation

The Hadamard theorem presented in Sect. 39.4 allows the renormalisation to be performed at the level of the propagator. To preserve the conservation of the SET, the following definition for the SET must be used [6]:

$$\langle \mathsf{T}_{\mu\nu}\rangle = \lim_{x'\to x} \operatorname{tr}\left(\frac{i}{2}\left[\gamma_{(\mu}D_{\nu)} - \gamma_{(\mu'}D_{\nu')}\right] + \frac{1}{6}g_{\mu\nu}\left[\frac{i}{2}(\not{D} - \not{D}') - m\right] S_F^{\text{reg}}(x,x')\right)$$
$$\times \Lambda(x',x), \tag{39.16}$$

where $S_F^{\text{reg}}(x,x') = (i\not{D}+m)(\mathscr{G}_F - \mathscr{G}_H)$ is the regularised propagator. The coefficient of $g_{\mu\nu}$ is proportional to the Lagrangian of the Dirac field and evaluates to zero when applied to a solution of (39.4). However, $S_F^{\text{reg}}(x,x')$ is not a solution of (39.4). The v.e.v. obtained from (39.16) matches perfectly the result obtained by Camporesi and Higuchi [3] using the zeta-function regularisation method (γ is Euler's constant):

$$\langle \mathsf{T}\rangle_{\text{vac}}^{\text{Had}} = -\frac{\omega^4}{4\pi^2}\left\{\frac{11}{60} + k - \frac{7k^2}{6} - k^3 + \frac{3k^4}{2} + 2k^2(k^2-1)\left[\ln\frac{\mu e^{-\gamma}\sqrt{2}}{\omega} - \psi(k)\right]\right\}. \tag{39.17}$$

Even though the results (39.15) and (39.17) are different for general values of the mass parameter k, they yield the same conformal anomaly. We would like to stress that the omission of the term proportional to $g_{\mu\nu}$ in (39.16) would increase the value of the conformal anomaly by a factor of 3.

39.6 Thermal Stress-Energy Tensor

The renormalised thermal expectation value (t.e.v.) of the SET can be written as:

$$\langle \mathsf{T}_{\mu\nu}\rangle_\beta^{\text{reg}} = \langle : \mathsf{T}_{\mu\nu} :\rangle_\beta + \langle \mathsf{T}_{\mu\nu}\rangle_{\text{vac}}^{\text{ren}}, \tag{39.18}$$

where $\beta = T^{-1}$ is the inverse temperature and the colons :: indicate that the operator enclosed is in normal order, i.e. with its v.e.v. subtracted. The bi-spinor of parallel transport can be used to show that

$$\langle : \mathsf{T}^{\mu}{}_{\nu} :\rangle_\beta = \operatorname{diag}(-\rho, \mathrm{p}, \mathrm{p}, \mathrm{p}), \tag{39.19}$$

where ρ is the energy density and p is the pressure. If $m = 0$, we have $p = \rho/3$ and:

$$\rho\rfloor_{m=0} = -\frac{3\omega^4}{4\pi^2}(\cos\omega r)^4 \sum_{j=1}^{\infty}(-1)^j \frac{\cosh(j\omega\beta/2)}{[\sinh(j\omega\beta/2)]^4}, \tag{39.20}$$

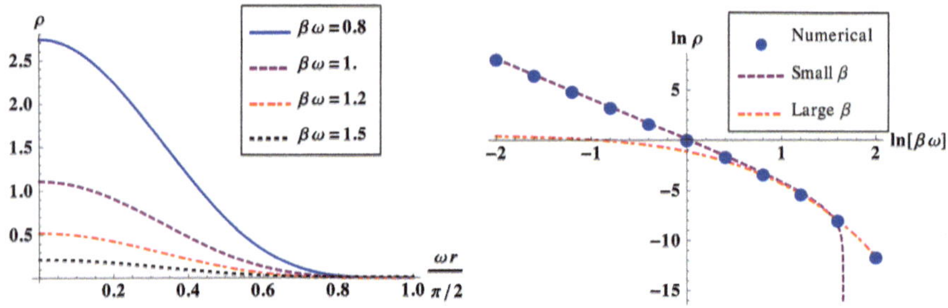

Fig. 39.1 **a** ρ between the origin ($r = 0$) and the boundary ($r\omega = \pi/2$) for $\beta\omega = 0.8$, 1.0, 1.2 and 1.4; **b** log-log plot of ρ in terms of $\beta\omega$; comparison with the asymptotic results in (39.21) and (39.22)

with the coordinate dependence fully contained in the $(\cos \omega r)^4$ prefactor. The first term in the sum over j is within 6 % of the sum, while the first two terms together are less than 1 % away, for all values of $\omega\beta$. The small and large $\omega\beta$ limits can be extracted:

$$\rho\rfloor_{m=0} = (\cos \omega r)^4 \left[\frac{7\pi^2}{60\beta^4} - \frac{\omega^2}{24\beta^2} + O(\omega^4) \right], \tag{39.21}$$

$$\rho\rfloor_{m=0} = \frac{6\omega^4}{\pi^2} \frac{(\cos \omega r)^4}{1 + e^{3\beta\omega/2}} \left[1 + 5e^{-\omega\beta} \frac{1 + e^{-3\omega\beta/2}}{1 + e^{-5\omega\beta/2}} + O(e^{-2\omega\beta}) \right]. \tag{39.22}$$

Figure 39.1 shows a graphical representation of the above results.

Acknowledgments This work is supported by the Lancaster-Manchester-Sheffield Consortium for Fundamental Physics under STFC grant ST/J000418/1, the School of Mathematics and Statistics at the University of Sheffield and European Cooperation in Science and Technology (COST) action MP0905 "Black Holes in a Violent Universe".

References

1. B. Allen, T. Jacobson, Vector two-point functions in maximally symmetric spaces. Commun. Math. Phys. **103**, 669–692 (1986)
2. V.E. Ambruş, E. Winstanley, Fermions on adS. Paper in preparation
3. R. Camporesi, A. Higuchi, Stress-energy tensors in anti-de Sitter spacetime. Phys. Rev. D **45**, 3591–3603 (1992)
4. S.M. Christensen, Regularization, renormalization, and covariant geodesic point separation. Phys. Rev. D **17**, 946–963 (1978)
5. I. Cotăescu, Dirac fermions in de Sitter and anti-de Sitter backgrounds. Rom. J. Phys. **52**, 895–940 (2007)
6. C. Dappiaggi, T.-P. Hack, N. Pinamonti, The extended algebra of observables for Dirac fields and the trace anomaly of their stress-energy tensor. Rev. Math. Phys. **21**, 1241–1312 (2009)
7. W. Mück, Spinor parallel propagator and Green's function in maximally symmetric spaces. J. Phys. A **33**, 3021–3026 (2000)
8. A.-H. Najmi, A.C. Ottewill, Quantum states and the Hadamard form. II. Energy minimization for spin-1/2 fields. Phys. Rev. D **30**, 2573–2578 (1984)

Chapter 40
Study on Rescaling Extrinsic Curvature in Gravitational Initial Data

Shan Bai and Niall Ó Murchadha

Abstract Vacuum solutions to the Einstein equations can be viewed as the interplay between the geometry and the gravitational wave energy content. The constraints on initial data reflect this interaction. We assume we are looking at cosmological solutions to the Einstein equations so we assume that the 3-space is compact, without boundary. In this article we investigate, using both analytic and numerical techniques, what happens when the extrinsic curvature is increased while the background geometry is held fixed. This is equivalent to trying to magnify the local gravitational wave kinetic energy on an unchanged background. We find that the physical intrinsic curvature does not blow up. Rather the local volume of space expands to accommodate this attempt to increase the kinetic energy.

40.1 Introduction

Initial data for the Einstein equations consists of two parts: the first part is a manifold equipped with a Riemannian 3-metric g_{ij}, and the second is a symmetric tensor K^{ij} on the same manifold. K^{ij} is the extrinsic curvature of the 3-slice, i.e., the time derivative of the 3-geometry. The metric and extrinsic curvature should satisfy the Hamiltonian constraint and the Momentum constraint. A comprehensive discussion of the constraints can be found in [1]. Interesting physics tends to occur at the boundaries of the space of free data: one gets at the very least some insight into the limitations of the conformal method. Parts of the boundary of the free data are easily accessible. We can scale any one of the three parts by multiplying it by a constant and letting the constant become either large or small. Such an action can be done easily in the Maxwell theory to increase the electromagnetic energy density $(E^2 + B^2)$ without limit. But it is not clear what happens with the Einstein equations because

S. Bai (✉)
Theoretisch-Physikalisches Institut, Friedrich-Schiller-Universität Jena, Jena, Germany
e-mail: shan.bai@kit.edu

N. Ó Murchadha
Physics Department, University College Cork, Cork, Ireland
e-mail: n.omurchadha@ucc.ie

P. Nicolini et al. (eds.), *1st Karl Schwarzschild Meeting on Gravitational Physics*,
Springer Proceedings in Physics 170, DOI 10.1007/978-3-319-20046-0_40

the gravitational wave energy interacts in a very nonlinear way with the geometry. One natural question about the recalling is what effect has this on the physics? For example, does the solution just cease to exist, does the volume of the spacetime blow up (or shrink to zero), do apparent horizons (which may be interpreted as cosmological horizons) appear?

In this article, we discuss one such rescaling of the gravitational free data. The work is a combination of analytic and numerical works. In Sect. 40.2 we describe the conformal method of solving the constraints. In Sect. 40.3 we show that if the extrinsic curvature vanishes nowhere on a compact manifold and we increase it, the conformal factor uniformly blows up. However, when the extrinsic curvature vanishes somewhere, the analysis in Sect. 40.3 is no longer valid. To investigate this special case, we revert to a spherically symmetric toy model, deriving some analytic results and showing the relevant numerical work in Sect. 40.4. Here we supply strong evidence that we do not get blow-up in regions of vanishing extrinsic curvature. We conclude with a summary and an outline of future work.

40.2 Solving the Einstein Constraints

Let us start on the Hamiltonian constraint,

$$R - K_{ij}K^{ij} + K^2 = 0\,, \tag{40.1}$$

where R is the 3-scalar curvature of g_{ij} and K is the trace of K^{ij}, i.e., $K = g_{ij}K^{ij}$. The terminology and notation comes from [2]. The standard way of generating solutions is by means of a conformal transformation which means $\bar{g}_{ij} = \phi^4 g_{ij}$ where ϕ is any function positive function. On any given manifold, it is easy to construct tracefree and divergencefree (TT) tensors. For example, any K^{ij} is the sum of a constant trace and a trace-free part, i.e., $K^{ij} = K^{ij}_{TT} + \frac{1}{3}Kg^{ij}$, where the constant K is conformal invariant and satisfies the momentum constraint. K^{ij}_{TT} is TT tensor [3]. TT tensors are conformally covariant. Now the (40.1) reduces to the famous Lichnerowicz-York equation [3, 4]

$$\nabla^2\phi - \frac{R}{8}\phi + \frac{1}{8}A^2\phi^{-7} - \frac{K^2}{12}\phi^5 = 0 \tag{40.2}$$

where $A^2 = K^{TT}_{ij}K^{ij}_{TT}$. This equation is very well behaved [5]. In this article we focus on the situation where the topology of the 3-manifold is compact and without boundary. It can be shown that (40.2) has a unique positive solution if $K \neq 0$ and if K^{ij}_{TT} is not identically zero [5]. Let us remind the reader that K is a constant, while K^{ij}_{TT} is a function. In the special cases, where either $K = 0$ or $K^{ij}_{TT} \equiv 0$, we have an extra condition related to the sign of the scalar curvature. However, in the general case no such restriction applies. This existence result does not depend either on the metric, other than it be uniformly elliptic, or on the topology of the 3-manifold.

To recapitulate: we start with a triplet, i.e., the free data, (g_{ij}, K^{ij}_{TT}, K), and construct a new set $(\bar{g}_{ij}, \bar{K}^{ij}_{TT}, K) = (\phi^4 g_{ij}, \phi^{-10} K^{ij}_{TT}, K)$ that satisfies the constraints.

40.3 Harnack Type Inequality for the Conformal Factor

In this section we consider such a rescaling. We pick one triplet (g_{ij}, K^{ij}_{TT}, K), and use it to construct a family of free data of the form $(g_{ij}, \alpha^{12} K^{ij}_{TT}, K)$, where α is a running parameter. We also wish to show that ϕ scales linearly with α as α becomes large, so we write $\tilde{\phi} = \phi/\alpha$ and then (40.2) becomes

$$\nabla^2\tilde{\phi} - \frac{R}{8}\tilde{\phi} + \alpha^4(\frac{1}{8}A^2\tilde{\phi}^{-7} - \frac{K^2}{12}\tilde{\phi}^5) = 0 . \tag{40.3}$$

We wish to solve the family of equations on a compact manifold without boundary. It turns out that the sign of the scalar curvature plays a minor role in the behaviour of the solutions. We can always set the scalar curvature to a constant value because of the Yamabe theorem [6], which tells us that any Riemannian metric on a compact manifold can be conformally transformed to a metric of constant scalar curvature. The key quantity is the Yamabe number

$$Y = \inf \frac{\int[(\nabla\theta)^2 + \frac{1}{8}R\theta^2]\mathrm{d}v}{[\int \theta^6 \mathrm{d}v]^{1/3}} , \tag{40.4}$$

where the infimum is taken over all smooth functions θ. The sign of the Yamabe number fixes the sign of the constant scalar curvature. Since (40.3) is conformally covariant, and since conformal transformations form a group under composition, we can set R to a constant value without losing any generality. However, we do need to handle the three separate cases, $Y > 0, R > 0$; $Y < 0, R < 0$ and $Y = 0, R = 0$ independently. In each case we will set the value of $K^2 = 9$. This choice does not change in any fundamental way the behaviour of the solution. To illuminate our treatments, we take the case of $Y > 0, R > 0$ as an example.

We assume that we are in the positive Yamabe class and set the scalar curvature $R = +24$; the specific number can be chosen freely. Now (40.3) reduces to

$$\nabla^2\tilde{\phi} - 3\tilde{\phi} + \alpha^4(\frac{1}{8}A^2\tilde{\phi}^{-7} - \frac{3}{4}\tilde{\phi}^5) = 0 \tag{40.5}$$

(40.5), because it is just a rescaled version of the original Lichnerowicz-York equation (40.2), which is extremely well behaved, has a regular positive solution. Let us look at what happens at the maximum of $\tilde{\phi}$, which we shall assume occurs at a point $r = r_{max}$. The first two terms in (40.5) will be negative at $r = r_{max}$ so we get

$$\left[\frac{1}{8}A^2\tilde{\phi}^{-7} - \frac{3}{4}\tilde{\phi}^5\right]_{r_{max}} > 0\,. \tag{40.6}$$

This becomes

$$[\max\tilde{\phi}]^{12} < \frac{1}{6}A^2|_{r_{max}} \leq \frac{1}{6}\max A^2\,. \tag{40.7}$$

We get a uniform lower bound by looking at (40.5) when $\tilde{\phi}$ is a minimum, which we shall assume occurs at $r = r_{min}$. At $r = r_{min}$ we get

$$\frac{1}{6}\min A^2 \leq \frac{1}{6}A^2|_{r_{min}} \leq 4\frac{[\min\tilde{\phi}]^8}{\alpha^4} + [\min\tilde{\phi}]^{12}\,. \tag{40.8}$$

Using the bounds on both $\min\phi$ and $\max\phi$, we have shown that there exists a universal constant C_0 independent of α such that

$$\frac{\min\phi}{\max\phi} > C_0\left[\frac{\min A^2}{\max A^2}\right]^{1/3}\,. \tag{40.9}$$

The maximum and minimum of ϕ both increase together proportional to α so that their ratio remains bounded independent of α. This can be regarded as a version of the Harnack inequality [7] for the non-linear equation (40.5).

The Harnack inequalities are clearly only valid when $\min(A^2) = \min(K^{ij}_{TT}K^{TT}_{ij})$ $\neq 0$. For the case that A^2 vanishes in a region, we will prove that the ϕ has minimum varying α and show the order of α proportional to the minimum ϕ numerically in the next section. Meanwhile, in the region where $A^2 \neq 0$, we can show numerically that ϕ has the standard linear scaling with α which is in consistent with the above statement.

40.4 Spherical Toy Model: Analytical Results and Numerical Results

When dealing with spherical symmetry, we are free to take advantage of the fact that a round 3-sphere can be decompactified to flat 3-space, and that conformal transformations form a group under composition of which the conformal factor is $\theta = \sqrt{b}/\sqrt{b^2 + r^2}$ for any $b > 0$. Denote the new conformed solution to be $\hat{\phi} = \theta^{-1}\phi$. We note that ϕ will be finite at the 'point at infinity' in the compact manifold, while $\hat{\phi} \approx \sqrt{b}/r$ at the corresponding infinity in R^3. A spherical region around the north pole corresponds to a disc $0 \leq r < r_1$ on which $A^2 = 0$. We can write down the conformed solution that $\hat{\phi} = \sqrt{a}/\sqrt{a^2 - r^2}$, where a is a parameter. These functions blow up at $r = a$, and since the total solution is regular, the blow-up must occur outside the range of validity of these functions. So it gives a lower bound for a, i.e., $a \geq r_1$. Clearly, the minimum value of $\hat{\phi}$ for these special solutions occurs at the origin,

$$\min \hat{\phi} = \hat{\phi}(r = 0) = \sqrt{\frac{1}{a}} \leq \sqrt{\frac{1}{r_1}} . \tag{40.10}$$

Therefore $\min \phi$ does not blow up like α but reaches some limit, while $\max \phi$ becomes unboundedly large. A similar argument holds when A^2 vanishes near the south pole. We can repeat this argument when A^2 vanishes on some belt $r_1 < r < r_2$. Fix the location of the minimum, in this case in the interval (r_1, r_2), and fix the value of $\hat{\phi}$ at the minimum. This uniquely determines the solution. The solution is 'U' shaped, blowing up twice at r_A and r_B. The bigger the $\min \hat{\phi}$, the narrower the 'U', i.e., $(r_B - r_A)(\min \hat{\phi})^2$ is bounded [8] where $r_A < r_1, r_B > r_2$ and $r_B - r_A > r_2 - r_1$, and so the value of $\min \hat{\phi}$ is bounded above.

In the numerical calculation, we assume A^2 to be vanishing in (r_1, r_2) where r_1 or r_2 could be zero. We solve the ordinary differential equation for a range of parameters α by pseudo-spectral method. We can observe that the value of $\hat{\phi}$, on the support of A^2, tends to a stationary limit, while $\hat{\phi}$ collapses off the support of A^2. This shows that ϕ scales linearly with α, on the support of A^2, while in the region where A^2 is zero, $\hat{\phi}$ continues to diminish so that ϕ approaches a stationary value. In the above analysis, we showed that $\min \phi = \alpha \min \hat{\phi}$ increases with α but approaches some fixed upper bound. The bound could be figured out here numerically in the Fig. 40.1.

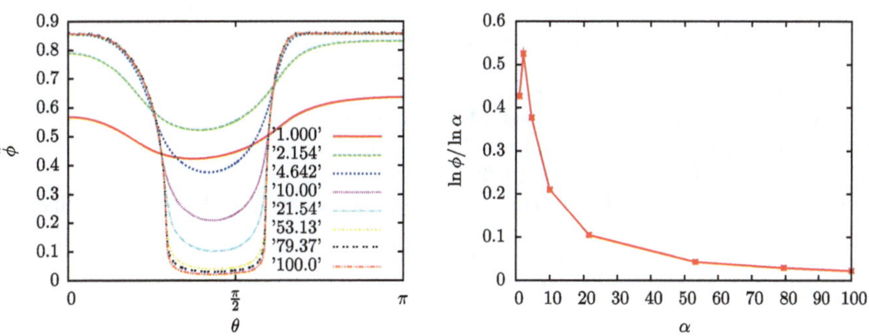

Fig. 40.1 The first figure is $\hat{\phi}$ on S^3. The parameters used are $\alpha = 1.000, 2.154, 4.642, 10.00, 21.54, 53.13, 79.37$, and 100.0. These correspond to the *red*, *green*, *blue*, *pink*, *light blue*, *yellow*, *dark blue*, and *mauve lines* respectively. The minimum ϕ occurs around $\theta = 1.35$. The second figure is to show the asymptotic behavior of the min ϕ when α is larger. It is shown that $\min \phi \sim \alpha^{0.03}$

40.5 Conclusions

We have shown that if we scale $K^{ij}_{TT}K^{TT}_{ij}$ by α^{12}, we find that the conformal factor, in general, scales like α. However, the physical $\bar{K}^{ij}_{TT}\bar{K}^{TT}_{ij} = \phi^{-12}\alpha^{12}K^{ij}_{TT}K^{TT}_{ij} = \tilde{\phi}^{-12}K^{ij}_{TT}K^{TT}_{ij}$, and as α becomes larger and larger $\tilde{\phi}$ remains finite. This means that $\bar{K}^{ij}_{TT}\bar{K}^{TT}_{ij}$ remains finite. Hence the velocity part of the gravitational wave energy density remains bounded even though the corresponding 'free' data blows up. On the other hand, the volume of space becomes unboundedly large, because $\sqrt{\bar{g}} = \phi^6\sqrt{g} = \alpha^6\tilde{\phi}^6\sqrt{g}$ blows up. Therefore the total gravitational wave energy in a coordinate sphere becomes larger and larger while the local energy density remains bounded.

It would be interesting to repeat this analysis in the asymptotically flat case. We would probably want to work with maximal initial data, i.e., $K = 0$, and just have a metric and a TT tensor as free data. It is clear that one can change the metric so that the ADM mass becomes unboundedly large and trapped surfaces appear [9]. What happens if we blow-up the extrinsic curvature on a fixed background metric? Will we get the same behaviour? Preliminary investigations indicate that we do: the ADM mass diverges and trapped surfaces appear. We intend to investigate this further.

Acknowledgments SBai and NÓM were supported by Grant 07/RFP/PHYF148 from Science Foundation Ireland

References

1. Y. Choquet-Bruhat, *General Relativity and the Einstein Equations* (OUP, Oxford, 2009)
2. R. Arnowitt, S. Deser, C. Misner, *Gravitation: An Introduction to Current Research* (Wiley, New York, 1962), pp. 227–265
3. J.W. York, doi:10.1063/1.1666338
4. A. Lichnerowicz, J. Math. Pures Appl. **23**, 3763 (1944)
5. N. Ó Murchadha, J.W. York, doi:10.1063/1.1666225
6. H. Yamabe, Osaka Math J. **12**, 21 (1960)
7. D. Gilbarg, N. Trudinger, *Elliptic Partial Differential Equations of the Second Order* (Springer, Berlin, 1983)
8. R. Conboye, N. Ó Murchadha, arXiv:1306.1363
9. R. Beig, N. Ó Murchadha, doi:10.1103/PhysRevLett.66.2421

Chapter 41
Massive Gravities

Dennis D. Dietrich

Abstract Massive gravities (here exclusively in 4 dimensions) could provide solutions for longstanding challenges, e.g. the vacuum- and dark-energy problems. They are, however, facing a multitude of constraints, which allow us to select viable approaches. Models with variable mass are the most promising candidates.

41.1 Introduction

Motivations for studying massive gravities are the vacuum- and dark-energy problems: Massive gravities can degravitate [1] the huge spacetime homogeneous energy-momentum source represented by the vacuum energy, which is predicted by quantum field theory. For intact equivalence principle it would imply a strongly curved universe in conflict with its observed approximate flatness. Moreover, massive gravities can provide self-accelerating solutions in the absence of dark energy.

41.1.1 Degravitation

In Maxwell theory with a constant source J^μ,

$$\mathscr{L}_\mathrm{M} = -\frac{1}{4}F_{\mu\nu}F^{\mu\nu} - A_\mu J^\mu \qquad \Rightarrow \qquad \Gamma^{\mu\kappa\lambda\nu}\partial_\kappa\partial_\lambda A_\nu = J^\mu \stackrel{!}{=} \mathrm{const.}$$

D.D. Dietrich (✉)
Institut für Theoretische Physik, Johann Wolfgang Goethe-Universität Frankfurt am Main, Max-von-Laue Str. 1, D-60438 Frankfurt am Main, Germany
e-mail: dietrich@th.physik.uni-frankfurt.de

D.D. Dietrich
Arnold-Sommerfeld-Zentrum für Theoretische Physik, Ludwig-Maximilians-Universität München, Theresienstr. 37, D-80333 Munich, Germany

P. Nicolini et al. (eds.), *1st Karl Schwarzschild Meeting on Gravitational Physics*, Springer Proceedings in Physics 170, DOI 10.1007/978-3-319-20046-0_41

we find that A_μ ($F_{\mu\nu}$) grows quadratically (linearly). To the contrary, Proca theory,

$$\mathscr{L}_{\rm P} = -\frac{1}{4}F_{\mu\nu}F^{\mu\nu} - \frac{m^2}{2}A_\mu A^\mu - A_\mu J^\mu \Rightarrow \Gamma^{\mu\kappa\lambda\nu}\partial_\kappa\partial_\lambda A_\nu - m^2 A^\mu = J^\mu \stackrel{!}{=} \text{const.},$$

allows for A_μ = constant and $F_{\mu\nu} = 0$. Analogously, when we expand general relativity (GR) with a cosmological constant (CC) Λ around a flat background,

$$\mathscr{L}_{\rm EH} = \sqrt{-g}(R + 2\Lambda) \qquad \Rightarrow \qquad G_{\mu\nu} = \Lambda g_{\mu\nu},$$

$$g_{\mu\nu} = \eta_{\mu\nu} + h_{\mu\nu} \quad \text{such that} \quad \Lambda = O(h_{\mu\nu}) \quad \text{and} \quad G_{\mu\nu}(\eta_{\mu\nu}) = 0,$$

an $h_{\mu\nu}$ growing quadratically ad infinitum solves the linearised equations of motion

$$\mathscr{E}_{\mu\nu}^{\kappa\lambda\alpha\beta}\partial_\alpha\partial_\beta h_{\kappa\lambda} = \Lambda\eta_{\mu\nu}.$$

At variance, in Fierz-Pauli theory (FP) [2] around a Minkowski background,

$$\mathscr{L}_{\rm FP} = \sqrt{-g}(R + 2\Lambda)|_{\rm quad} - \sqrt{-\eta}\frac{m^2}{4}(\eta^{\mu\kappa}\eta^{\nu\lambda} - \eta^{\mu\nu}\eta^{\kappa\lambda})h_{\mu\nu}h_{\kappa\lambda}$$

$$\Rightarrow \quad \mathscr{E}_{\mu\nu}^{\kappa\lambda\alpha\beta}\partial_\alpha\partial_\beta h_{\kappa\lambda} - m^2(\delta_\mu^\kappa\delta_\nu^\lambda - \eta_{\mu\nu}\eta^{\kappa\lambda})h_{\kappa\lambda} = \Lambda\eta_{\mu\nu},$$

$h_{\mu\nu} \propto \eta_{\mu\nu}$ and thus flat space is a solution. In both cases the infrared modification due to the mass screens the constant source and makes it physically ineffective.

41.1.2 Self-Acceleration

Matter slows down the expansion of the universe at late times. If the matter's long-range effect is lessened by a graviton mass this can mimic the opposite effect arising from a CC. Indeed, the difference between the potentials in the Newtonian limits of GR and in FP, respectively, is as between the Coulomb and the Yukawa potentials,

$$V_{\rm C} = \frac{\#}{r} \quad \rightarrow \quad V_{\rm Y} = \#\frac{e^{-mr}}{r} \qquad \text{versus} \qquad V_{\rm N} = \frac{\sharp}{r} \quad \rightarrow \quad V_{\rm FP} = \frac{4}{3}\sharp\frac{e^{-mr}}{r}.$$

Interestingly, it is this aspect of degravitation and self-acceleration that originally motivated Einstein [3] to introduce the cosmological constant.

41.2 Challenges

Challenges for massive gravities arise at the linear and the non-linear levels.

41.2.1 Linear Level

In the limit $m \to 0$ the $m = 0$ result is not recovered, as the scalar degree of freedom (dof) in the decomposition over a flat background,

$$h_{\mu\nu} = h^{TT}_{\mu\nu} + \partial_{(\mu} V^T_{\nu)} + \partial_\mu \partial_\nu \phi,$$

does not decouple and, e.g. gives rise to the factor 4/3 above. ($h^{TT}_{\mu\nu}$ is transverse and traceless, V^T_ν traceless.) This is known as van Dam-Veltman-Zakharov (vDVZ) discontinuity [4, 5], which need not be present over a curved background [6, 7]. The factor affects the gravitational interaction of macroscopic bodies, but not the bending of light in a gravitational field, which makes it detectable.

Furthermore, there can be unitarity violations and classical instabilities for massive graviton theories formulated relative to curved backgrounds, $G_{\mu\nu}(\eta_{\mu\nu}) \neq 0$,

$$h_{\mu\nu} = h^{TT}_{\mu\nu} + \nabla_{(\mu} V^T_{\nu)} + \nabla_{(\mu} \partial_{\nu)} \phi.$$

(Mass as Casimir of the Poincaré group is of course only strictly defined over a Minkowski background, but we will address FP-related theories as massive gravities. They all have additional dofs like a massive vector has a longitudinal dof.)

According to the Goldstone-boson equivalence theorem ϕ dominates at large graviton momenta. The ϕ Lagrangian reads ($i, j \in \{1; 2; 3\}$, $\dot{\phi} = \partial\phi/\partial t$)

$$\mathscr{L} = \sqrt{-\eta}[A\dot{\phi}^2 + B^{ij}(\partial_i\phi)(\partial_j\phi) + \dot{\phi} D^i \partial_i \phi],$$

where the functions A, B^{ij}, and D^i depend on the background. The Hamiltonian

$$\mathscr{H} = \frac{\pi^2}{4\sqrt{-\eta}A} - \sqrt{-\eta} B^{ij}(\partial_i\phi)(\partial_j\phi)$$

signals unitarity violation by a negative A and a classical instability by a positive definite B^{ij}. Here $\pi = \partial\mathscr{L}/\partial\dot{\phi}$. The unitarity violation arises from the wrong sign (negative norm) in the canonical commutation relations for the generators/annihilators

$$[a(\mathbf{k}), a^\dagger(\mathbf{k}')] = \text{sgn}(A)\delta^{(3)}(\mathbf{k} - \mathbf{k}').$$

The classical instability implies that the condition $||h_{\mu\nu}|| \ll ||\eta_{\mu\nu}||$ is violated and perturbation theory breaks down. The latter, however, does not signal a fundamental problem of the theory, but only the failure of the analysis tool.

On a Friedmann-Robertson-Walker (FRW) background $\mathscr{L}$ simplifies to

$$\mathscr{L}_{\mathrm{FRW}} = \sqrt{-\eta}[A\dot{\phi}^2 + B(\nabla\phi/a)^2].$$

$A < 0$ leads to unitarity violation, $B > 0$ to a classical instability. In an expanding FRW background the unitarity violation happens at $m^2 > H^2 + \dot{H}$, which can never be reached from our observation point, i.e. from late times, without first violating the stability requirement $m^2 > H^2 + \dot{H}/3$, i.e. within the range of applicability of perturbation theory, as $\dot{H} < 0$. (Exceptions are a pure CC [7], where $\dot{H} \equiv 0$ and phantom matter with $\dot{H} > 0$.) We call this mechanism self-protection [8]. It is an example for the phenomenon of classicalisation [9, 10]. The result persists to low graviton momenta [11], where another stability bound comes from the vector sector,

$$m^2 > -\tfrac{3}{2}\dot{H}.$$

While avoiding theoretical inconsistencies, a self-protected theory does not allow for an approximately homogeneous early universe. This requires a model that is stable and unitary at all times. Therefore, we extend the mass term beyond the naive FP term [12]. The requirement that only second-order equations of motion arise necessitates that $\mathscr{M}^{\mu\nu\alpha\beta}$ in the mass term $h_{\mu\nu}\mathscr{M}^{\mu\nu\alpha\beta}h_{\alpha\beta}$ must have the symmetries of $R^{\mu\alpha\nu\beta}$. Then the unique choice to second adiabatic order reads [12]

$$\mathscr{M}^{\mu\nu\alpha\beta} = (m_0^2 + \alpha R_0)g_0^{\mu[\nu}g_0^{\beta]\alpha} + \beta(g_0^{\mu[\nu}R_0^{\beta]\alpha} + R_0^{\mu[\nu}g_0^{\beta]\alpha}) + \gamma R_0^{\mu\alpha\nu\beta}.$$

On FRW backgrounds, which are conformally flat, the γ term can be expressed by a combination of the α and β terms. Hence, we set $\gamma = 0$ without loss of generality. Then absolute stability is achieved for $\beta = 0$ and α either equal to m_0^2/Ω_Λ or smaller than a constant $\alpha_{\max} < 0$ without a simple analytical expression [12]. (See Fig. 41.1.) Ω_Λ is the current relative density parameter of the CC. What this difference means concretely for the backward evolution of a scalar mode is shown in Fig. 41.2. Taking stock, this generalisation cures the instability issues at least over FRW spacetimes. Next, after further checks of the linear model, e.g. for other backgrounds, a non-linear completion of the model must be found and subjected to further tests.

41.2.2 Nonlinear Level

The vDVZ discontinuity is cured at the nonlinear level by the Vainshtein mechanism [13]. The solution of the nonlinear equations of motion belonging to the Lagrangian

$$\mathscr{L}_{\mathrm{FP,nlin}} = \sqrt{-g}R|_{\not{\mathrm{quad}}} - \sqrt{-\eta}\frac{m^2}{4}(\eta^{\mu\kappa}\eta^{\nu\lambda} - \eta^{\mu\nu}\eta^{\kappa\lambda})h_{\mu\nu}h_{\kappa\lambda}$$

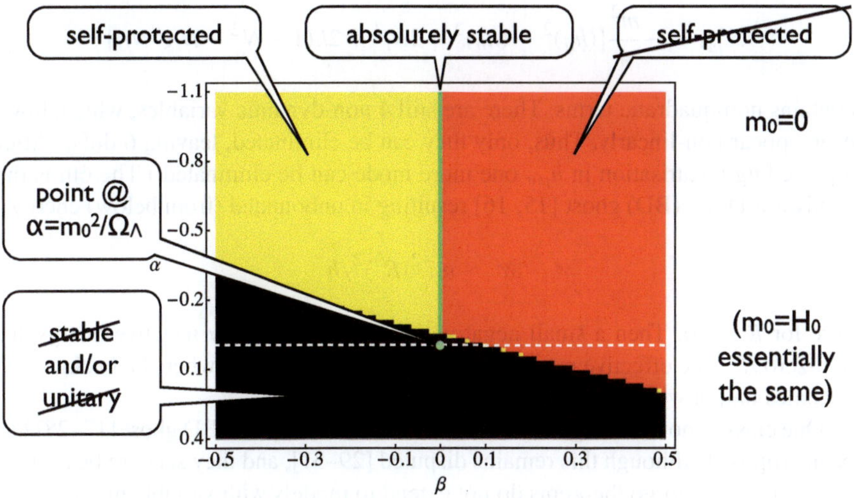

Fig. 41.1 Adapted from [12]. Parameter plot in the α–β plane, for $m_0 = 0$ and $\gamma = 0$

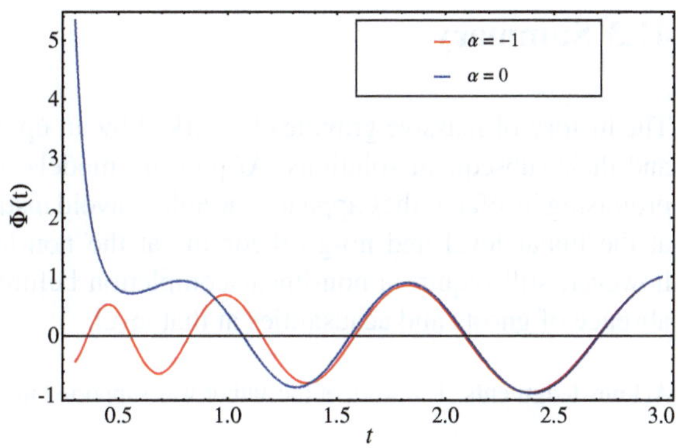

Fig. 41.2 Taken from [12]. Scalar mode on a matter dominated background, $m_0 = H_0$, $\beta = 0 = \gamma$. The *solid red line* for $\alpha = -1$ represents an absolutely stable case, while the *dashed blue line* for $\alpha = 0$ represents an unstable one

are non-analytic in r_V/r (around a spherically symmetric source), where $r_V = \sqrt[5]{r_S m^{-4}}$ is the Vainshtein and r_S the Schwarzschild radius. GR is recovered for $r < r_V$, although the perturbative solution is a power series in r_V/r, which diverges in the limit $m \to 0$.

Beyond the linear level 6 instead of 5 dofs propagate: In ADM variables [14]

$$\mathscr{L}_{\text{EH}} = \pi^{ij}\dot{g}_{ij} - NR^0 - N_i R^i.$$

The variation wrt. the non-dynamic lapse N and shift N_i variables yields 4 constraints eliminating a total of 8 variables. This leaves the 2 transverse traceless modes of the 10 independent components of a 4d symmetric tensor. The mass term

$$\mathscr{L}_{m^2} = -\frac{m^2}{4}[(h_{ij})^2 - (h_{ii})^2 - 2N_i{}^2 + 2h(1 - N^2 - N_i g^{ij} N_j)]$$

contains non-quadratic terms. There are still 4 non-dynamic variables, which, however, appear non-linearly. Thus, only they can be eliminated, leaving 6 dofs. (After a preceding linearisation in $h_{\mu\nu}$ one more mode can be eliminated.) The 6th is the Boulware-Deser (BD) ghost [15, 16] resulting in unbounded (from below) energy,

$$\mathscr{H} \sim m^{-2}(R^0)^2/h$$

here for $R^i = 0$. Then a small negative h leads to arbitrarily negative energy for non-zero R^0. The effective mass of the ghost becomes infinite when the background becomes flat, in which case the ghost cannot be excited.

One class of non-linear generalisations of FP that avoid the BD ghost [17–28] has been proposed, although this remains disputed [29–31], and they seem to be acausal [32, 33]. These no-go theorems do not extend to models with variable mass.

41.3 Summary

The history of massive gravities is marked by an up-and-down between challenges and their subsequent solutions. At present, models with variable mass seem most promising insofar as they appear to be able to avoid instabilities and unitarity violation at the linear level and no-go theorems at the non-linear. The models from [12], however, still require a non-linear completion before they can be checked for the absence of ghosts and acausalities at that level.

Acknowledgments The work of the author was supported in part by the Humboldt foundation.

References

1. G. Dvali, S. Hofmann, J. Khoury, Phys. Rev. D **76**, 084006 (2007)
2. M. Fierz, W. Pauli, Proc. Roy. Soc. Lond. A **173**, 211 (1939)
3. A. Einstein, Sitzungsber. Preuss. Akad. Wiss. Berlin (Math. Phys.) **1917**, 142 (1917)
4. H. van Dam, M.J.G. Veltman, Nucl. Phys. B **22**, 397 (1970)
5. V.I. Zakharov, JETP Lett. **12**, 312 (1970)
6. C. Deffayet, Phys. Rev. D **71**, 103501 (2005)
7. A. Higuchi, Nucl. Phys. B **282**, 397 (1987)
8. F. Berkhahn, D.D. Dietrich, S. Hofmann, JCAP **1011**, 018 (2010)
9. F. Berkhahn, D.D. Dietrich, S. Hofmann, Phys. Rev. Lett. **106**, 191102 (2011)
10. G. Dvali, G.F. Giudice, C. Gomez, A. Kehagias, JHEP **1108**, 108 (2011)
11. F. Berkhahn, D.D. Dietrich, S. Hofmann, JCAP **1109**, 024 (2011)
12. F. Berkhahn, D.D. Dietrich, S. Hofmann, F. Kuhnel, P. Moyassari, Phys. Rev. Lett. **108**, 131102 (2012)

13. A.I. Vainshtein, Phys. Lett. B **39**, 393 (1972)
14. R.L. Arnowitt, S. Deser, C.W. Misner, Gen. Rel. Grav. **40**, 1997 (2008)
15. D.G. Boulware, S. Deser, Phys. Rev. D **6**, 3368 (1972)
16. D.G. Boulware, S. Deser, Phys. Lett. B **40**, 227 (1972)
17. S.F. Hassan, R.A. Rosen, Phys. Rev. Lett. **108**, 041101 (2012)
18. S.F. Hassan, R.A. Rosen, JHEP **04**, 123 (2012)
19. S.F. Hassan, R.A. Rosen, A. Schmidt-May, JHEP **02**, 026 (2012)
20. S.F. Hassan, A. Schmidt-May, M. von Strauss, Phys. Lett. B **715**, 335 (2012)
21. C.J. Isham, A. Salam, J.A. Strathdee, Phys. Lett. B **31**, 300 (1970)
22. C. de Rham, G. Gabadadze, Phys. Rev. D **82**, 044020 (2010)
23. C. de Rham, G. Gabadadze, A.J. Tolley, Phys. Rev. Lett. **106**, 231101 (2011)
24. C. de Rham, G. Gabadadze, A.J. Tolley, Phys. Lett. B **711**, 190 (2012)
25. C. de Rham, G. Gabadadze, A.J. Tolley, JHEP **1111**, 093 (2011)
26. A. Salam, J.A. Strathdee, Phys. Rev. **184**, 1750 (1969)
27. A. Salam, J.A. Strathdee, Phys. Rev. **184**, 1760 (1969)
28. B. Zumino, Effective Lagrangians, broken symmetries, in *Brandeis University 1970, Lectures On Elementary Particles and Quantum Field Theory*, vol. 2 (Cambridge, 1970), pp. 437–500
29. L. Alberte, A.H. Chamseddine, V. Mukhanov, JHEP **1104**, 004 (2011)
30. A.H. Chamseddine, V. Mukhanov, JHEP **1108**, 091 (2011)
31. A.H. Chamseddine, V. Mukhanov, JHEP **1303**, 092 (2013)
32. S. Deser, A. Waldron, Phys. Rev. Lett. **110**, 111101 (2013)
33. S. Deser, K. Izumi, Y.C. Ong, A. Waldron, Phys. Lett. B **726**, 544 (2013)

Chapter 42
Self Sustained Traversable Wormholes and Topology Change Induced by Gravity's Rainbow

Remo Garattini

Abstract We consider the effects of Gravity's Rainbow on the self-sustained equation which is responsible to find new traversable wormholes configurations which are sustained by their own gravitational quantum fluctuations. The same self-sustained equation is also used to discover if topology change is possible. In this contribution, we will show that in both uses, the self-sustained equation will produce a Wheeler wormhole, namely a wormhole of Planckian size. This means that, from the point of view of traversability, the wormhole will be traversable in principle, but not in practice. From the topology change point of view, the background metric will be fixed to be Minkowskian in the equation governing the quantum fluctuations, which behaves essentially as a backreaction equation, and the quantum fluctuations are let to evolve. Analyzing this procedure, we will show that the self-sustained equation, endowed with a Gravity's Rainbow distortion, will be responsible of a topology change with the appearance of a Planckian wormhole.

42.1 Introduction

A wormhole is often termed Einstein-Rosen bridge because a "*bridge*" connecting two "sheets" was the result obtained by Einstein and Rosen in attempting to build a geometrical model of a physical elementary "particle" that was everywhere finite and singularity free [1]. It was Wheeler who introduced the term wormhole [2], although his wormholes were at the quantum scale. We have to wait for Morris and Thorne [3] to see the subject of wormholes seriously considered by the scientific community. In practice a traversable wormhole is a solution of the Einstein's Field equations, represented by two asymptotically flat regions joined by a bridge or, in other word, it is a short-cut in space and time. To exist, traversable

R. Garattini (✉)
Dipartimento di Ingegneria, Università degli Studi di Bergamo,
Viale Marconi, 5, 24044 Dalmine, Italy
e-mail: remo.garattini@unibg.it

R. Garattini
Istituto Nazionale di Fisica Nucleare, Sezione di Milano, Milan, Italy

P. Nicolini et al. (eds.), *1st Karl Schwarzschild Meeting on Gravitational Physics*,
Springer Proceedings in Physics 170, DOI 10.1007/978-3-319-20046-0_42

wormholes must violate the null energy conditions, which means that the matter threading the wormhole's throat has to be "*exotic*". Classical matter satisfies the usual energy conditions. Therefore, it is likely that wormholes must belong to the realm of semiclassical or perhaps a possible quantum theory of the gravitational field. Since a complete theory of quantum gravity has yet to come, it is important to approach this problem semiclassically. On this ground, the Casimir energy on a fixed background has the correct properties to substitute the exotic matter: indeed, it is known that, for different physical systems, Casimir energy is negative. Usually one considers some matter or gauge fields which contribute to the Casimir energy necessary to the traversability of the wormholes, nevertheless nothing forbids to use the Casimir energy of the graviton on a background of a traversable wormhole. In this way, one can think that the quantum fluctuations of the gravitational field of a traversable wormhole are the same ones which are responsible to sustain traversability. Nevertheless, Casimir energy is a form of Zero Point Energy (ZPE) which, usually manifests Ultra Violet (UV) divergences. To keep under control the UV divergences, usually one invokes a standard regularization/renormalization process. However, an alternative procedure can be taken under consideration by distorting spacetime since the beginning. This distortion is better known as Gravity's Rainbow. Since Gravity's Rainbow switches on at the Planck scale it is likely that ZPE can be used as a tool to produce a topology change. Note that in [4], the ZPE was used as an indicator for a topology change without a Gravity's Rainbow scheme. In this contribution we will explicitly show how Gravity's Rainbow comes into play to produce a topology change as a ZPE consequence.

42.2 Self-sustained Traversable Wormholes

In this section we shall consider the formalism outlined in detail in [5–7], where the graviton one loop contribution to a classical energy in a wormhole background is used. The spacetime metric representing a spherically symmetric and static wormhole is given by

$$ds^2 = -e^{2\Phi(r)}\, dt^2 + \frac{dr^2}{1 - b(r)/r} + r^2\,(d\theta^2 + \sin^2\theta\, d\phi^2)\,, \tag{42.1}$$

where $\Phi(r)$ and $b(r)$ are arbitrary functions of the radial coordinate, r, denoted as the redshift function, and the shape function, respectively [3]. The radial coordinate has a range that increases from a minimum value at r_0, corresponding to the wormhole throat, to infinity. A fundamental property of a wormhole is that a flaring out condition of the throat, given by $(b - b'r)/b^2 > 0$, is imposed [3, 8], and at the throat $b(r_0) = r = r_0$, the condition $b'(r_0) < 1$ is imposed to have wormhole solutions. Another condition that needs to be satisfied is $1 - b(r)/r > 0$. For the wormhole to be traversable, one must demand that there are no horizons present, which are identified as the surfaces with $e^{2\Phi} \to 0$, so that $\Phi(r)$ must be finite everywhere. The classical energy is given by

$$H_{\Sigma}^{(0)} = \int_{\Sigma} d^3x \, \mathcal{H}^{(0)} = -\frac{1}{16\pi G} \int_{\Sigma} d^3x \sqrt{g} \, R \,,$$

where the background field super-hamiltonian, $\mathcal{H}^{(0)}$, is integrated on a constant time hypersurface. R is the curvature scalar, and using metric (42.1), is given by

$$R = -2\left(1 - \frac{b}{r}\right)\left[\Phi'' + (\Phi')^2 - \frac{b'}{r(r-b)} - \frac{b'r + 3b - 4r}{2r(r-b)}\Phi'\right].$$

We shall henceforth consider a constant redshift function, $\Phi'(r) = 0$, which provides interestingly enough results, so that the curvature scalar reduces to $R = 2b'/r^2$. Thus, the classical energy reduces to

$$H_{\Sigma}^{(0)} = -\frac{1}{2G} \int_{r_0}^{\infty} \frac{dr \, r^2}{\sqrt{1 - b(r)/r}} \frac{b'(r)}{r^2} \,. \tag{42.2}$$

A traversable wormhole is said to be "*self sustained*" if

$$H_{\Sigma}^{(0)} = -E^{TT} , \tag{42.3}$$

where E^{TT} is the total regularized graviton one loop energy. Basically this is given by

$$E^{TT} = -\frac{1}{2} \sum_{\tau} \left[\sqrt{E_1^2(\tau)} + \sqrt{E_2^2(\tau)}\right] , \tag{42.4}$$

where τ denotes a complete set of indices and $E_i^2(\tau) > 0, i = 1, 2$ are the eigenvalues of the modified Lichnerowicz operator

$$\left(\hat{\triangle}_L^m h^{\perp}\right)_{ij} = \left(\triangle_L h^{\perp}\right)_{ij} - 4R_i^k h_{kj}^{\perp} + {}^3R \, h_{ij}^{\perp} \,, \tag{42.5}$$

acting on traceless-transverse tensors of the perturbation and where $\triangle_L$ is the Lichnerowicz operator defined by

$$(\triangle_L \, h)_{ij} = \triangle h_{ij} - 2R_{ikjl} \, h^{kl} + R_{ik} \, h_j^k + R_{jk} \, h_i^k \,, \tag{42.6}$$

with $\triangle = -\nabla^a \nabla_a$. For the background (42.1), one can define two r-dependent radial wave numbers

$$k_i^2\left(r, l, \omega_{i,nl}\right) = \omega_{i,nl}^2 - \frac{l\,(l+1)}{r^2} - m_i^2(r) \quad i = 1, 2 \quad , \tag{42.7}$$

where

$$\begin{cases} m_1^2(r) = \frac{6}{r^2}\left(1 - \frac{b(r)}{r}\right) + \frac{3}{2r^2}b'(r) - \frac{3}{2r^3}b(r) \\ \\ m_2^2(r) = \frac{6}{r^2}\left(1 - \frac{b(r)}{r}\right) + \frac{1}{2r^2}b'(r) + \frac{3}{2r^3}b(r) \end{cases} \tag{42.8}$$

are two r-dependent effective masses $m_1^2(r)$ and $m_2^2(r)$. When we perform the sum over all modes, E^{TT} is usually divergent. In [5–7] a standard regularization/renormalization scheme has been adopted to handle the divergences. In this contribution, we will consider the effect of Gravity's Rainbow on the graviton to one loop. One advantage in using such a scheme is to avoid the renormalization process and to use only one scale: the Planck scale.

42.3 Gravity's Rainbow at Work and Topology Change

One of the purposes of (42.3) is the possible discovery of a traversable wormhole with the determination of the shape function. When Gravity's Rainbow is taken under consideration, spacetime is endowed with two arbitrary functions $g_1(E/E_P)$ and $g_2(E/E_P)$ having the following properties

$$\lim_{E/E_P \to 0} g_1(E/E_P) = 1 \quad \text{and} \quad \lim_{E/E_P \to 0} g_2(E/E_P) = 1. \tag{42.9}$$

$g_1(E/E_P)$ and $g_2(E/E_P)$ appear into the solutions of the modified Einstein's Field Equations [9]

$$G_{\mu\nu}(E/E_P) = 8\pi G(E/E_P)\, T_{\mu\nu}(E/E_P) + g_{\mu\nu}\Lambda(E/E_P), \tag{42.10}$$

where $G(E/E_P)$ is an energy dependent Newton's constant, defined so that $G(0)$ is the low-energy Newton's constant and $\Lambda(E/E_P)$ is an energy dependent cosmological constant. Usually E is the energy associated to the particles deforming the spacetime geometry. Since the scale of deformation involved is the Planck scale, it is likely that spacetime itself fluctuates in such a way to produce a ZPE. However the deformed Einstein's gravity has only one particle available: the graviton. As shown in [10], the self sustained equation (42.3) becomes

$$\frac{b'(r)}{2Gg_2(E/E_P)\,r^2} = \frac{2}{3\pi^2}(I_1 + I_2)\,. \tag{42.11}$$

Equation (42.11) is finite for appropriate choices of the Rainbow's functions $g_1(E/E_P)$ and $g_2(E/E_P)$. We assume that

$$g_1(E/E_P) = \exp\left(-\alpha E^2/E_P^2\right) \qquad g_2(E/E_P) = 1\,, \tag{42.12}$$

where $\alpha \in \mathbb{R}$ and $g_2(E/E_P) = 1$, to avoid Planckian distortions in the classical term. We find

$$I_1 = 3\int_{\sqrt{m_1^2(r)}}^{\infty} \exp(-\alpha\frac{E^2}{E_P^2})E^2\sqrt{E^2 - m_1^2(r)}dE\,, \tag{42.13}$$

and

$$I_2 = 3\int_{\sqrt{m_2^2(r)}}^{\infty} \exp(-\alpha\frac{E^2}{E_P^2})E^2\sqrt{E^2 - m_2^2(r)}dE\,. \tag{42.14}$$

Following [10], after the integration one finds that (42.11) can be rearranged in the following way

$$\frac{b'(r)}{2Gr^2} = \frac{E_P^4}{2\pi^2}\left[\frac{x_1^2}{\alpha}\exp\left(-\frac{\alpha x_1^2}{2}\right)K_1\left(\frac{\alpha x_1^2}{2}\right) + \frac{x_2^2}{\alpha}\exp\left(\frac{\alpha x_2^2}{2}\right)K_1\left(\frac{\alpha x_2^2}{2}\right)\right], \tag{42.15}$$

where $x_1 = \sqrt{m_1^2(r)/E_P^2}$, $x_2 = \sqrt{m_2^2(r)/E_P^2}$ and $K_1(x)$ is a modified Bessel function of order 1. Note that it is extremely difficult to extract any useful information from this relationship, so that in the following we consider two regimes, namely the cis-planckian regime, where $x_i \ll 1$ $(i = 1, 2)$, and the trans-planckian régime, where $x_i \gg 1$. In [10], it has been shown that the cis-planckian regime does not produce solutions compatible with traversability. On the other hand when we fix our attention on the trans-planckian regime, i.e., $x_1 \gg 1$ and $x_2 \gg 1$, we obtain the following approximation

$$\frac{1}{2G}\frac{b'(r)}{r^2} \simeq \frac{E_P^4}{8\sqrt{\alpha^3\pi^3}}\left[\exp\left(-\alpha x_1^2\right)x_1 + O\left(\frac{1}{x_1}\right) + \exp\left(-\alpha x_2^2\right)x_2 + O\left(\frac{1}{x_2}\right)\right]. \tag{42.16}$$

Note that in this regime, the asymptotic expansion is dominated by the Gaussian exponential so that the quantum correction vanishes. Thus, the only solution is $b'(r) = 0$ and consequently we have a constant shape function, namely, $b(r) = r_t$. It is interesting to observe that (42.11) can be interpreted also in a different way. Indeed, if we fix the background on the r.h.s. of (42.11) and consequently let the quantum fluctuations evolve, one can verify what kind of solutions it is possible to extract from the l.h.s. in a recursive way. In this way, if we discover that the l.h.s. has solutions which topologically differ from the fixed background of the r.h.s., we can conclude that a topology change has been induced from quantum fluctuations of the graviton for any spherically symmetric background on the r.h.s of (42.11). Of course, this is not a trivial task, therefore the simplest way to see if a topology change is realized, we fix the Minkowski background on the r.h.s. of (42.11). This means that $b(r) = 0$ $\forall r$ and the effective masses become $m_1^2(r) = m_2^2(r) = 6/r^2$. Then (42.15) reduces to

$$\frac{1}{2G}\frac{b'(r)}{r^2} = \frac{E_P^4}{\pi^2}\left[\frac{6}{\alpha\,(rE_P)^2}\exp\left(-\frac{3\alpha}{(rE_P)^2}\right)K_1\left(\frac{3\alpha}{(rE_P)^2}\right)\right]. \tag{42.17}$$

Let us fix our attention on the trans-planckian regime, i.e., $rE_P \ll 1$, where we can write

$$\frac{b'(r)}{r^2} \simeq \frac{E_P^2}{2\sqrt{\alpha^3\pi^3}}\left[\exp\left(-\alpha\frac{6}{(rE_P)^2}\right)\frac{\sqrt{6}}{rE_P} + O\left(rE_P\right)\right]. \tag{42.18}$$

Since this is a particular case of (42.16), we conclude that the only solution is $b'(r) = 0$ and consequently we have a constant shape function, namely, $b(r) = r_t$. A comment to the result (42.18) is in order. One could think that (42.18) is only a special case of (42.16). Of course this is not true, because (42.16) uses a different initial condition with respect to (42.18). Indeed, in (42.16) the background is arbitrary, while in (42.18) one considers a Minkowski line element and the solution is obtained with an iterative process. One can observe that this procedure could be approached also distorting the one loop graviton by means of a Noncommutative geometry like in [11–13], where the classical Liouville measure is modified into [11]

$$dn_i = \frac{d^3\mathbf{x}d^3\mathbf{k}}{(2\pi)^3}\exp\left(-\frac{\theta}{4}\left(\omega_{i,nl}^2 - m_i^2\left(r\right)\right)\right), \quad i = 1, 2. \tag{42.19}$$

$m_i^2\left(r\right)$ are the effective masses described in (42.8) and θ is the Noncommutative parameter. While nothing can be said about the effect of Noncommutative geometry on topology change, a result can be extracted from the traversability of the wormhole. Indeed, if one fixes the form of the shape function to be $b\left(r\right) = r_0^2/r$, which is the prototype of the traversable wormholes [3], one gets $r_0 = 0.28l_P$, with θ fixed at $\theta = 7.43 \times 10^{-2}l_P^2$. If we compare the result obtained in [10] using Gravity's Rainbow, one finds the following value for the radius $r_t = 1.46l_P$, which is slightly larger than r_0. The conclusion is that Gravity's Rainbow and Noncommutative geometry keep under control the UV divergences in this ZPE calculation connected with the self-sustained equation [14]. In both cases we find that the result is a Wheeler wormhole. This means that, from the point of view of traversability, the wormhole will be traversable in principle, but not in practice.

References

1. A. Einstein, N. Rosen, Phys. Rev. **48**, 73 (1935)
2. J.A. Wheeler, Phys. Rev. **97**, 511–536 (1955)
3. M.S. Morris, K.S. Thorne, Am. J. Phys. **56**, 395 (1988)
4. A. DeBenedictis, R. Garattini, F.S.N. Lobo, Phys. Rev. D **78**, 104003 (2008). arXiv:0808.0839 [gr-qc]
5. R. Garattini, Class. Quant. Grav. **22**, 1105 (2005). arXiv:gr-qc/0501105
6. R. Garattini, Class. Quant. Grav. **24**, 1189 (2007). arXiv:gr-qc/0701019

7. R. Garattini, F.S.N. Lobo, Class. Quant. Grav. **24**, 2401 (2007). arXiv:gr-qc/0701020
8. M. Visser, *Lorentzian Wormholes: From Einstein to Hawking* (American Institute of Physics, New York, 1995)
9. J. Magueijo, L. Smolin, Class. Quant. Grav. **21**, 1725 (2004). arXiv:gr-qc/0305055
10. R. Garattini, F.S.N. Lobo, Phys. Rev. D **85**, 024043 (2012). arXiv:1111.5729 [gr-qc]
11. R. Garattini, P. Nicolini, Phys. Rev. D **83**, 064021 (2011). arXiv:1006.5418 [gr-qc]
12. R. Garattini, EPJ Web Conf. **58** 01007 (2013). arXiv:1212.4311 [gr-qc]
13. R. Garattini, F.S.N. Lobo, Phys. Lett. B **671**, 146 (2009). arXiv:0811.0919 [gr-qc]
14. R. Garattini, G. Mandanici, Phys. Rev. D **83**, 084021 (2011). arXiv:1102.3803 [gr-qc]

Chapter 43
A General Maximum Entropy Principle for Self-Gravitating Perfect Fluid

Sijie Gao

Abstract We consider a self-gravitating system consisting of perfect fluid in a spherically symmetric spacetime. Using the Gibbs-Duhem relation, we extremize the total entropy S under the constraints that the total number and energy of particles are fixed. We show that the extrema of S coincides precisely with the relativistic Tolman-Oppenheimer-Volkoff (TOV) equation of hydrostatic equilibrium. Furthermore, we show that the maximum entropy principle is also valid for a charged perfect fluid. Our work provides a strong evidence for the fundamental relationship between general relativity and ordinary thermodynamics.

43.1 Introduction

The four laws of black hole mechanics were originally derived from the Einstein equation at the purely classical level [1, 2]. The discovery of the Hawking radiation [3] confirms that black holes are thermodynamic objects with Hawking temperature. In fact, the relations between gravity and ordinary thermodynamics had been studied even before the establishment of black hole mechanics. In 1965, Cocke [4] proposed a maximum entropy principle for self-gravitating fluid spheres. Cocke showed that the requirement that the total entropy S be an extremum leads to the equation of hydrostatic equilibrium which was originally derived from the Einstein equation. However, a critical assumption in Cocke's derivation is that the fluid is in adiabatic motion so that the total entropy is invariant. In relation to Cocke's work, Sorkin et al. (SWZ) [5] proved an entropy principle for radiation, where the adiabatic condition was not needed in SWZ's derivation. SWZ's discussion was restricted to radiation. In this paper, we prove a maximum entropy principle for a general self-gravitating perfect fluid with spherical symmetry. In addition to the Einstein constraint equation, we only make use of the ordinary thermodynamic laws such as Gibbs-Duhem relation

S. Gao (✉)
Department of Physics, Beijing Normal University, Beijing 100875, China
e-mail: sijie@bnu.edu.cn

P. Nicolini et al. (eds.), *1st Karl Schwarzschild Meeting on Gravitational Physics*,
Springer Proceedings in Physics 170, DOI 10.1007/978-3-319-20046-0_43

to derive the Tolman-Oppenheimer-Volkoff (TOV) equation. Finally, we extend our treatment to a general charged fluid and derive the generalized TOV equation for a charged fluid.

43.2 Maximum Entropy Principle for Perfect Fluid

We consider a static spacetime with spherical symmetry, which can be described by the metric

$$ds^2 = g_{tt}(r)dt^2 + \left[1 - \frac{2m(r)}{r}\right]^{-1} dr^2 + r^2 d\Omega^2 . \tag{43.1}$$

We assume the spacetime is filled with a perfect fluid with stress-energy tensor

$$T_{ab} = \rho u_a u_b + \frac{1}{3}\rho(g_{ab} + u_a u_b) , \tag{43.2}$$

It is not difficult to find that the constraint Einstein equation, which is the time-time component of the Einstein equation, yields

$$\rho = \frac{m'(r)}{4\pi r^2} \tag{43.3}$$

Our purpose is to show that the TOV equation can be derived from the extrema of total entropy and (43.3).

We start with the familiar first law of thermodynamics

$$dS = \frac{1}{T}dE + \frac{p}{T}dV - \frac{\mu}{T}dN , \tag{43.4}$$

where S, E, N represent the total entropy, energy and particle number within the volume V. Rewrite (43.4) in terms of densities s, ρ, n

$$d(sV) = \frac{1}{T}d(\rho V) + \frac{p}{T}dV - \frac{\mu}{T}d(nV) , \tag{43.5}$$

which gives

$$sdV + Vds = \frac{1}{T}\rho dV + Vd\rho + \frac{p}{T}dV - \frac{\mu}{T}ndV - \frac{\mu}{T}Vdn . \tag{43.6}$$

Applying (43.4) to a unit volume, we find

$$ds = \frac{1}{T}d\rho - \frac{\mu}{T}dn . \tag{43.7}$$

Combining (43.6) and (43.7), we arrive at the integrated form of the Gibbs-Duhem relation [6]

$$s = \frac{1}{T}(\rho + p - \mu n)\,. \tag{43.8}$$

Note that this is a general expression for perfect fluid. We shall treat (ρ, n) as two independent variables, e.g.,

$$s = s(\rho, n),\ \ \mu = \mu(\rho, n),\ \ \ p = p(\rho, n)\,. \tag{43.9}$$

Our task is to extremize the total entropy

$$S = 4\pi \int_0^R s(r)\left[1 - \frac{2m(r)}{r}\right]^{-1/2} r^2 dr\,. \tag{43.10}$$

Now S can be treated as a functional of $m(r)$. It is natural to require that

$$\delta m(0) = \delta m(R) = 0 \tag{43.11}$$

for all variations.

In addition to the constraint (43.11), we also impose that the total number of particles

$$N = 4\pi \int_0^R n(r)\left[1 - \frac{2m(r)}{r}\right]^{-1/2} r^2 dr \tag{43.12}$$

be fixed, i.e.,

$$\delta N = 0\,. \tag{43.13}$$

Following the standard method of Lagrange multipliers, the extrema of S leads to

$$\delta S + \lambda \delta N = 0\,. \tag{43.14}$$

Define the "total Lagrangian" by

$$L(m, m', n) = s(\rho(m'), n)\left[1 - \frac{2m(r)}{r}\right]^{-1/2} r^2 + \lambda n(r)\left[1 - \frac{2m(r)}{r}\right]^{-1/2} r^2\,. \tag{43.15}$$

Now the constrained Euler-Lagrange equation is given by

$$\frac{\partial L}{\partial n} = 0\,, \tag{43.16}$$

$$\frac{d}{dr}\frac{\partial L}{\partial m'} + \frac{\partial L}{\partial m} = 0\,. \tag{43.17}$$

Thus, (43.16) yields

$$\frac{\partial s}{\partial n} + \lambda = 0\,. \tag{43.18}$$

Using (43.7), we have

$$-\frac{\mu}{T} + \lambda = 0\,, \tag{43.19}$$

which shows that $\frac{\mu}{T}$ must be a constant for self-gravitating fluid.

From (43.15), we have

$$\frac{\partial L}{\partial m} = r\left(1 - \frac{2m}{r}\right)^{-3/2}(n\lambda + s)\,, \tag{43.20}$$

and

$$\frac{\partial L}{\partial m'} = \frac{\partial s}{\partial m'} r^2 \left(1 - \frac{2m}{r}\right)^{-1/2}. \tag{43.21}$$

Here

$$\frac{\partial s}{\partial m'} = \frac{\partial s}{\partial \rho}\frac{\partial \rho}{\partial m'} = \frac{1}{T}\frac{1}{4\pi r^2}\,, \tag{43.22}$$

where (43.3) and (43.7) have been used. Hence

$$\frac{\partial L}{\partial m'} = \frac{1}{4\pi T}\left(1 - \frac{2m}{r}\right)^{-1/2}, \tag{43.23}$$

and

$$\frac{d}{dr}\frac{\partial L}{\partial m'} = \frac{T(m'r - m) - r(r - 2m)T'}{4\pi T^2 (r - 2m)^{3/2} r^2}\,. \tag{43.24}$$

Using (43.8) and (43.19), (43.20) becomes

$$\frac{\partial L}{\partial m} = r\left(1 - \frac{2m}{r}\right)^{-3/2}\left(\frac{\rho + p}{T}\right). \tag{43.25}$$

So the Euler-Lagrange (43.17) yields

$$(4\pi p r^3 + m)T + (r - 2m)rT' = 0. \tag{43.26}$$

The constraint (43.19) yields

$$\mu' = \lambda T'. \tag{43.27}$$

Rewrite (43.8) as

$$p = Ts + \mu n - \rho, \tag{43.28}$$

and then

$$dp = Tds + sdT + \mu dn + nd\mu - d\rho. \tag{43.29}$$

By substituting (43.7), we have

$$dp = sdT + nd\mu. \tag{43.30}$$

It follows immediately that

$$p'(r) = sT'(r) + n\mu'(r). \tag{43.31}$$

Substituting (43.8), (43.19) and (43.27) into (43.31), we have

$$T' = \frac{T}{p + \rho} p'(r). \tag{43.32}$$

Substituting (43.32) into (43.26), we obtain the desired TOV equation

$$p' = -\frac{(p + \rho)(4\pi r^3 p + m)}{r(r - 2m)}. \tag{43.33}$$

Note that this equation was originally derived from Einstein's equation. The above result suggests that the TOV equation is consistent with the entropy principle for perfect fluid.

43.3 Maximum Entropy Principle for Charged Fluid

In this section, we sketch the proof of the entropy principle for a charged self-gravitating fluid. Detailed proof can be found in [7, 8]. In coordinates (t, r, θ, ϕ), assume that a spherically symmetric charged fluid is associated with the metric

$$ds^2 = g_{tt}(r)dt^2 + \left[1 - \frac{2m(r)}{r} + \frac{Q^2(r)}{r^2}\right]^{-1} dr^2 + r^2 d\theta^2 + r^2 \sin^2\theta d\phi^2 . \tag{43.34}$$

One can show from Maxwell's equation that $Q(r)$ is the total charge up to the radius r.

Then the time-time component of Einstein's equation is modified as

$$m'(r) = 4\pi r^2 \rho + \frac{QQ'}{r} . \tag{43.35}$$

Now we derive the hydroelectrostatic equation by extremizing the total entropy

$$S = \int_0^R s(r)\left[1 - \frac{2m}{r} + \frac{Q^2}{r^2}\right]^{-1/2} r^2 dr . \tag{43.36}$$

For simplicity, we assume that all the particles have the same charge q. Thus, the charge density is proportional to the particle number density n

$$\rho_e = qn . \tag{43.37}$$

Then the conservation of particle number N is equivalent to the conservation of charge with the radius R.

In addition to m, m', we can also treat $Q(r)$, $Q'(r)$ as independent variables in the Lagrangian formalism. So the Lagrangian is written as

$$L(m, m', Q, Q') = s\left[1 - \frac{2m}{r} + \frac{Q^2}{r^2}\right]^{-1/2} r^2 . \tag{43.38}$$

with the constraints

$$m(0) = Q(0) = 0, \quad m(R) = constant, \quad Q(R) = constant . \tag{43.39}$$

The extrema of S leads to the following Euler-Lagrange equations

$$\frac{d}{dr}\frac{\partial L}{\partial Q'} + \frac{\partial L}{\partial Q} = 0 \tag{43.40}$$

$$\frac{d}{dr}\frac{\partial L}{\partial m'} + \frac{\partial L}{\partial m} = 0 \tag{43.41}$$

Using these equations, together with thermodynamic relations used in last section, one can derive

$$p' = \frac{QQ'}{4\pi r^4} - (\rho + p)\left(4\pi r p + \frac{m}{r^2} - \frac{Q^2}{r^3}\right)\left(1 - \frac{2m}{r} + \frac{Q^2}{r^2}\right)^{-1} . \quad (43.42)$$

This is exactly the generalized Oppenheimer-Volkoff equation for charged fluid [9].

43.4 Conclusions

We have discussed a general self-gravitating fluid with spherical symmetry. We have derived the TOV equation of hydrostatic equilibrium by applying the maximum entropy principle. The TOV equation is an important equation for self-gravitating system which was originally derived from the Einstein equation. Our results show that the Einstein equation is consistent with the ordinary thermodynamic laws.

Acknowledgments This research was supported by NSFC Grants No. 11235003, 11375026 and NCET-12-0054.

References

1. J.D. Bekenstein, Phys. Rev. D **7**, 2333 (1973)
2. J.M. Bardeen, B. Carter, S.W. Hawking, Commun. Math. Phys. **31**, 161 (1973)
3. S.W. Hawking, Commun. Math. Phys. **43**, 199 (1975)
4. W.J. Cocke, Ann. Inst. Henri Poincaré **2**, 283 (1965)
5. R.D. Sorkin, R.M. Wald, Z.J. Zhang, Gen. Rel. Grav. **13**, 1127 (1981)
6. S. Gao, R.M. Wald, Phys. Rev. D **64**, 084020 (2001)
7. S. Gao, Phys. Rev. D **84**, 104023 (2011)
8. S. Gao, Phys. Rev. D **85**, 027503 (2012)
9. J.D. Bekenstein, Phys. Rev. D **4**, 2185 (1971)

Chapter 44
Dynamical Holographic QCD Model: Resembling Renormalization Group from Ultraviolet to Infrared

Mei Huang and Danning Li

Abstract Resembling the renormalization group from ultraviolet (UV) to infrared (IR), we construct a dynamical holographic model in the graviton-dilaton-scalar framework, where the dilaton background field Φ and scalar field X are responsible for the gluodynamics and chiral dynamics, respectively. At the UV boundary, the dilaton field is dual to the dimension-4 gluon operator, and the scalar field is dual to the dimension-3 quark-antiquark operator. The metric structure at IR is automatically deformed by the nonperturbative gluon condensation and chiral condensation in the vacuum. The produced scalar glueball spectra in the graviton-dilaton framework agree well with lattice data, and the light-flavor meson spectra generated in the graviton-dilaton-scalar framework are in good agreement with experimental data. Both the chiral symmetry breaking and linear confinement are realized in this dynamical holographic QCD model.

44.1 Introduction

Quantum chromodynamics (QCD) is accepted as the fundamental theory of the strong interaction. In the ultraviolet (UV) or weak coupling regime of QCD, the perturbative calculations agree well with experiment. However, in the infrared (IR) regime, the description of QCD vacuum as well as hadron properties and processes in terms of quarks and gluons still remains an outstanding challenge in the formulation of QCD as a local quantum field theory.

In order to derive the low-energy hadron physics and understand the deep-infrared sector of QCD from first principle, various non-perturbative methods have been employed, in particular lattice QCD, Dyson-Schwinger equations (DSEs), and functional renormalization group equations (FRGs). In recent decades, an entirely

M. Huang (✉) · D. Li
Institute of High Energy Physics, Chinese Academy of Sciences,
Beijing 100049, China
e-mail: huangm@mail.ihep.ac.cn

D. Li
e-mail: lidn@mail.ihep.ac.cn

P. Nicolini et al. (eds.), *1st Karl Schwarzschild Meeting on Gravitational Physics*,
Springer Proceedings in Physics 170, DOI 10.1007/978-3-319-20046-0_44

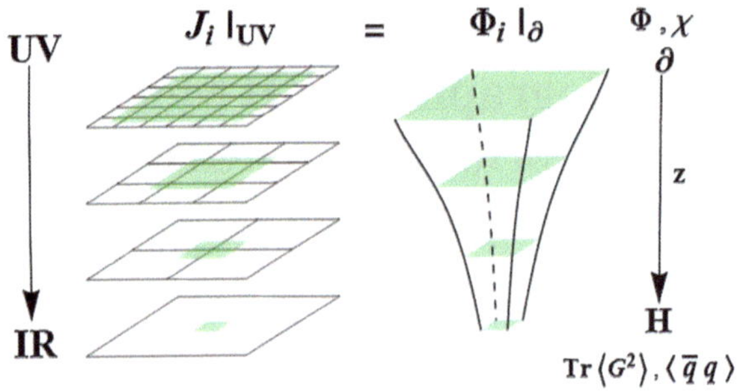

Fig. 44.1 Duality between d-dimensional QFT and $(d+1)$-dimensional gravity as shown in [9] (*Left-hand side*). Dynamical holographic QCD model resembles RG from UV to IR (*Right-hand side*): in the UV boundary the dilaton bulk field $\Phi(z)$ and scalar field $X(z)$ are dual to the dimension-4 gluon operator and dimension-3 quark-antiquark operator, which develop condensates in the IR

new method based on the anti-de Sitter/conformal field theory (AdS/CFT) correspondence and the conjecture of the gravity/gauge duality [1–3] provides a revolutionary method to tackle the problem of strongly coupled gauge theories. Though the original discovery of holographic duality requires supersymmetry and conformality, the holographic duality has been widely used in investigating hadron physics [4–8], strongly coupled quark gluon plasma and condensed matter. It is widely believed that the duality between the quantum field theory and quantum gravity is an unproven but true fact. In general, holography relates quantum field theory (QFT) in d-dimensions to quantum gravity in (d + 1)-dimensions, with the gravitational description becoming classical when the QFT is strongly coupled. The extra dimension can be interpreted as an energy scale or renormalization group (RG) flow in the QFT [9] as shown in Fig. 44.1.

In this talk, we introduce our recently developed dynamical holographic QCD model [10, 11], which resembles the renormalization group from ultraviolet (UV) to infrared (IR). The dynamical holographic model is constructed in the graviton-dilaton-scalar framework, where the dilaton background field $\Phi(z)$ and scalar field $X(z)$ are responsible for the gluodynamics and chiral dynamics, respectively. At the UV boundary, the dilaton field $\Phi(z)$ is dual to the dimension-4 gluon operator, and the scalar field $X(z)$ is dual to the dimension-3 quark-antiquark operator. The metric structure at IR is automatically deformed by the nonperturbative gluon condensation and chiral condensation in the vacuum. In Fig. 44.1, we show the dynamical holographic QCD model, which resembles the renormalization group from UV to IR.

44.2 Pure Gluon System: Graviton-Dilaton Framework

For the pure gluon system, we construct the quenched dynamical holographic QCD model in the graviton-dilaton framework by introducing one scalar dilaton field $\Phi(z)$ in the bulk. The 5D graviton-dilaton coupled action in the string frame is given below:

$$S_G = \frac{1}{16\pi G_5}\int d^5x\sqrt{g_s}e^{-2\Phi}\left(R_s + 4\partial_M\Phi\partial^M\Phi - V_G^s(\Phi)\right), \tag{44.1}$$

where G_5 is the 5D Newton constant, g_s, Φ and V_G^s are the 5D metric, the dilaton field and dilaton potential in the string frame, respectively. The metric ansatz in the string frame is often chosen to be

$$g_{MN}^s = b_s^2(z)(dz^2 + \eta_{\mu\nu}dx^\mu dx^\nu), \quad b_s(z) \equiv e^{A_s(z)}. \tag{44.2}$$

To avoid the gauge non-invariant problem and to meet the requirement of gauge/gravity duality, we take the dilaton field in the form of

$$\Phi(z) = \mu_G^2 z^2 \tanh(\mu_{G^2}^4 z^2/\mu_G^2). \tag{44.3}$$

In this way, the dilaton field in the UV behaves $\Phi(z) \overset{z\to 0}{\to} \mu_{G^2}^4 z^4$, and is dual to the dimension-4 gauge invariant gluon operator $\mathrm{Tr}G^2$, while in the IR it takes the quadratic form $\Phi(z) \overset{z\to\infty}{\to} \mu_G^2 z^2$. By self-consistently solving the Einstein equations, the metric structure will be automatically deformed in the IR by the dilaton background field, for details, please refer to [10, 11].

We assume the glueball can be excited from the QCD vacuum described by the quenched dynamical holographic model, and the 5D action for the scalar glueball $\mathcal{G}(x, z)$ in the string frame takes the form as

$$S_{\mathcal{G}} = \int d^5x\sqrt{g_s}\frac{1}{2}e^{-\Phi}\left[\partial_M\mathcal{G}\partial^M\mathcal{G} + M_{\mathcal{G},5}^2\mathcal{G}^2\right]. \tag{44.4}$$

The Equation of motion for $\mathcal{G}$ has the form of

$$-e^{-(3A_s-\Phi)}\partial_z(e^{3A_s-\Phi}\partial_z\mathcal{G}_n) = m_{\mathcal{G},n}^2\mathcal{G}_n, \tag{44.5}$$

where n the excitation number. After the transformation $\mathcal{G}_n \to e^{-\frac{1}{2}(3A_s-\Phi)}\mathcal{G}_n$, we get the Schroedinger like equation of motion for the scalar glueball

$$-\mathcal{G}_n'' + V_{\mathcal{G}}\mathcal{G}_n = m_{\mathcal{G},n}^2\mathcal{G}_n, \tag{44.6}$$

with the 5D effective Schroedinger potential

$$V_{\mathcal{G}} = \frac{3A_s'' - \Phi''}{2} + \frac{(3A_s' - \Phi')^2}{4}. \tag{44.7}$$

Then from (44.6), we can solve the scalar glueball spectra and the result is shown in Fig. 44.2. It is a surprising result that if one self-consistently solves the metric background under the dynamical dilaton field, it gives the correct ground state and at the same time gives the correct Regge slope.

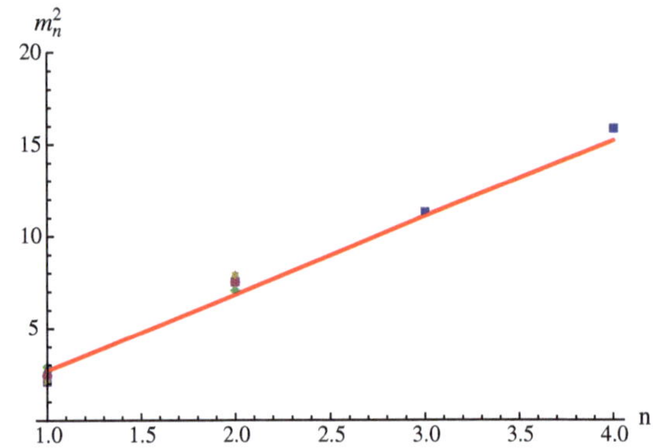

Fig. 44.2 The scalar glueball spectra for the dilaton field $\Phi(z) = \mu_G^2 z^2 \tanh(\mu_{G^2}^4 z^2/\mu_G^2)$ with $\mu_G = \mu_{G^2} = 1\text{GeV}$. The *dots* are lattice data taken from [12–15]

44.3 Dynamical Holographic QCD Model for Meson Spectra

We then add light flavors in terms of meson fields on the gluodynamical background. The total 5D action for the graviton-dilaton-scalar system takes the following form:

$$S = S_G + \frac{N_f}{N_c} S_{KKSS}, \tag{44.8}$$

with

$$S_G = \frac{1}{16\pi G_5} \int d^5x \sqrt{g_s} e^{-2\Phi} \left(R + 4\partial_M \Phi \partial^M \Phi - V_G(\Phi)\right), \tag{44.9}$$

$$S_{KKSS} = -\int d^5x \sqrt{g_s} e^{-\Phi} Tr(|DX|^2 + V_X(X^\dagger X, \Phi) + \frac{1}{4g_5^2}(F_L^2 + F_R^2)), \tag{44.10}$$

where S_{KKSS} takes the same form as in [4].

In the vacuum, it is assumed that there are both gluon condensate and chiral condensate. The dilaton background field Φ is supposed to be dual to some kind of gluodynamics in QCD vacuum. We take the dilaton background field $\Phi(z) = \mu_G^2 z^2 \tanh(\mu_{G^2}^4 z^2/\mu_G^2)$. The scalar field $X(z)$ is dual to dimension-3 quark-antiquark operator, and $\chi(z)$ is the vacuum expectation value (VEV) of the scalar field $X(z)$. For detailed analysis please refer to [10, 11]. The equations of motion of the vector, axial-vector, scalar and pseudo-scalar mesons take the form of:

$$-\rho_n'' + V_\rho \rho_n = m_n^2 \rho_n, \tag{44.11}$$

$$-a_n'' + V_a a_n = m_n^2 a_n, \tag{44.12}$$

$$-s_n'' + V_s s_n = m_n^2 s_n, \tag{44.13}$$

Table 44.1 Two sets of parameters

	G_5/L^3	m_q (MeV)	$\sigma^{1/3}$(MeV)	$\mu_G = \mu_{G^2}$(GeV)
Mod A	0.75	8.4	165	0.43
Mod B	0.75	6.2	226	0.43

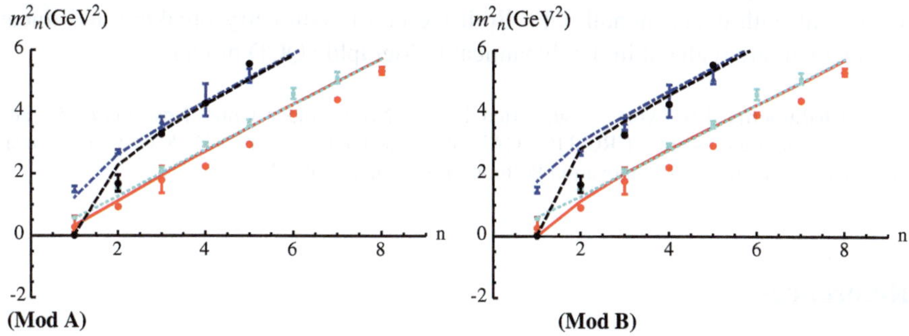

Fig. 44.3 Meson spectra in the dynamical soft-wall model with two sets of parameters in Table 44.1 comparing with experimental data. The *red* and *black lines* are for scalars and pseudoscalars, the *green* and *blue lines* are for vectors and axial-vectors

$$-\pi_n'' + V_{\pi,\varphi}\pi_n = m_n^2(\pi_n - e^{A_s}\chi\varphi_n),$$
$$-\varphi_n'' + V_\varphi\varphi_n = g_5^2 e^{A_s}\chi(\pi_n - e^{A_s}\chi\varphi_n). \tag{44.14}$$

with V_ρ, V_a, V_s, $V_{\pi,\varphi}$, V_φ are Schroedinger like potentials given in [10, 11]. For our numerical calculations, we take two sets of parameters for G_5/L^3 with L the AdS_5 radius, the current quark mass m_q, chiral condensate $\sigma^{1/3}$, μ_G and μ_{G^2} in Table 44.1. The parameters in Mod A have a smaller chiral condensate, which gives a smaller pion decay constant $f_\pi = 65.7$ MeV, and the parameters in Mod B have a larger chiral condensate, which gives a reasonable pion decay constant $f_\pi = 87.4$ MeV. The meson spectra are shown in Fig. 44.3. It is observed from Fig. 44.3 that in our graviton-dilaton-scalar system, with two sets of parameters, the generated meson spectra agree well with experimental data.

44.4 Discussion and Summary

In this work, we construct a quenched dynamical holographic QCD (hQCD) model in the graviton-dilaton framework for the pure gluon system, and develop a dynamical hQCD model for the two flavor system in the graviton-dilaton-scalar framework by adding light flavors on the gluodynamical background. The dynamical holographic model resembles the renormalization group from UV to IR. The dilaton background field Φ and scalar field X are responsible for the gluodynamics and chiral dynamics, respectively. At the UV boundary, the dilaton field is dual to the dimension-4 gluon operator, and the scalar field is dual to the dimension-3 quark-antiquark operator. The metric structure at IR is automatically deformed by the nonperturbative gluon condensation and chiral condensation in the vacuum. The produced scalar glueball spectra in the graviton-dilaton framework agree well with lattice data, and the light-flavor meson spectra generated in the graviton-dilaton-scalar framework are in well agreement with experimental data. Both the chiral symmetry breaking and linear confinement are realized in the dynamical holographic QCD model.

Acknowledgments This work is supported by the NSFC under Grant Nos. 11175251 and 11275213, DFG and NSFC (CRC 110), CAS key project KJCX2-EW-N01, K.C. Wong Education Foundation, and Youth Innovation Promotion Association of CAS.

References

1. J.M. Maldacena, Adv. Theor. Math. Phys. **2**, 231 (1998)
2. S.S. Gubser, I.R. Klebanov, A.M. Polyakov, Phys. Lett. B **428**, 105 (1998)
3. E. Witten, Adv. Theor. Math. Phys. **2**, 253 (1998)
4. A. Karch, E. Katz, D.T. Son, M.A. Stephanov, Phys. Rev. D **74**, 015005 (2006)
5. C. Csaki, M. Reece, JHEP **0705**, 062 (2007)
6. T. Gherghetta, J.I. Kapusta, T.M. Kelley, Phys. Rev. D **79**, 076003 (2009)
7. Y.-Q. Sui, Y.-L. Wu, Z.-F. Xie, Y.-B. Yang, Phys. Rev. D **81**, 014024 (2010)
8. Y.-Q. Sui, Y.-L. Wu, Y.-B. Yang, Phys. Rev. D **83**, 065030 (2011)
9. A. Adams, L.D. Carr, T. Schaefer, P. Steinberg, J.E. Thomas, New J. Phys. **14**, 115009 (2012)
10. D.N. Li, M. Huang, Q.-S. Yan, Eur. Phys. J. C **73** 2615 (2013)
11. D.N. Li, M. Huang, JHEP **1311**, 088 (2013)
12. H.B. Meyer, arXiv:hep-lat/0508002
13. B. Lucini, M. Teper, JHEP **0106**, 050 (2001)
14. C.J. Morningstar, M.J. Peardon, Phys. Rev. D **60**, 034509 (1999)
15. Y. Chen, A. Alexandru, S.J. Dong, T. Draper et al., Phys. Rev. D **73**, 014516 (2006)

Chapter 45
Modified Theories of Gravity with Nonminimal Coupling and the Faint Young Sun Paradox

Lorenzo Iorio

Abstract A certain general class of modified gravitational theories with nonminimal coupling predicts a "pressure"-type, non-geodesic acceleration for a non-rotating, massive test particle. The resulting orbital perturbations for a two-body system consist of secular rates of change of all the standard orbital elements. The resulting variation of the mutual distance yields a physical mechanism which has the potential capability to explain, in principle, the Faint Young Sun Paradox in terms of a recession of the Earth from the Sun during the Archean.

45.1 The Faint Young Sun Paradox

The so-called "Faint Young Sun Paradox" (FYSP) [20] consists in the fact that, according to consolidated models of the Sun's evolution history, the energy output of our star during the Archean, from 3.8 to 2.5 Ga ago, would have been too low to keep liquid water on the Earth's surface. Instead, there are compelling and independent evidences that, actually, our planet was mostly covered by liquid water oceans, hosting also forms of life, during that eon. For a recent review of the FYSP, see [4] and references therein.

Setting the origin of the time t at the Zero-Age Main Sequence (ZAMS) epoch, i.e. when the nuclear fusion ignited in the core of the Sun, a formula which accounts for the temporal evolution of the solar luminosity $L(t)$ reasonably well over the eons, with the possible exception of the first $\approx$0.2 Ga in the life of the young Sun, is [6]

$$\frac{L(t)}{L_0} = \frac{1}{1 + \frac{2}{5}\left(1 - \frac{t}{t_0}\right)}, \tag{45.1}$$

L. Iorio (✉)
Ministero dell'Istruzione, dell'Università e della Ricerca,
Viale Unitàdi, Italia 68, 70125 Bari, Italy
e-mail: lorenzo.iorio@libero.it

P. Nicolini et al. (eds.), *1st Karl Schwarzschild Meeting on Gravitational Physics*,
Springer Proceedings in Physics 170, DOI 10.1007/978-3-319-20046-0_45

where $t_0 = 4.57$ Ga is the present epoch, and L_0 is the current Sun's luminosity. The formula of (45.1) is in good agreement with recent standard solar models such as, e.g., [1].

According to (45.1), at the beginning of the Archean 3.8 Ga ago, corresponding to $t_{Ar} = 0.77$ Ga in our ZAMS-based temporal scale, the solar luminosity was just

$$L_{Ar} = 0.75 L_0. \tag{45.2}$$

Thus, if the heliocentric distance of the Earth was the same as today, (45.2) implies a solar irradiance I as little as

$$I_{Ar} = 0.75 I_0. \tag{45.3}$$

As extensively reviewed in [4], there is ample and compelling evidence that the Earth hosted liquid water, and even life, during the entire Archean eon spanning about 1.3 Ga. Thus, our planet could not be entirely frozen during such a remote eon, as, instead, it would have necessarily been if it really received only ≈75 % of the current solar irradiance, as it results from (45.3).

Although intense efforts by several researchers in the last decades to find a satisfactory solution to the FYSP involving multidisciplinary investigations, it not only refuses to go away [4, 5, 9], but rather it becomes even more severe [10] in view of some recent studies.

45.2 Was the Earth Closer to the Sun Than Now?

As a working hypothesis, let us provisionally assume that, at t_{Ar}, the solar irradiance I_{Ar} was approximately equal to a fraction of the present one I_0 large enough to allow for a global liquid ocean on the Earth. As noticed in [4], earlier studies required an Archean luminosity as large as 98–85 % of the present-day value to have liquid water. Some more recent models have lowered the critical luminosity threshold down to about 90–86 % [8, 11], with a lower limit as little as [11]

$$L_{oc} \approx 0.82 L_0. \tag{45.4}$$

Since the same heliocentric distance as the present-day one was assumed in the literature, (45.4) is equivalent to the following condition for the irradiance required to keep liquid ocean

$$I_{oc} \approx 0.82 I_0. \tag{45.5}$$

By assuming $I_{Ar} = I_{oc}$, together with (45.2), implies

$$r_{Ar} = 0.956 r_0, \tag{45.6}$$

i.e. the Earth should have been closer to the Sun by about ≈4.4 % with respect to the present epoch. As a consequence, if one assumes that the FYSP could be resolved, to some extent, by a closer Earth, some physical mechanism should have subsequently displaced out planet to roughly its current heliocentric distance by keeping the irradiance equal to at least $I_{\rm oc}$ over the next 1.3 Ga until the beginning of the Proterozoic era 2.5 Ga ago, corresponding to $t_{\rm Pr} = 2.07$ Ga with respect to the ZAMS epoch, when the luminosity of the Sun was

$$L_{\rm Pr} = 0.82 L_0, \tag{45.7}$$

according to (45.1). Thus, by imposing

$$I(t) = 0.82 I_0,\ 0.77\ {\rm Ga} \leq t \leq 2.07\ {\rm Ga}, \tag{45.8}$$

one gets

$$\frac{r(t)}{r_0} = \frac{1}{\sqrt{0.82\left[1 + \frac{2}{5}\left(1 - \frac{t}{t_0}\right)\right]}}, \tag{45.9}$$

$$\frac{\dot{r}(t)}{r(t)} = \frac{1}{7 t_0 \left(1 - \frac{2}{7}\frac{t}{t_0}\right)}. \tag{45.10}$$

The plots of (45.9) and (45.10) are depicted in Figs. 45.1 and 45.2.

It can be noticed that a percent distance rate as large as

$$\frac{\dot{r}}{r} \approx 3.4 \times 10^{-11}\ {\rm a}^{-1} \tag{45.11}$$

is enough to keep the irradiance equal to about 82 % of the present one during the entire Archean by displacing the Earth towards its current location.

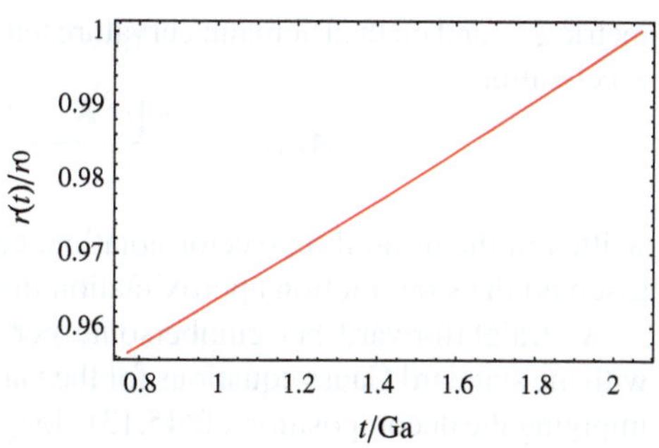

Fig. 45.1 Temporal evolution of the Earth-Sun distance $r(t)$, normalized to its present-day value r_0, over the Archean according to (45.9). The constraint $I(t) = 0.82 I_0$ throughout the Archean was adopted

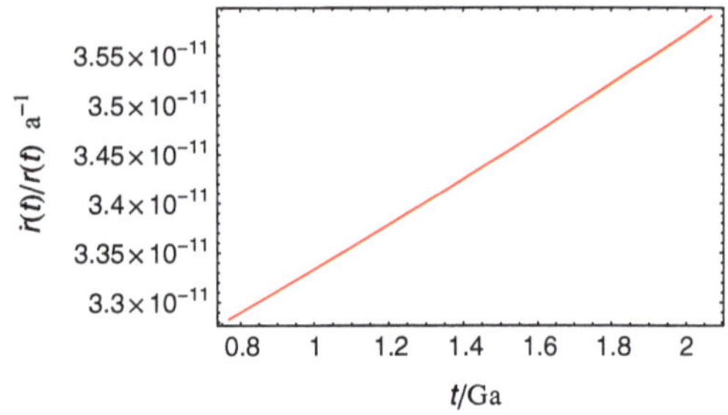

Fig. 45.2 Temporal evolution of $\dot{r}(t)/r(t)$ over the Archean according to (45.10). The constraint $I(t) = 0.82 I_0$ throughout the Archean was adopted

45.3 Modified Gravitational Theories with Nonminimal Coupling

Standard general relativity does not predict notable cosmological effects able to expand the orbit of a localized two-body system [3]. Instead, it can potentially occur in a certain class [17] of modified gravitational theories with nonminimal coupling between the matter and the gravitational field. This is not the place to delve into the technical details of such alternative theories of gravitation predicting a violation of the equivalence principle [2, 17]. Suffice it to say that a class of them, recently investigated in [17], yields an extra-acceleration $\boldsymbol{A}_{\mathrm{nmc}}$ for a test particle orbiting a central body which, in principle, implies a long-term impact on its distance.

In the usual four-dimensional spacetime language, a non-geodesic four-acceleration of a non-rotating test particle [17]

$$A^{\mu}_{\mathrm{nmc}} = \frac{c\xi}{\mathfrak{m}}\left(\delta^{\mu}_{\nu} - \frac{v^{\mu}v_{\nu}}{c^2}\right)K^{\nu},\ \mu = 0, 1, 2, 3 \tag{45.12}$$

occurs. In (45.12), $\mathfrak{m}$ is the mass of the test particle as defined in multipolar schemes in the context of general relativity [18, 19], ξ is an integrated quantity depending on the matter distribution of the system, $K^{\mu} \doteq \nabla^{\mu} \ln F$, where ∇^{μ} denotes the covariant derivative, and the nonminimal function F depends arbitrarily on the spacetime metric $g_{\mu\nu}$ and on the Riemann curvature tensor $R_{\mu\nu\alpha}^{\ \ \ \ \beta}$. From (45.12), the test particle acceleration

$$\boldsymbol{A}_{\mathrm{nmc}} = -\frac{\xi\left[c^2\boldsymbol{K} - cK_0\boldsymbol{v} + (\boldsymbol{K}\cdot\boldsymbol{v})\,\boldsymbol{v}\right]}{c\mathfrak{m}}, \tag{45.13}$$

written in the usual three-vector notation, can be extracted. In deriving (45.13), we assumed the slow-motion approximation in such a way that $v^{\mu} \approx \{c, \boldsymbol{v}\}$.

A straightforward but cumbersome perturbative calculation can be performed with the standard Gauss equations for the variation of the Keplerian orbital elements, implying the decomposition of (45.13) along the radial, transverse and normal directions of an orthonormal trihedron comoving with the particle and their evaluation onto a Keplerian ellipse, usually adopted as unperturbed reference trajectory. Such a procedure, which has the advantage of being applicable to whatsoever perturbing acceleration, yields, to zero order in the eccentricity e of the test particle, the following percent secular variation of its semimajor axis

$$\frac{\dot{a}}{a} = \frac{2\xi K_0}{\mathfrak{m}} + \mathcal{O}(e) . \tag{45.14}$$

It must be stressed that, for the quite general class of theories covered in [17], $\mathfrak{m}, \xi, K_0$ are, in general, not constant. As a working hypothesis, in obtaining (45.14) we assumed that they can be considered unchanging over the period of the test particle. Thus, there is still room for a slow temporal dependence with characteristic time scales quite larger that the test particle's period. Such a feature is important to explain the fact that, at present, there is no evidence for any anomalous increase of the Sun-Earth distance as large as a few meters per year, as it would be required by (45.11). Indeed, it can always be postulated that, in the last $\approx$2 Ga, $\mathfrak{m}, \xi, K_0$ became smaller enough to yield effects below the current threshold of detectability which, on the basis of the results in [15], was evaluated to be of the order of [16] $\approx 1.5 \times 10^{-2}$ m a^{-1} for the Earth. The rate of change of (45.14) is an important result since it yields an effect which is rooted in a well defined theoretical framework. It also envisages the possibility that a modification of the currently accepted laws of the gravitational interaction can, in principle, have an impact on the ancient history of our planet and, indirectly, even on the evolution of the life on it.

45.4 Competing Classical Orbital Effects

According to [12], a gravitational billiard involving planet-planet scattering between the Earth itself and a rogue rocky protoplanetesimal X, with $m_X \approx 0.75 m_\oplus$, which would have impacted on Venus, might have affected the Earth in the desired way during the Archean. However, compelling evidences for it are still missing.

Another classical physical mechanism which, in principle, may lead to the desired orbit expansion is a steady mass loss from either the Sun [7] or the Earth itself [13, 14]. However, it presents some difficulties both in terms of the magnitude of the mass loss rate(s) required, especially as far as the Earth's hydrosphere is concerned, and of the timescale itself. Suffice it to say that the Earth should have lost about 2 % of its current mass during the Archean. Moreover, it is generally accepted that a higher mass loss rate for the Sun due to an enhanced solar wind in the past could last for just (0.2–0.3) Ga at most.

45.5 Conclusions

In conclusion, it is entirely possible that the Faint Young Sun paradox can be solved by a stronger greenhouse effect on the early Earth; nonetheless, the quest for alternative explanations should definitely be supported and pursued.

Acknowledgments I thank the organizers of the Karl Schwarzschild Meeting 2013 for their kind invitation and for their financial support.

References

1. J.N. Bahcall, M.H. Pinsonneault, S. Basu, Solar models: current epoch and time dependences, neutrinos, and helioseismological properties. Astrophys. J. **555**, 990–1012 (2001)
2. O. Bertolami, J. Páramos, Mimicking the cosmological constant: constant curvature spherical solutions in a nonminimally coupled model. Phys. Rev. D **84**, 064022 (2011)
3. M. Carrera, D. Giulini, Influence of global cosmological expansion on local dynamics and kinematics. Rev. Mod. Phys. **82**, 169–208 (2010)
4. Feulner, G.: The faint young Sun problem. Rev. Geophys. **50**, RG2006 (2012)
5. C. Goldblatt, K.J. Zahnle, Faint young Sun paradox remains. Nature **474**, E1 (2011)
6. D.O. Gough, Solar interior structure and luminosity variations. Sol. Phys. **74**, 21–34 (1981)
7. V. Holzwarth, M. Jardine, Theoretical mass loss rates of cool main-sequence stars. Astron. Astrophys. **463**, 11–21 (2007)
8. G.S. Jenkins, A general circulation model study of the effects of faster rotation rate, enhanced CO_2 concentration, and reduced solar forcing: Implications for the faint young sun paradox. J. Geophys. Res. **98**, 20803 (1993)
9. J.F. Kasting, Early Earth: faint young Sun redux. Nature **464**, 687–689 (2010)
10. H. Kienert, G. Feulner, V. Petoukhov, Faint young Sun problem more severe due to ice-albedo feedback and higher rotation rate of the early Earth. Geophys. Res. Lett. **39**, 23710 (2012)
11. B. Longdoz, L.M. Francois, The faint young sun climatic paradox: influence of the continental configuration and of the seasonal cycle on the climatic stability. Global Planet. Change **14**, 97–112 (1997)
12. D. Minton, Was the Earth always at 1 AU (and was the Sun always one solar mass)? http://realserver4.stsci.edu/t/data/2012/04/2983/Minton.pdf 2012. Talk delivered at the conference *2012 The Faint Early Sun* at the Space Telescope Science Institute in Baltimore on 10 April 2012
13. J.E. Owen, Y. Wu, Kepler planets: a tale of evaporation. Astrophys. J. **775**, 105 (2013)
14. D. Perez-Becker, E. Chiang, Catastrophic evaporation of rocky planets. Mon. Not. Roy. Astron. Soc. **433**, 2294–2309 (2013)
15. E. Pitjeva, Use of optical and radio astrometric observations of planets, satellites and spacecraft for ephemeris astronomy, in *A Giant Step: from Milli- to Micro-arcsecond Astrometry, Proceedings IAU Symposium*, vol. 248, eds. by W. Jin, I. Platais, M. Perryman (Cambridge University Press, Cambridge, 2007), pp. 20–22
16. E.V. Pitjeva, N.P. Pitjev, Relativistic effects and dark matter in the Solar system from observations of planets and spacecraft. Mon. Not. Roy. Astron. Soc. **432**, 3431–3437 (2013)
17. D. Puetzfeld, Y.N. Obukhov, Covariant equations of motion for test bodies in gravitational theories with general nonminimal coupling. Phys. Rev. D **87**, 044045 (2013)
18. W.G. Dixon, A covariant multipole formalism for extended test bodies in general relativity. Il Nuovo Cimento **34**, 317–339 (1964)

19. J.L. Synge, *Relativity: The General Theory* (North Holland, Amsterdam, 1960)
20. C. Sagan, G. Mullen, Earth and Mars: evolution of atmospheres and surface temperatures. Science **177**, 52–56 (1972)

Chapter 46
A Practical Look at Regge Calculus

Dimitri Marinelli and Giorgio Immirzi

Abstract Regge calculus is the classical starting point for a bunch of different models of quantum gravity. Moreover, it is often considered a finite-element discretization of general relativity, providing so a potential practical scheme for numerical relativity. Despite these important roles, not many sample calculations have been studied. I will present one detailed example which, step-by-step, will show the power and the limits of this model. Many interesting aspects and open problems related to the Lorentzian structure of the discrete system will emerge.

Regge Calculus is a dynamical theory of space-time introduced in 1961 by Regge as a discrete approximation of the Einstein theory of gravity [1]. The basic idea is to replace a smooth space-time with a collection of simplices. The collective dynamics of these geometric objects is driven by the Regge action and the dynamical variables are their edge lengths (which play the role of the metric tensor of General Relativity). Simplices are generalizations of triangles and tetrahedra to arbitrary dimensions. In this context, they are convex pieces of space and they can be suitably glued together to build an extended geometric object *e.g.* triangles (or 2−simplices) glued together can form a 2-dimensional surface. An important point is that even if simplices are flat (so are "pieces" of space with zero curvature) the spaces they may generate, in general, are not (Fig. 46.1).

Coming back to General Relativity, from a heuristic point of view, we can think to build a (discrete) space-time gluing together pieces of 4−dimensional Minkowski space getting thus in general a globally non-flat space-time. Mathematically we are associating to a space-time a piecewise-linear simplicial manifold (PL-manifold in short).

A very illustrative example is the study of the discretized version of the Freedman-Robertson-Walker closed universe. The topology of this space-time is $S^3 \times I$

D. Marinelli (✉)
Dipartimento di Fisica, Università degli Studi di Pavia, via A Bassi 6, 27100 Pavia, Italy
e-mail: dimitri.marinelli@gmail.com

G. Immirzi
Colle Ballone, Montopoli di Sabina, Italy
e-mail: giorgio.immirzi@pg.infn.it

P. Nicolini et al. (eds.), *1st Karl Schwarzschild Meeting on Gravitational Physics*,
Springer Proceedings in Physics 170, DOI 10.1007/978-3-319-20046-0_46

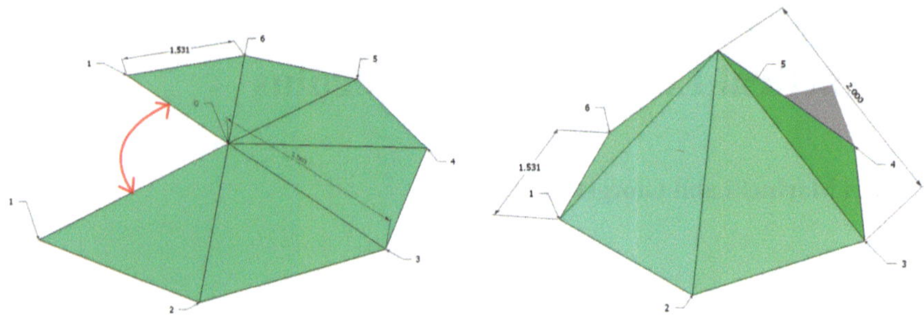

Fig. 46.1 On the *left*, geometric view of a PL-manifold with a non-null *deficit angle* at the central vertex (the two edges pointed by the *arrow* are identified). On the *right*, The PL-manifold of Fig. 46.1 embedded in $\mathbb{R}^3$, the curvature is concentrated in the vertex 0

(a 4-dimensional cylinder), that is particularly convenient, in fact, there is a clear distinction between time and space, moreover, its spatial foliation (S^3) is compact, so it can be described with a finite number of simplices making the calculation easier. This example corresponds to the 33th Problem listed by Wheeler in 1964 [2].

46.1 Triangulations

The first step to implement the Regge Calculus is to find a suitable PL-manifold to be associated with the space-time we want to approximate.

In this example, we want a foliated space-time. We can build a triangulation for a 3-dimensional hypersurface and then "evolve" it in a 4-dimensional PL-manifold.

We can use the boundaries of convex regular 4-polytopes with tetrahedral faces as triangulations for S^3 (in the same way as boundaries of tetrahedra, octahedra and icosahedra can be triangulations of S^2). 4-simplex (5-cell), 16-cell and 600-cell are the 4-polytopes with respectively 5, 16 and 600 boundary tetrahedra and they can be thought as triangulation refinements: 600-cell will better approximate the 3-sphere then the 5-cell. Here we choose 5-cell for the initial triangulation.

Now we have to build a triangulation that fills up the portion of space-time between two 3-dimensional compact foliations in a sort of "sandwich". Once fixed the arrow of time, namely choosing a future direction, and normalizing the interval $I := [0, 1]$, we can call one hypersurface "future" the one with $t = 1$ and "past" with $t = 0$.

We have implemented two different triangulations. One is found following an algorithm, the tent-like triangulation, proposed in [3] (See Fig. 46.2). The second triangulation is found combinatorially and is completely represented in the graph in Fig. 46.3.

This topology allows to compose "wider" space-times (longer evolutions) piling up several copies of these triangulations, identifying the past 3-dimensional foliation of one with the future of another one, and constructing thus a triangulation for a potentially infinitely long cylinder

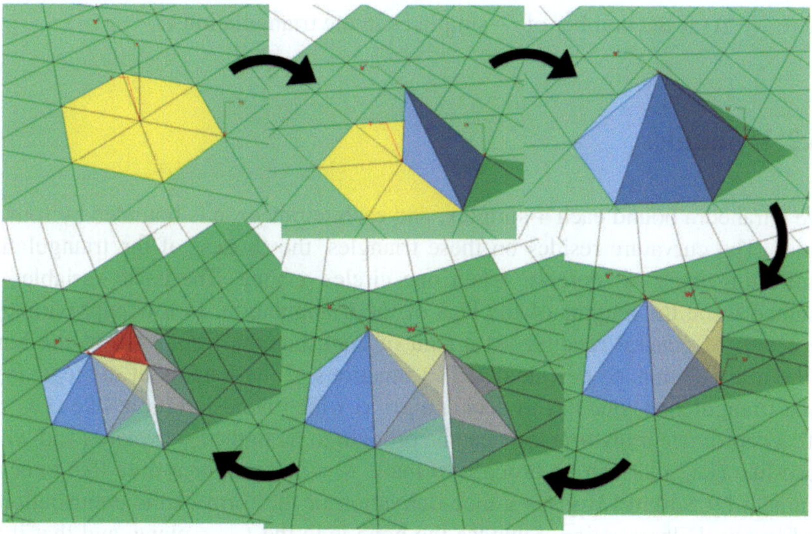

Fig. 46.2 An illustration of the tent-like evolution algorithm for the $(2+1)$ dimensional case

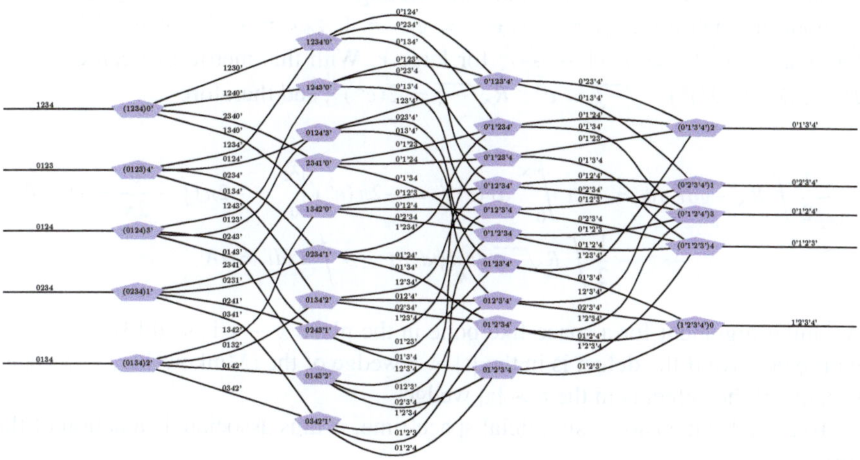

Fig. 46.3 Dual graph of the symmetric evolution: each vertex is a 4-simplex, each edge a tetrahedron

Some combinatorial relations must be fulfilled to check that the triangulation Δ is the result of an appropriate gluing process. Those relations are the Dehn-Sommerville relations. In cases where the triangulation has a boundary $\partial\Delta$, they read

$$N_k(\Delta) - N_k(\partial\Delta) = \sum_{i=0}^{4}(-1)^{i+4}\binom{i+1}{k+1} N_i(\Delta), \quad k = 0, \ldots, 4$$

where $N_i(\Delta)$ is the number of $i-$simplices in the triangulation Δ.

46.2 Dynamics

Five tetrahedra bound each 4-simplex, and each tetrahedron is bounded by four triangles. The curvature resides on these triangles, the 'bones' of the triangulation, manifesting itself as 'defects' when one circles a bone. The basic variables are assumed to be the lengths of the legs of the triangulation. In this approximation, the action of general relativity is a sum of contributions from each of the bones of the triangulation, to be calculated as functions of the lengths of the legs. For a given triangulation, one has to decide a priori which bones are time-like (i.e., that with a coordinate transformation can be brought to lie in the $t-z$ plane) and which space-like (i.e., that can be brought to the $x-y$ plane). For a time-like bone, an argument by Sorkin [3] gives a nice derivation of the form of the contribution to the Einstein-Hilbert action. Suppose the bone is in the $t-z$ plane, and that it has a defect θ; in the space-time around it is possible to introduce the cylindrical coordinates $x = r\cos\phi$, $y = r\sin\phi$, but deleting the wedge $2\pi-\theta \le \phi \le 2\pi$. Smooth the metric to: $g_{tt} = -1$, $g_{zz} = g_{rr} = 1$, $g_{\phi\phi} = e^{2\lambda(r)}$, with $e^{2\lambda(r)} = r^2$ for small r, $e^{2\lambda(r)} = r^2(1-\frac{\theta}{2\pi})^2$ for large r. With this metric one can calculate $R = 2(\lambda'' + (\lambda')^2)$, $\sqrt{-g} = e^{\lambda}$, $R\sqrt{-g} = 2(e^{\lambda})''$, and therefore

$$-\frac{1}{2}\int R\sqrt{-g}dr\,d\phi = -2\pi\int_0^{\infty} dr(e^{\lambda})'' = -2\pi(e^{\lambda})'\Big|_0^{\infty} = -2\pi(1-\frac{\theta}{2\pi}-1) = \theta$$
$$\rightarrow \quad -\frac{1}{2}\int R\sqrt{-g}\,dr\,d\phi dz dt = \theta\int dz dt = \theta A$$

A similar argument for a space-like bone in the plane $x-y$ is said [3] to give ηA with η positive if the defect is in the $|t| < z$ wedge of the (Minkowski) $t-z$ plane, negative if the defect is in the $t > |z|$ wedge.

To each 4-dimensional simplicial space-time, is thus associated an action of the form

$$S_R\left[\left\{l_{ij}^2\right\}\right] := \sum_f A_f\,\theta_f, \tag{46.1}$$

where the sum is over all the triangles (the "bones") and the A_f and θ_f are the area and "defect" associated with the triangle f. The action (46.1) can diverge (e.g., in a non-compact curved space-time), but its variation with respect to the dynamical variables (the square of the lengths) is well defined, and leads to a set of implicit equations: The discrete counterpart of the Einstein equations.

To evaluate this action it is useful to define two 4-vectors as shown in the paper [4]: Each tetrahedron in a 4-simplex has a unique outward pointing vector associated with it (up to a normalization factor) V^μ orthogonal to all its edges; each triangle inside this tetrahedron can have a vector orthogonal to V^μ and orthogonal to all of three its edges N^μ. If we normalize these vectors to +1 for time-like vectrs and -1 for space-like ones, we can give a neat definition of defect angle following the prescription of [5]:

- time-like bone:

$$\theta_f = 2\pi - \sum_{\sigma_4(f)} (\phi_f)_{\sigma_4}, \quad \cos\phi_f = \hat{N}\cdot\hat{N}' \tag{46.2}$$

- space-like bone:

$$\theta_f = -\sum_{\sigma_4(f)} (\phi_f)_{\sigma_4}, \sinh\phi_f = \epsilon(\hat{N}^2)\epsilon(\hat{V}\cdot\hat{N}')\hat{N}\cdot\hat{N}' \qquad \textbf{if } |\hat{N}\cdot\hat{N}'| < |\hat{V}\cdot\hat{N}'|$$

$$\sinh\phi_f = \epsilon(\hat{N}^2)\epsilon(\hat{N}\cdot\hat{N}')\hat{V}\cdot\hat{N}' \qquad \text{otherwise}| \tag{46.3}$$

the first case will occur if $\hat{V}^2 = -\hat{V}'^2$, the second if $\hat{V}^2 = \hat{V}'^2 = -1$.

Here we have denoted with a hat the normalized 4-vectors, $(\phi_f)_{\sigma_4}$ is the dihedral angle of the 4-simplex σ_4 in the triangular face f, $\sigma_4(f)$ is all the 4-simplices sharing the triangle f and $\epsilon(a)$ is the sign of the scalar a. Edge lengths of the simplices are scalar quantities, so they are invariant under coordinate changes

$$(l_{ij}^2) = \eta_{\mu\nu}(v_i - v_j)^\mu (v_i - v_j)^\nu = g_{\mu'\nu'}(v_i' - v_j')^{\mu'}(v_i' - v_j')^{\nu'} \tag{46.4}$$

where v_i^μ are the coordinates of the ith vertex of a simplex in the usual orthogonal coordinates where the metric is $\eta = \mathrm{diag}(-1, +1, +1, +1)$. It is possible [6] to associate with each 4-simplex a metric such that a vertex v_0 lies in the origin of the Minkowski space while vectors to the other vertices build an orthonormal base, namely $v_i^\mu = \delta_{i-1}^\mu$ and the metric reads

$$g_{\mu\nu} = -\frac{1}{2}\delta_\mu^{i+1}\delta_\nu^{j+1}(l_{ij}^2 - l_{i0}^2 - l_{j0}^2).. \tag{46.5}$$

Using this coordinate system, the coordinates 4-vectors N, V are simply linear combination of Kronecker deltas and the scalar products in (46.2) and (46.3) are just functions of the metric (46.5) and its inverse.

The variation of the action w.r.t. the edge lengths, gives a set of implicit equations. A matter part have to be considered in the non-vacuum case. These equations can be solved numerically as described in [7]. A numerical study of the evolution, will be published in a paper in preparation by the same authors of this work.

46.3 Discussion

The two PL-manifolds presented in Sect. 46.1 are not the best choice if one is interested in Numerical Relativity. They are both a coarse-grained approximation of a cylindrical space-time, which for the homogeneous and isotropic case can even be solved analytically. But, the interest in this particular triangulations resides on the fact that they are composed of a very small number of simplices (20 and 30 4-simplices). In particular, the "tent-like triangulation", can be generalized to initial surface with different topologies, and initial triangulations with different number of simplices, making the evolution local and even parallelizable [7], providing thus a practical framework for Numerical Regge Calculus.

The PL-manifold described by the graph in Fig. 46.3, instead, is not easily generalizable to different initial simplicial surfaces (combinatorial constrain apply). On the other hand, it provides a triangulation which is homogeneous: All the 4-simplices are treated on the same footing. Moreover, due to its homogeneity, it has only four different kind of simplices (only two if we considered the symmetry with respect to time inversion). These features make this triangulation an ideal candidate to be promoted as a base triangulation for a spinfoam model (the path integral approach to quantum gravity) as proposed in [4].

Dynamical evolution, in General Relativity, is driven by equations of motion of hyperbolic type, namely we are dealing with a causal evolution. In Regge Calculus, on the other hand, the role of causality is not well understood with consequences also in those quantum gravity model based on a discrete space–time. Looking at Regge Calculus as a finite element approximation of space-time, causality have to be connected to the Courant–Friedrichs–Lewy condition [8] and the model we presented can be a useful tool to study this issue.

Acknowledgments DM acknowledges partial support from PRIN 2010-11 Geometrical and analytical theories of finite and infinite dimensional Hamiltonian systems.

References

1. T. Regge, Il Nuovo Cimento **19**(3), 558 (1961)
2. J.A. Wheeler, in *Relativity, Groups and Topology*, eds. by B. DeWitt, C. DeWitt (Gordon and Breach, London, 1964), pp. 317–522
3. R. Sorkin, Phys. Rev. D **12**(2), 385 (1975)
4. G. Immirzi (2013). arXiv:1311.6942 [gr-qc]
5. L. Brewin, Class. Quant. Grav. **28**(18), 185005 (2010). arXiv:1011.1885
6. J. Fröhlich, in *Advanced Series in Mathematical Physics*, vol. 15 (World Scientific, Singapore, 1992), pp. 523–545 (1992)
7. J.W. Barrett, M. Galassi, W.A. Miller, R.D. Sorkin, P.A. Tuckey, R.M. Williams, Int. J. Theor. Phys. **36**(4), 815 (1997). arXiv:gr-qc/9411008
8. P.D. Lax, IBM J. Res. Dev. **11**(2), 235 (1967)

Chapter 47
Boundary States of the Potts Model on Random Planar Maps

Benjamin Niedner, Max R. Atkin and John F. Wheater

Abstract We revisit the 3-states Potts model on random planar triangulations as a Hermitian matrix model. As a novelty, we obtain an algebraic curve which encodes the partition function on the disc with both fixed and mixed spin boundary conditions. We investigate the critical behaviour of this model and find scaling exponents consistent with previous literature. We argue that the conformal field theory that describes the double scaling limit is Liouville quantum gravity coupled to the (A_4, D_4) minimal model with extended $\mathcal{W}_3$-symmetry.

47.1 Introduction

Matrix models provide a powerful tool for defining quantum gravity partition functions in two dimensions. The conformal field theories describing their critical points generically correspond to Euclidean quantum gravity interacting with conformal matter [1]. The spectrum of the theory is then determined by the consistent boundary states and as a result, the computation of such boundary states using matrix models has been of continued interest [2–5]. In this contribution, we shall consider this problem for the 3-states Potts model coupled to gravity, whose partition function can be written as the $U(N) \times \mathbb{Z}_3$-invariant matrix integral

B. Niedner (✉) · J.F. Wheater
Rudolf Peierls Centre for Theoretical Physics, Oxford University, 1 Keble Road, Oxford OX1 3NP, UK
e-mail: benjamin.niedner@physics.ox.ac.uk

J.F. Wheater
e-mail: j.wheater@physics.ox.ac.uk

M.R. Atkin
Université Catholique de Louvain, Chemin du Cyclotron 2, 1348 Louvain-la-Neuve, Belgium
e-mail: max.atkin@uclouvain.be

M.R. Atkin
Fakultät für Physik, Universität Bielefeld, Postfach 100131, 33501 Bielefeld, Germany

P. Nicolini et al. (eds.), *1st Karl Schwarzschild Meeting on Gravitational Physics*, Springer Proceedings in Physics 170, DOI 10.1007/978-3-319-20046-0_47

$$Z(N, c, g) = \int \prod_{i=1}^{3} \mathrm{d}X_i \; \mathrm{e}^{-N \mathrm{tr}[\sum_{i=1}^{3} V(X_i) - \sum_{\langle ij \rangle} X_i X_j]}\,, \tag{47.1}$$

where we have set $V(x) = cx^2/2 + gx^3/3$ and the X_i are $N \times N$ Hermitian matrices. This matrix model was first partially solved in [6], followed by the more detailed investigations [7–9]; related results have been obtained from combinatorial approaches [10, 11].

To find the relevant observables in this matrix model, we note that in flat space, the critical point of the 3-states Potts model is described by the (A_4, D_4) minimal model, the simplest model in which the presence of an additional higher-spin current extends the Virasoro algebra to the $\mathscr{W}_3$-algebra. Its conformally invariant boundary states were classified in [12, 13] and are listed in Table 47.1 together with their discrete counterparts. Accordingly, we introduce the generating function for triangulations of the disc with fixed color i on the boundary,

$$w_i(x) = \frac{1}{N} \sum_{k=0}^{\infty} \langle \mathrm{tr}\, X_i^k \rangle x^{-k-1} = \frac{1}{N} \left\langle \mathrm{tr} \frac{1}{x - X_i} \right\rangle . \tag{47.2}$$

We see from Table 47.1 that the continuum limit of these observables should correspond to the 3 states $\mathbb{1}$, ψ and $\psi^\dagger$. The mixed-color boundary conditions ε, σ and $\sigma^\dagger$ may be imposed by considering

$$w_+(x) = \frac{1}{N} \left\langle \mathrm{tr} \frac{1}{x - (X_i + X_j)} \right\rangle , \qquad i \neq j\,. \tag{47.3}$$

The remaining boundary conditions F and N correspond to operators not contained in the bulk spectrum of (A_4, D_4); these will not be discussed herein. The purpose of

Table 47.1 Boundary states in the (A_4, D_4) minimal model, their decomposition in (A_4, A_5) Virasoro modules, their conformal weights h, their $\mathbb{Z}_3$ charge q, and the corresponding microscopic boundary conditions

	(A_4, D_4)	(A_4, A_5)	h	q	
$\mathbb{1}$	$(1, 1)$	$(1, 1) \oplus (1, 5)$	$0, 3$	0	X_1
F	$(1, 2)$	$(1, 2) \oplus (1, 4)$	$1/8, 13/8$	–	$X_1 + X_2 + X_3$
ψ	$(1, 3)$	$(1, 3)$	$2/3$	$+1$	X_2
$\psi^\dagger$	$(1, 4)$	$(1, 3)$	$2/3$	-1	X_3
ε	$(2, 1)$	$(2, 1) \oplus (3, 1)$	$2/5, 7/5$	0	$X_2 + X_3$
N	$(2, 2)$	$(2, 2) \oplus (2, 4)$	$1/40, 21/40$	–	
σ	$(2, 3)$	$(2, 3)$	$1/15$	$+1$	$X_1 + X_3$
$\sigma^\dagger$	$(2, 4)$	$(2, 3)$	$1/15$	-1	$X_1 + X_2$

this note is then to compute the functions (47.2) and (47.3) at large N.[1] In the next section, we shall demonstrate that this can be achieved by requiring the saddle point equations to have a consistent analytic continuation in this limit.

47.2 Discrete Solution

In this section, we determine the disc partition functions with boundary conditions corresponding to the orbits $\{\mathbb{1}, \psi, \psi^\dagger\}$ and $\{\varepsilon, \sigma, \sigma^\dagger\}$ under the action of $\mathbb{Z}_3$ in Table 47.1 for generic values of the couplings c and g. Consider the change of variables

$$X_1 = \frac{1}{2}(X_+ + X_-) - \frac{c+1}{2g}\,, \qquad X_2 = \frac{1}{2}(X_+ - X_-) - \frac{c+1}{2g}\,, \tag{47.4}$$

so that the resolvent of X_+ coincides with the expression (47.3) for the disc partition function with boundary condition $\sigma^\dagger$ up to a shift in X_+ which leaves Z invariant. We may pick $w_+(x)$ and $w_3(x)$ as representatives of each $\mathbb{Z}_3$ orbit. In these variables, the integrand in (47.1) is Gaussian in X_-, which can hence be integrated out, leaving us with an integral over just two matrices. Upon gauge-fixing the $U(N)$-symmetry, we can then carry out the integral over the unitary group using the well-known result [14]. Denoting the respective eigenvalues of X_+ and X_3 by x_+^i and x_3^i, $i = 1 \dots N$, the resulting saddle point equations read

$$\frac{\partial U_+(x_+^i)}{\partial x_+^i} = \frac{1}{N}\left(\frac{\partial}{\partial x_+^i}\ln \det_{k,l}\, e^{N x_+^k x_3^l} + \sum_{j<i}\frac{1}{x_+^i - x_+^j} - \sum_{j}\frac{1}{x_+^i + x_+^j}\right), \tag{47.5a}$$

$$\frac{\partial U_3(x_3^i)}{\partial x_3^i} = \frac{1}{N}\left(\frac{\partial}{\partial x_3^i}\ln \det_{k,l}\, e^{N x_+^k x_3^l} + \sum_{j<i}\frac{1}{x_3^i - x_3^j}\right). \tag{47.5b}$$

Here, U_+ and U_3 are polynomials of degree 3 with coefficients determined by $V(x)$ and (47.4). Since the left-hand side of (47.5) is holomorphic, the above equations can be analytically continued to the complex plane for any N. On the other hand, for $N \to \infty$, individual expressions on the right-hand side will develop branch cuts located at the support of the densities of eigenvalues. To determine their analytic continuations, we follow [15] in introducing the functions

$$x_3(x) = \lim_{N\to\infty}\frac{1}{N}\frac{\partial}{\partial x_+^i}\ln \det_{k,l} e^{N x_+^k x_3^l}\Bigg|_{x_+^i = x}\,, \qquad x_+(x) = \lim_{N\to\infty}\frac{1}{N}\frac{\partial}{\partial x_3^i}\ln \det_{k,l} e^{N x_+^k x_3^l}\Bigg|_{x_3^i = x}\,. \tag{47.6}$$

[1] Note that whilst on the disc, the individual boundary conditions that form an orbit under the $\mathbb{Z}_3$ action are indistinguishable, this will generally not be the case for higher topologies.

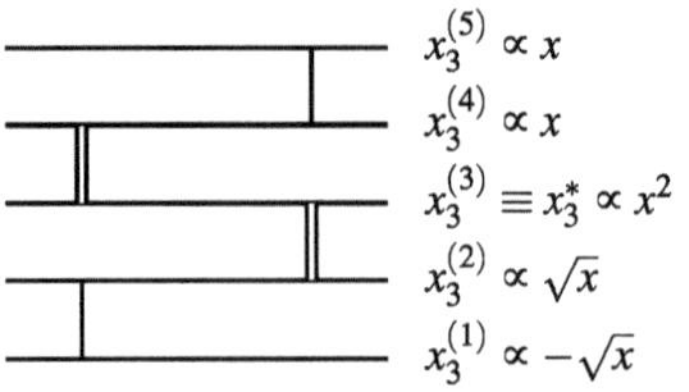

Fig. 47.1 Analytic structure of the function $x_3(x)$. *Horizontal lines* depict sheets, *vertical lines* cuts. Of the latter, *double lines* correspond to finite cuts on the real axis and *single lines* to cuts that extend to ∞

Let us denote the spectral density of X_+ by ρ_+, whose support we assume connected. It was shown in [15] that for $N \to \infty$,

1. $\exists \gamma \in \mathbb{R}$ such that the function $x_3(x)$ is analytic on $\mathbb{C} \setminus (-\infty, \gamma] \cup \operatorname{supp} \rho_+$,
2. its discontinuity across $\operatorname{supp} \rho_+$ coincides with that of $w_+(x)$ and
3. it is the functional inverse of $x_+(x)$.

By the symmetry of the definitions (47.6), the analogous statements apply to $x_+(x)$. Furthermore, let $x_3^*(x)$ (resp. $x_+^*(x)$) be the function obtained by analytic continuation through $\operatorname{supp} \rho_+$ onto the next sheet. Using properties 1 and 2 above, the large N limit of equations (47.5) for $x \notin (-\infty, \gamma] \cup \operatorname{supp} \rho_+$ can then be written as

$$U'_+(x) = x_3^*(x) + w_+(x) + w_+(-x) \,, \tag{47.7a}$$
$$U'_3(x) = x_+^*(x) + w_3(x) \,. \tag{47.7b}$$

These equations determine the desired disc partition functions in terms of the multi-valued function $x_3(x)$ and its inverse: $w_3(x)$ follows straightforwardly from (47.7b), and $w_+(x)$ may be obtained as the solution to the Riemann-Hilbert problem defined by properties 1 and 2, with the condition that $w_+(x) = x^{-1} + \mathcal{O}(x^{-2})$ as $x \to \infty$ as a consequence of the definition (47.3).

It remains to determine the function $x_3(x)$. To this purpose, note first that property 3 allows us to write (47.7b) as $U'_3(x_3(x)) = w_3(x_3(x)) + x$, which when expanded about $x = \infty$ gives the asymptotic behaviour $\pm\sqrt{x}$ of $x_3(x)$ on the initial sheet (47.6) and another sheet connected to it through $(-\infty, \gamma]$. We may then use (47.7a) to determine the cut structure and asymptotic behaviour on all other sheets by circling around the various branch points. The result of this procedure is depicted in Fig. 47.1. We thus make an ansatz that the function $x_3(x)$ takes values on an algebraic curve[2] $\mathcal{C}$,

$$\mathcal{C} = \{(x_3, x_+) | Q(x_3, x_+) = 0\} \tag{47.8}$$

$$Q(x_3, x_+) = \prod_{k=1}^{5} (x_3 - x_3^{(k)}(x_+)) \,. \tag{47.9}$$

---

[2]Closely related generating functions were proven to satisfy algebraic equations in [10]. We expect similar theorems to hold in our case.

Demanding that the coefficients of x_+ themselves be polynomials in x_3 fixes $Q(x_3, x_+)$ up to 10 unknown functions of the coupling constants. These unknowns are fully determined if we restrict to solutions for which the genus of the curve vanishes: in that case, we may parametrise $\mathscr{C}$ by two rational functions $\mathbb{CP}^1 \setminus \{z_i\} \to \mathscr{C}$. The poles z_i of these functions will correspond to the asymptotic regions on each sheet of $x_3(x)$. For concreteness, we may position these poles at the canonical points 0, 1 and ∞ using a conformal transformation of z. A possible parametrisation then reads

$$x_+(z) = \frac{\sum_{k=0}^{5} \alpha_k z^k}{z^2(z-1)}, \qquad x_3(z) = \frac{\sum_{k=0}^{5} \beta_k z^k}{z(z-1)^2}. \tag{47.10}$$

These functions are single-valued on the punctured Riemann sphere and cover $\mathscr{C}$ exactly once. Demanding that their Laurent expansion about each pole reproduce the appropriate asymptotic behaviour, we obtain 4 conditions per pole allowing us to solve for the 12 unknowns $\alpha_k, \beta_k, k = 0 \ldots 5$. This completes our solution for the disc partition functions with fixed and mixed boundary conditions.

47.3 Critical Behaviour

To explore the phase diagram of the model, we do not need to solve for the coefficients in (47.10) explicitly. Instead, we note that the possible critical exponents are determined by the multiplicity of the singularity at the left edge of the spectral density, which controls the large-order behaviour of the generating function $w_+(x)$. As we let g and c approach their critical values, the various branch points on the curve merge so that at the highest critical point $\mathscr{C}$ will exhibit a single singularity. Its multiplicity is fixed to 5 if the genus of $\mathscr{C}$ vanishes. This results in the conditions

$$\frac{\partial^{m+n}}{\partial x_+^m \partial x_3^n} Q(x_3, x_+) = 0, \qquad m + n < 5, \tag{47.11}$$

giving the critical values $c_c = 2 + \sqrt{47}$, $g_c = \sqrt{105}/2$ in agreement with [6, 9]; the singularity is located at $x_{+,c} = 0$, $x_{3,c} = -g_c^{-1}(4 + c_c)/2$. To find the scaling behaviour of $w_+(x)$ near this point, it is useful to resolve the branch points at γ via the change of variables $x(\zeta) = \gamma(1 - 2\zeta^2)$. We also introduce the auxiliary function

$$f(\zeta) = P(\zeta) - \int_{\zeta(\mathrm{supp}\,\rho_+)} \mathrm{d}\zeta' \frac{\rho_+(x(\zeta'))}{\zeta - \zeta'}, \qquad \zeta \notin \zeta(\mathrm{supp}\,\rho_+), \tag{47.12}$$

where $P(\zeta)$ is a polynomial of degree 4 and we integrate over the image of the support of ρ_+ obtained by selecting the positive branch of $\zeta(x)$. For a suitable choice of coefficients in P, (47.7a) then has the equivalent homogeneous form

$$2\,\mathrm{Re}\, f(\zeta) + f(-\zeta) + f(\sqrt{1-\zeta^2}) + f(-\sqrt{1-\zeta^2}) = 0\,, \quad \zeta \in \zeta(\mathrm{supp}\, \rho_+)\,. \tag{47.13}$$

If we parametrise the vicinity of the singularity as $\zeta \propto \cosh\phi$ so that $f(\zeta) \propto \cosh(\mu\phi)$, we find that (47.13) implies $5\mu = \pm 4n + 20m$ with $n \in \{1, 2\}$ and $m \in \mathbb{Z}$. Taking into account the upper bound on μ implied by the degree of $\mathscr{C}$ leads to the conclusion that $\mu = 12/5$ for $g = g_c$, $c = c_c$ and hence that at the critical point, $w_+(x)$ has scaling exponent $6/5$. Indeed, the corresponding value of the string susceptibility $\gamma_s = -1/5$ agrees with that found for fixed boundary conditions [6, 8, 9].

47.4 Discussion

In this contribution, we computed the partition function for the 3-states Potts model on the randomly triangulated disc with both fixed and mixed boundary conditions. By requiring that the large N saddle point equations possess a consistent analytic continuation, we found that both boundary conditions can be encoded in a single algebraic curve and determined the scaling behaviour of the disc partition functions at the highest critical point from its singularities. As expected, the resulting value of the string susceptibility $\gamma_s = -1/5$ is independent of the chosen boundary conditions and consistent with Liouville quantum gravity coupled to conformal matter with central charge $c = 4/5$. Given the symmetries of the model, the identification of the latter sector with the (A_4, D_4) modular invariant hence appears justified.

The algebraic curve (47.8) forms part of the initial data for topological recursion [16] which enables systematic computation of finite N corrections to various observables. Of particular interest would be the computation of cylinder amplitudes between different boundary states to probe the spectrum of the continuum theory beyond the planar limit. This would allow a check of the conjecture made in [17] that in the presence of gravity, not all states in Table 47.1 are independent.[3] We intend to address these and related issues in a forthcoming publication.

Acknowledgments BN would like to acknowledge support by the German National Academic Foundation and STFC grant ST/J500641/1.

References

1. P. Di Francesco, P.H. Ginsparg, J. Zinn-Justin, Phys. Rept. **254**, 1 (1995).arXiv:9306153
2. M.R. Atkin, J.F. Wheater, JHEP **1102**, 084 (2011). arXiv:1011.5989 [hep-th]
3. J.H. Oh, J. Park, C. Rim (2011). arXiv:1109.5465 [hep-th]
4. C.T. Chan, H. Irie, C.H. Yeh, Nucl. Phys. B **855**, 46 (2012). arXiv:1109.2598 [hep-th]
5. M.R. Atkin, S. Zohren, JHEP **1211**, 163 (2012). arXiv:1204.4482 [hep-th]

[3] Similar observations were made for the (A_p, A_q) minimal models coupled to gravity [2, 5, 18].

6. J.M. Daul (1994). arXiv:hep-th/9502014
7. G. Bonnet, Phys. Lett. B **459**, 575 (1999). arXiv:hep-th/9904058
8. B. Eynard, G. Bonnet, Phys. Lett. B **463**, 273 (1999). arXiv:hep-th/9906130
9. P. Zinn-Justin, J. Stat. Phys. **98**(1–2), 245 (2000).arXiv:cond-mat/9903385
10. O. Bernardi, M. Bousquet-Melou, J. Comb. Theor. B **101**(5), 315 (2011). arXiv:0909.1695
11. G. Borot, J. Bouttier, E. Guitter, J. Phys. A **45**(49), 494017 (2012). arXiv:1207.4878 [math-ph]
12. J.L. Cardy, Nucl. Phys. B **324**(3), 581 (1989)
13. I. Affleck, M. Oshikawa, H. Saleur, J. Phys. A **31**(28), 5827 (1998). arXiv:cond-mat/9804117
14. C. Itzykson, J. Zuber, J. Math. Phys. **21**(3) (1980)
15. P. Zinn-Justin, Commun. Math. Phys. **194**(3), 631 (1998).arXiv:cond-mat/9705044
16. B. Eynard, N. Orantin, J. Phys. A **42**(29), 293001 (2009)
17. S. Kawamoto, J.F. Wheater, S. Wilshin, Int. J. Mod. Phys. A **23**, 2257 (2008)
18. N. Seiberg, D. Shih, JHEP **0402**, 021 (2004). arXiv:hep-th/0312170

Chapter 48
One-Loop Effective Action in Quantum Gravitation

Leslaw Rachwal, Alessandro Codello and Roberto Percacci

Abstract We present the formalism of computing one-loop effective action for Quantum Gravitation using non-local heat kernel methods. We found agreement with previous old results. In main part of my presentation I considered the system of E-H gravitation and scalar fields. We were able to derive nonlocal quantum effective action up to the second order in heat kernel generalized curvatures. By going to flat spacetime expressions for gravitational formfactors are possible to construct and compare with the results from effective field theory for gravity.

In this work we will review the results of computation of 1-loop effective action in a system, where we have standard Einstein-Hilbert gravitation and a minimally coupled scalar field. Standard computation, known in the literature, are mainly based on perturbative quantization methods and they exploit Feynman diagrams techniques [1]. Here we will follow a different route. Namely we will obtain 1-loop quantum effective action as the effect of integrating average effective action along the RG flow trajectory from UV down to IR limit. Moreover in the core of our calculation we will use non-local heat kernel techniques to evaluate some functional traces. We will pay special attention to the appearance of nonlocal terms in the quantum effective action. All the calculations will be performed in Euclidean spacetimes and later we will specify to four spacetime dimensions. One of the goal of such calculation is the quantum effective action per se. Another is related to gravitational formfactors of simple interactions with scalars.

L. Rachwal (✉)
Abdus Salam International Centre for Theoretical Physics, Strada Costiera 11, 34014 Trieste, Italy
e-mail: grzerach@gmail.com

A. Codello · R. Percacci · L. Rachwal
Scuola Internazionale Superiore di Studi Avanzati, via Bonomea 265, 34136 Trieste, Italy

A. Codello
CP3–Origins and Danish Institute for Advanced Study, University of Southern Denmark, Odense, Denmark

P. Nicolini et al. (eds.), *1st Karl Schwarzschild Meeting on Gravitational Physics*, Springer Proceedings in Physics 170, DOI 10.1007/978-3-319-20046-0_48

48.1 Effective Average Action

First we want to introduce the notion of the average effective action (EAA). The EAA is a scale-dependent generalisation of the standard effective action, that interpolates smoothly between the bare action for $k \to \infty$ and the standard quantum effective action for $k \to 0$. In this way, we avoid the problems of performing the functional integral. Instead they are converted into the problem of integrating the exact flow of the EAA from the UV to the IR. The EAA formalism deals naturally with several different aspects of quantum field theories. One aspect is related to the discovery of non-Gaussian fixed points of the RG flow. In particular, the EAA framework is a useful setting to search for Asymptotically Safe theories, i.e. theories valid up to arbitrarily high energy scales. A second aspect, in which the EAA reveals its big usefulness, is the domain of nonperturbative calculations. In fact, the exact flow, that EAA satisfies is a valuable starting point for inventing new approximation schemes.

In EAA the crucial point is the separation between high and small energy modes of quantum fields. The elimination of higher energy modes is performed by separating the low energy modes, to be integrated out, from the high modes in a covariant way. To do this we introduce a cutoff action constructed using the covariant d'Alembertian, that respects the symmetries of the underlying theory. In full generality in order to construct EAA we add to the bare action S an infrared (IR) "cutoff" or regulator term ΔS_k of the form:

$$\Delta S_k = \frac{1}{2}\int \mathrm{d}^d x \, \sqrt{g}\, \phi R_k(\Box)\phi \, . \tag{48.1}$$

In above formula the operator kernel R_k is chosen in such a way to suppress the field modes ϕ_n, eigenfunctions of the covariant second order differential operator $\Box$, with eigenvalues smaller than the cutoff scale $\nu_n < k^2$. Generic fields of our quantum field theory are denoted here by ϕ. We will call ΔS_k the cutoff action. The functional form of the cutoff kernels $R_k(z)$ is arbitrary except for the requirements that they should be monotonically decreasing functions in both z and k arguments, i.e. rigorously that $R_k(z) \to 0$ for $z \gg k^2$ and that $R_k(z) \to k^2$ for $z \ll k^2$. It is important to recall two limits of EAA. First in the IR limit ($k = 0$) quantum effective action is obtained. On the other hand, when $k \to \infty$, then EAA equals to the bare action of considered quantum theory. In this way we obtain the scale dependent generalisation of the standard effective action, which interpolates between the two.

48.2 Truncation Ansatz and 'Inverse Propagator'

Quantum gravity gives unambiguous predictions at low energy in the framework of effective field theories. The low energetic action contains only the simplest Einstein-Hilbert term (with a possibility of adding a cosmological constant, which we however neglect here). In this effective theory there exist observables, which do not depend on the particular way of UV completion. They are genuine predictions of quantum gravity. The quantum divergences, which must be absorbed during the renormalization procedure, are contained in local, but not universal terms in the quantum effective action. We are mainly interested in nonlocal terms in quantum effective action. The reason for this is that they are universal terms in low-energetic effective field theory of quantum gravity [2, 3]. They do not depend on any specific way of UV completion of gravity. There exist different ways, by which, one can obtain quantum effective action in the infrared limit. However it is without any doubt that low-energetic predictions of quantum gravity are calculable and solid, regardless of any complicated dynamics, which saves the theory in UV. In our method for integration RG flows we will use exact (also known as functional) Renormalization Group equations. In integration of RG flow of scale-dependent effective action such nonlocal terms originate from the part of integration done for the lowest momentum scales.

We will use the following ansatz for the form of the action of our system

$$S = \int \mathrm{d}^d x \sqrt{g} \left[\frac{1}{K^2} R - \frac{1}{2} (\partial\phi)^2 - V(\phi) \right] - \frac{1}{2K^2\alpha} \int \mathrm{d}^d x \sqrt{g}\, \chi^2 + \int \mathrm{d}^d x \sqrt{g}\, \bar{C}_\mu \left(-\Box \delta^\mu_\nu - R^\mu_\nu \right) C^\nu , \tag{48.2}$$

where d'Alembertian is given by $\Box = \nabla_\mu \nabla^\mu$. Due to the gauge diffeomorphism symmetry present in the system we are forced to introduce gauge fixing conditions necessary for perturbative quantization of the system: $\chi_\mu = \nabla^\nu h_{\mu\nu} - \frac{1}{2} \nabla_\mu h$. Moreover another consequence of this gauge redundancy is that for consistency, we also had to add vector ghosts denoted by C_μ in the second line of (48.2). In our computation we use the background field method and we take the metric perturbations in the form $h_{\mu\nu} = \delta g_{\mu\nu}$ and in contracted version $h = g^{\mu\nu} h_{\mu\nu}$. All covariant derivatives are with respect to the background metric. As we can see in the action (48.2) we included minimally coupled scalar field ϕ and we allowed for the existence of potential $V(\phi)$ for it. Gravitational coupling appears there as K, which has the inverse energy dimension. In the gravitational part of the action R is the only present curvature invariant built out of the full metric $g_{\mu\nu}$. Additionally constant α is a gauge parameter in our gauge fixing condition.

When we have the explicit form of the action, then the next step is to compute the second variational derivative w.r. to all fluctuating quantum fields like in [4]. Usually this takes the form of second order differential operator, which is of fundamental importance in our construction of the cutoff kernels in the EAA.

48.3 Exact RG Flows

Using the methods of nonlocal heat kernel we will now exploit the power of Exact RG formalism applied to the EAA. At the beginning we need to know flows of which terms to consider and for this reason we first look for simple task related to local terms.

48.3.1 Local Terms of One-Loop Effective Action

Firstly we will look for local terms in 1-loop effective action for our system. They are related to UV divergences of the theory. In general these divergences give rise to the renormalization of couplings in front of local terms. They are not universal and depend on the precise way of UV completion. However we assume, that the bare action (in UV) is given by (48.2). At one loop order the quantum effective action is given by the integral

$$\Gamma[\phi, g] = -\frac{1}{2}\int_0^\infty \frac{\mathrm{d}s}{s}\mathrm{Tr}\,\mathrm{e}^{-s\hat{S}^{(2)}}\,, \tag{48.3}$$

where $\mathrm{Tr}\,\mathrm{e}^{-s\hat{S}^{(2)}}$ is the functional trace of some differential operator, which we are going to compute with the heat kernel techniques. For our applications in the exponent of heat kernel we use inverse propagator, spoken about in the previous section, denoted here by $\hat{S}^{(2)}$ (second variational derivative of the action S with respect to all fluctuating fields). This operator, as other quantities with a hat over, is a matrix in field space of gravitons and scalar field perturbations. In order to find logarithmically divergent part of one-loop effective action to second order in curvature we can use the Schwinger-DeWitt method for quadratic operators:

$$\begin{aligned}\mathrm{Tr}\,\mathrm{e}^{-s\hat{S}^{(2)}} &= \frac{1}{(4\pi s)^{d/2}}\int \mathrm{d}^d x\,\sqrt{g}\,\mathrm{tr}\left\{\hat{1} + s\hat{P} + s^2\left[\frac{1}{2}\hat{P}^2 + \frac{1}{12}\hat{\mathcal{R}}_{\mu\nu}\hat{\mathcal{R}}^{\mu\nu}\right.\right.\\ &\left.\left.+\frac{1}{180}\mathrm{Riem}^2\hat{1} - \frac{1}{180}R_{\mu\nu}R^{\mu\nu}\hat{1}\right]\right\}\,. \end{aligned}\tag{48.4}$$

We will restrict ourselves to second order contribution in operators $\hat{P}$, $\hat{\mathcal{R}}_{\mu\nu}$ and gravitational curvatures. (We don't consider here application of this method to the ghost part of the action, because we are mainly interested in terms with nonminimally coupled matter.) Using Schwinger-DeWitt technique we reduced the functional trace to matrix traces. The terms, which appear explicitly in the above expression are the basis for consideration of RG flows for nonlocal operators.

48.3.2 Nonlocal Terms and Their Exact RG Flows

In order to go beyond Schwinger-DeWitt technique and find form of nonlocal part of one-loop action we insert nonlocal structure functions. They are functions of s parameter and box operator $\Box = \nabla^\mu\nabla_\mu$ (acting under the integral). We insert these structure functions between two matrix operators present at the second order as in the detailed formula below

$$\frac{1}{(4\pi s)^{d/2}}\int \mathrm{d}^d x\,\sqrt{g}\,s^2\,\mathrm{tr}\left\{\hat{P}\,f_P\,(-s\Box)\,\hat{P}+\hat{\mathscr{R}}_{\mu\nu}\,f_{\mathscr{R}}\,(-s\Box)\,\hat{\mathscr{R}}^{\mu\nu}\right.$$
$$\left.+\hat{P}\,f_{PR}\,(-s\Box)\,R+R\,f_R\,(-s\Box)\,R\,\hat{1}+R_{\mu\nu}\,f_{\mathrm{Ric}}\,(-s\Box)\,R^{\mu\nu}\,\hat{1}\right\}. \quad (48.5)$$

It must be emphasised, that the leading order in s contribution is equal to constants, which were written in the formula (48.4) in section above (for $\hat{P}R$ operator this constant vanishes). Moreover we have used the Euler identity here. The traces of matrix terms of order curvature square are modified with respect to expressions given in previous section by the appearance of structure functions f_P, $f_{\mathscr{R}}$, f_{PR}, f_R and f_{Ric}.

Now we want to consider the exact RG flow of EAA, which will be denoted here by $\bar{\Gamma}_k$. As the ansatz for it we choose the nonlocal expression above, understood that all the couplings and structure functions now acquire dependence on the momentum scale k. The exact RG flow equation for the background effective average action (bEAA) is the following

$$\partial_t\bar{\Gamma}_k[\phi,g]=\frac{1}{2}\mathrm{Tr}\frac{\partial_t R_k(-\mathscr{D}^2))}{-\mathscr{D}^2+R_k(-\mathscr{D}^2)}-\mathrm{Tr}\frac{\partial_t R_k(\Delta_{gh})}{\Delta_{gh}+R_k(\Delta_{gh})}. \quad (48.6)$$

In the above formula $\mathscr{D}$ is a general operator of the covariant derivative and R_k are cutoff kernels (suitably chosen functions of momenta to suppress the contributions from high energy modes in the path integral. We denote by Δ_{gh} the kinetic operator in the vector ghost sector.). The r.h.s. of this equation expresses itself by functional traces of some differential operators and the RG time derivatives of cutoff kernels where $t=\log k/k_0$. We note that in the denominator we have differential part $\mathscr{D}^2$ of our inverse propagator operator. The r.h.s. of the flow equation is then schematically given as

$$\partial_t\bar{\Gamma}_k[\phi,g]=\frac{1}{(4\pi)^{d/2}}\int \mathrm{d}^d x\,\sqrt{g}\,\left\{\mathscr{O}_{i,1}\left[\int_0^\infty \mathrm{d}s\,\tilde{h}_k(s)\,s^{2-\frac{d}{2}}\,\tilde{f}_i(-s\Box)\right]\mathscr{O}_{i,2}\right\}, \quad (48.7)$$

where the structure functions $\tilde{f}_i(x)$ were derived combining non-local heat kernel structure functions and $\mathscr{O}_{i,1,2}$ stand for operators in between which we insert these structure functions. In this derivation we follow [5].

48.4 Effective Action and Formfactors

Finally integrating the flow (48.7) and putting some boundary conditions in UV, we arrive to the following explicit form for one-loop quantum effective action in our model of scalar field interacting minimally with Quantum Gravitation:

$$
\begin{aligned}
\bar{\Gamma}_0\big|_{\mathscr{R}^2} = \frac{1}{32\pi^2}\int \mathrm{d}^4x\sqrt{g}\Bigg\{&\frac{43}{30}R_{\mu\nu}\log\left(\frac{-\Box}{k_0^2}\right)R^{\mu\nu}+\frac{1}{20}R\log\left(\frac{-\Box}{k_0^2}\right)R \\
&+\frac{5}{2}K^4m^4\phi^2\log\left(\frac{-\Box}{k_0^2}\right)\phi^2-2K^2m^4\phi\log\left(\frac{-\Box}{k_0^2}\right)\phi \\
&-\frac{13}{3}K^2m^2R\log\left(\frac{-\Box}{k_0^2}\right)\phi^2-\frac{1}{6}m^2R+\frac{1}{2}m^4+\frac{5}{2}K^4\left(\nabla\phi\right)^2\log\left(\frac{-\Box}{k_0^2}\right)(\nabla\phi)^2 \\
&+K^4m^2\phi^2\log\left(\frac{-\Box}{k_0^2}\right)(\nabla\phi)^2-\frac{2}{3}K^2R\log\left(\frac{-\Box}{k_0^2}\right)(\nabla\phi)^2-K^2m^2\left(\nabla\phi\right)^2\Bigg\}.
\end{aligned}
\tag{48.8}
$$

This 1-loop quantum effective action is the main, solid result of this work.

The goal of this section would be to compute one-loop corrections to three-point vertex from above action. Above we have computed it to the second order in operators of heat kernel and we arrived at a nonanalytic expression with low-energetic logarithms. We want to consider the simplest vertex of interaction within our theory—with one gravitons and two scalar fields. That's why we shall compute the third variational derivative with respect to mentioned fluctuations. At the end we specify flat gravitational background and vanishing background scalar field. We also prefer to write the expression for the vertex in the momentum space and in such way we can compare to the perturbative results. Such comparison and more details of this computation and these techniques can be found in [6]. However in this short contribution we shed light only on the most important aspects of the lengthy calculations.

Acknowledgments L.R. is supported by the European Programme Unification in the LHC Era (UNILHC), under the contract PITN-GA-2009-237920.

References

1. J.F. Donoghue, in Glasgow 1994, Proceedings, High Energy Physics, vol. 2, pp. 1061–1063, Massachusetts U. Amherst - UMHEP-413 (rec.Oct.94) 6 p. arXiv:hep-th/9409143
2. J.F. Donoghue. arXiv:gr-qc/9512024
3. J.F. Donoghue, AIP Conf. Proc. **1483**, 73 (2012). arXiv:1209.3511 [gr-qc]
4. A.O. Barvinsky, A.Y. Kamenshchik, I.P. Karmazin, Phys. Rev. D **48**, 3677 (1993). arXiv:gr-qc/9302007
5. A. Satz, A. Codello, F.D. Mazzitelli, Phys. Rev. D **82**, 084011 (2010). arXiv:1006.3808 [hep-th]
6. A. Codello, R. Percacci, L. Rachwal, A. Tonero, arXiv:hep-th/1505.03119
7. L. Rachwal, 'Models for RG running for gravitational couplings and applications. Ph.D. thesis SISSA, Trieste (2013) http://hdl.handle.net/1963/7061

Chapter 49
Heavy Probes in Strongly Coupled Plasmas with Chemical Potential

Andreas Samberg and Carlo Ewerz

Abstract We study the properties of heavy probes moving in strongly coupled plasmas at finite chemical potential. Using the gauge/gravity duality we consider large classes of gravity models consisting in deformed AdS_5 spacetimes endowed with Reissner–Nordström-type black holes. We report on our analysis of the screening distance of a quark–antiquark pair, its free energy, and the running coupling. These observables show a certain insensitivity as to which model and deformation is used, pointing to strong-coupling universal behavior. Thus, the results may be relevant for modeling heavy quarkonia traversing a quark–gluon plasma at finite net baryon density, and their suppression by melting.

49.1 Introduction

Over the past years, gauge/gravity duality ([1–3]; see e.g. [4] for a review) has been successfully applied to the physics of the QCD medium created in heavy ion collisions at RHIC and LHC. One of the most prominent theoretical results is the computation of a universal value for the ratio of the shear viscosity η to the entropy density s in a large class of strongly coupled deconfined plasmas [5, 6] and the conjecture that this value ($\eta/s = 1/(4\pi)$) may be a lower bound for all physical substances [7]. On the experimental side, it was found that the Quark–Gluon Plasma (QGP) created at RHIC and LHC has a value of η/s in the ballpark of the holographic result, indicating strong coupling at the accessible temperatures.

A. Samberg (✉) · C. Ewerz
Institut für Theoretische Physik, Ruprecht-Karls-Universität Heidelberg,
Philosophenweg 16, D-69120 Heidelberg, Germany
e-mail: a.samberg@thphys.uni-heidelberg.de

C. Ewerz · A. Samberg
ExtreMe Matter Institute,
GSI Helmholtzzentrum für Schwerionenforschung,
Planckstraße 1, D-64291 Darmstadt, Germany
e-mail: c.ewerz@thphys.uni-heidelberg.de

P. Nicolini et al. (eds.), *1st Karl Schwarzschild Meeting on Gravitational Physics*,
Springer Proceedings in Physics 170, DOI 10.1007/978-3-319-20046-0_49

Beyond current high-energy heavy ion programs, which mainly study the high-temperature, low-chemical potential regime of QCD, a major open question is the phase structure of QCD at nonzero baryon chemical potential. Experimentally, this will be addressed for example at future FAIR experiments. Theoretically, the ability to handle nonzero chemical potential in QCD or at least QCD-like theories is crucial.

We explore, by means of the gauge/gravity duality, the phase structure of strongly coupled non-conformal theories similar to QCD by investigating the physics of probes in the thermal plasmas of these theories with nonzero chemical potential. In a spirit similar to the holographic computation of η/s we search for universality in the behavior of heavy quark–antiquark ($Q\bar{Q}$) bound states in large classes of holographic theories. This may yield insight into certain QCD processes relevant for the QGP produced in heavy ion collisions, e.g. suppression of charmonia [8] or bottomonia.

Specifically, we analyze the $Q\bar{Q}$ screening distance, the $Q\bar{Q}$ free energy in the medium (roughly speaking the interaction potential), and the running coupling extracted from the free energy. Previous work on similar problems includes [9, 10] for vanishing chemical potential, and [11, 12] for nonzero chemical potential in $\mathcal{N} = 4$ supersymmetric Yang–Mills theory. A more detailed account of our findings will be published elsewhere.

We start with the prototype of gauge/gravity duality between classical supergravity on AdS_5 and conformal $\mathcal{N} = 4$ supersymmetric Yang–Mills theory (SYM) with gauge group $\mathrm{SU}(N_c)$ in the limit of infinite number of colors, $N_c \to \infty$, and large 't Hooft coupling $\lambda \equiv g_{\mathrm{YM}}^2 N_c$. A thermal bath for the gauge theory is dual to a black hole in AdS_5. Putting charge on the black hole, we can induce a chemical potential in the dual theory. Therefore, as a starting point we consider AdS_5 with a Reissner–Nordström black hole (AdS-RN), in Poincaré coordinates,

$$\mathrm{d}s^2 = \frac{R^2}{z^2}\left(-h(z)\mathrm{d}t^2 + \mathrm{d}\mathbf{x}^2 + \frac{\mathrm{d}z^2}{h(z)}\right) \tag{49.1}$$

$$\text{with} \quad h(z) = 1 - \left(1 + Q^2\right)\left(\frac{z}{z_\mathrm{h}}\right)^4 + Q^2\left(\frac{z}{z_\mathrm{h}}\right)^6 .$$

Here, R is the curvature scale and the black hole horizon is at $z = z_\mathrm{h}$. The dual theory is in a thermal state with temperature $T = (1 - \frac{1}{2}Q^2)/(\pi z_\mathrm{h})$ and chemical potential $\mu = \sqrt{3}Q/z_\mathrm{h}$. We have $0 \leq Q \leq \sqrt{2}$.

To come closer to real-world physics, we study models in which conformality is explicitly broken by deforming the AdS spacetime. On the one hand, we consider the *CGN model* proposed by Colangelo et al. [13]. It is specified by the metric (49.1) with an additional overall warp factor $\mathrm{e}^{c^2 z^2}$ with deformation parameter c. On the other hand, we study a family of *1-parameter models* which we derive from the action used in [14], which adds to 5-dimensional gravity with metric $g_{\mu\nu}$ and negative cosmological constant a scalar field ϕ and a U(1) gauge field A_μ whose boundary value equals the chemical potential in the dual gauge theory. Our ansatz with deformation parameter κ is

$$g_{\mu\nu}\mathrm{d}x^{\mu}\mathrm{d}x^{\nu} = \mathrm{e}^{2A(z)}\left(-h(z)\mathrm{d}t^2 + \mathrm{d}\mathbf{x}^2\right) + \frac{\mathrm{e}^{2B(z)}}{h(z)}\mathrm{d}z^2\,, \tag{49.2}$$

$$A(z) = \log\left(\frac{R}{z}\right), \qquad \phi(z) = \sqrt{\frac{3}{2}}\kappa z^2, \qquad A_{\mu}\mathrm{d}x^{\mu} = \Phi(z)\mathrm{d}t\,, \tag{49.3}$$

where R is a constant and $h(z)$ is the redshift factor induced by the black hole. We derive two classes of models from this ansatz, treating ϕ as the dilaton or not, called 'string frame' and 'Einstein frame' models, respectively. The solutions can be given in closed form [15]. At fixed (μ, T), a maximal deformation $\kappa_{\max}$ exists that still allows a black hole solution representing (μ, T). Since these models solve gravity equations of motion (EOMs) they are expected to be thermodynamically consistent, as opposed to models in which the metric is deformed 'by hand'.

In the CGN model we find unusual behavior in some observables at low temperatures and chemical potentials. Such behavior occurs when we consider moving probes, for example in the case of the drag force. We believe that these artifacts are unphysical. Indeed, they do no longer occur in models obtained as solutions of gravity EOMs. In particular, they are absent in our 1-parameter models. We will report on this in detail elsewhere [16].

49.2 Screening Distance

We study a dipole of an infinitely heavy quark and its antiquark, separated by a distance L, in the deconfined plasma of the gauge theory. The quarks are situated at the 4-dimensional boundary ($z = 0$ in our coordinates) and are connected by a macroscopic string in the bulk (see e.g. [4]). We accommodate a finite velocity v of the $Q\bar{Q}$ system with respect to the surrounding medium by boosting the bulk metric with rapidity $\eta = \operatorname{artanh}(v)$. In order to find the classical string configuration we have to extremize the Nambu–Goto action in the given gravity background.

There is a distance L_s, such that for $L < L_\mathrm{s}$ there are two string configurations connecting the dipole, while no such solution exists for $L > L_\mathrm{s}$. Thus, L_s is called the *screening distance* of the $Q\bar{Q}$ interaction in the thermal medium. In the studies [17–19] it was found that at any temperature T the screening distance is bounded from below by $L_\mathrm{s}^{\mathcal{N}=4\,\mathrm{SYM}}(T)$ under consistent deformations of AdS-Schwarzschild, the dual of $\mathcal{N} = 4$ SYM at $\mu = 0$ (where we understand consistency in the sense of solving equations of motion of a suitable 5-dimensional gravity action).

The question arises whether this bound holds under the inclusion of a chemical potential. We find that in the CGN and similar models, depending on (sign) choices in the metric, L_s can change in both directions. More interesting are consistent deformations which in our case are the two classes of 1-parameter models.

As can be seen from Fig. 49.1, the bound on L_s is violated in the Einstein frame models at large chemical potential, approximately when $\mu \gtrsim \sqrt{\kappa}$. In the string frame models it is only violated for $Q\bar{Q}$ pairs moving sufficiently fast at large $\mu \gtrsim \sqrt{\kappa}$.

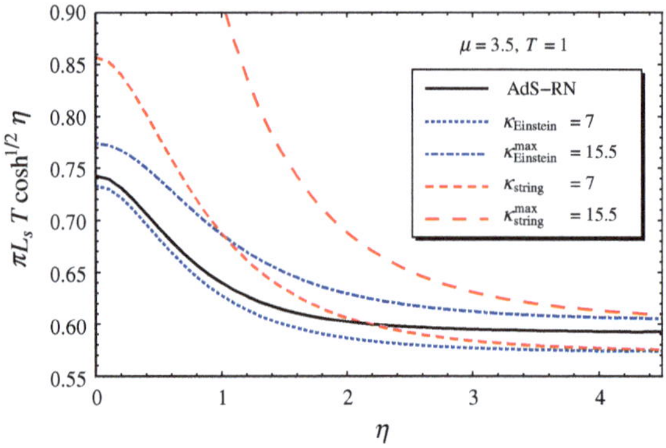

Fig. 49.1 Rapidity dependence of the screening distance L_s at finite temperature T and chemical potential μ in the 1-parameter models, evolving from the conformal case (dual to AdS-RN) to the maximal deformation κ_{max}. The asymptotic behavior $L_s \propto \cosh^{-1/2}(\eta)$ is scaled out. At large η, the differences between the two 1-parameter models vanish. L_s, μ, and κ are measured in units of temperature

However, for all deformations, the amount of violation of the bound is relatively small such that there might exist a slightly lower, improved bound. Moreover, in the Einstein frame models, the screening distance is a robust observable, depending only weakly on the deformation.

We also study the dependence of L_s on the $Q\bar{Q}$ rapidity η. Figure 49.1 illustrates that the ultrarelativistic scaling of the screening distance $L_s \propto \cosh^{-1/2}(\eta)$ is robust and remains valid in all models, at all chemical potentials. It is interesting to note that this robustness against deformation is different from what was found in [11] for other explicitly non-conformal models.

49.3 Free Energy and Running Coupling

The free energy $F(L)$ of the $Q\bar{Q}$ system can be extracted from the extremal classical string action following a well-known procedure (see e.g. [4]). The typical features in the consistently deformed models can be seen in Fig. 49.2. Here, $F(L)$ is normalized such that a configuration having $F < 0$ has less free energy than the non-interacting, unbound $Q\bar{Q}$ system. Thus, we see that the effect of increasing the chemical potential is a decrease in binding energy. Taking this together with our findings concerning the screening distance, we see that in these holographic models an increased net density around the $Q\bar{Q}$ dipole weakens its binding by screening the interaction. We find this effect regardless of the specific model under consideration.

To explore the impact of non-conformality on the interaction in more detail, we study the running coupling defined via the derivative of the free energy,

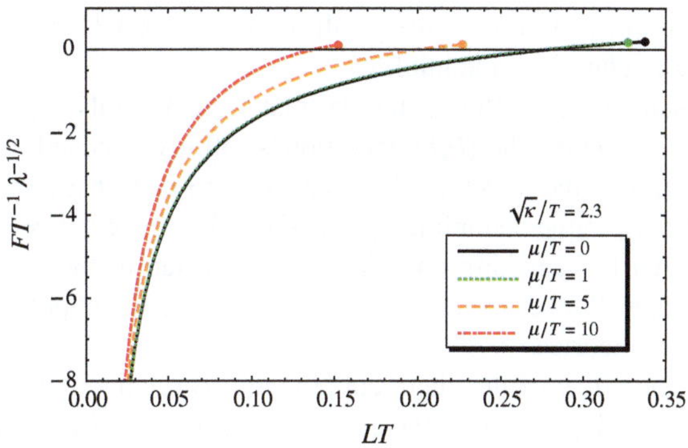

Fig. 49.2 Typical $Q\bar{Q}$ free energy $F(L)$ in a non-conformal 1-parameter string frame model for varying values of the chemical potential μ. The behavior in the Einstein frame models is qualitatively similar. $\sqrt{\lambda}$ is a constant. The endpoints of the *curves* are located at the respective screening distances

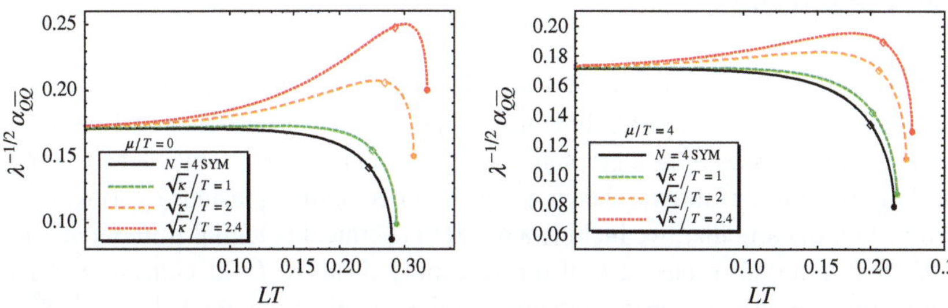

Fig. 49.3 Running coupling $\alpha_{Q\bar{Q}}(L)$ in $\mathcal{N} = 4$ SYM and 1-parameter string frame models with increasing non-conformality parameter κ, for zero (*left*) and nonzero (*right*) chemical potential μ

$$\alpha_{Q\bar{Q}}(L) \equiv \frac{3}{4}L^2\frac{\mathrm{d}F(L)}{\mathrm{d}L}\,. \tag{49.4}$$

In the conformal case, where $F(L) \propto 1/L$ is Coulombic, $\alpha_{Q\bar{Q}} = \text{const}$. Hence, any non-trivial dependence on L measures the deviation from conformality.

In Fig. 49.3 we show $\alpha_{Q\bar{Q}}(L)$ for the 1-parameter string frame models. The qualitative picture is the same in the Einstein frame models; however, in these models, $\alpha_{Q\bar{Q}}$ is very robust under deformations, so that the curves for different κ deviate only very little from each other.

From Fig. 49.3 we see that $\alpha_{Q\bar{Q}}$ is constant in the UV, i.e. at small distances L. One can also see this from the free energy itself in Fig. 49.2 which approaches Coulombic form in the UV. This is due to the restoration of conformality in the UV, the bulk realization of which is the condition on the metric to be asymptotically AdS.

At larger distance, both the explicit non-conformality and the thermal medium affect the $Q\bar{Q}$ interaction. In effect, the running coupling starts to deviate from its asymptotic value. We find a robust increase above the UV value at intermediate length scales due to non-conformality in all deformed models, at vanishing μ [17] but also at nonzero chemical potential.

The plasma starts to take effect at the thermal scale, roughly $L_{\mathrm{th}} \sim 1/T$, leading to a drop-off of $\alpha_{Q\bar{Q}}$, before the $Q\bar{Q}$ interaction is entirely screened. (The endpoint of the curves $\alpha_{Q\bar{Q}}(L)$ is at the screening distance L_{s}.) This pattern is also found in lattice QCD studies of $\alpha_{Q\bar{Q}}$ in the deconfined phase [20]. While the lattice calculations are presently restricted to vanishing chemical potential due to the sign problem, the holographic models allow us to explore the effect of a chemical potential in strongly coupled QCD-like theories.

We find that the effect of the chemical potential is relatively weak: the drop-off scale is only weakly dependent on the chemical potential while it strongly depends on the temperature.

49.4 Conclusion

We have reported on some results of our studies of deformed, i.e. non-conformal, gauge/gravity models for strongly coupled plasmas at nonzero chemical potential. Studying large classes of holographic models, we address the problem of the strong coupling dynamics of moving heavy mesons in deconfined plasmas by looking for universality. In particular, we include a nonzero chemical potential μ in our studies.

We find a certain robustness of the screening distance L_{s} at nonzero μ under deformations. However, when switching on the chemical potential, $L_{\mathrm{s}}^{\mathcal{N}=4\ \mathrm{SYM}}$ no longer is a lower bound on the screening distance under deformations, unlike in the case of $\mu = 0$.

Furthermore, we observe a weak impact of the chemical potential on the qualitative features of the quark–antiquark interaction. Even quantitatively, the dependence of characteristic scales on the chemical potential is generally significantly weaker than their dependence on temperature

Acknowledgments We thank K. Schade for many helpful discussions. This work was supported by the ExtreMe Matter Institute EMMI.

References

1. J.M. Maldacena, Adv. Theor. Math. Phys. **2**, 231 (1998)
2. S.S. Gubser, I.R. Klebanov, A.M. Polyakov, Phys. Lett. B **428**, 105 (1998)
3. E. Witten, Adv. Theor. Math. Phys. **2**, 253 (1998)
4. J. Casalderrey-Solana, H. Liu, D. Mateos, K. Rajagopal, U.A. Wiedemann (2011). arXiv:1101.0618
5. G. Policastro, D.T. Son, A.O. Starinets, Phys. Rev. Lett. **87**, 081601 (2001)
6. P. Kovtun, D.T. Son, A.O. Starinets, JHEP **0310**, 064 (2003)
7. P. Kovtun, D.T. Son, A.O. Starinets, Phys. Rev. Lett. **94**, 111601 (2005)
8. T. Matsui, H. Satz, Phys. Lett. B **178**, 416 (1986)

9. H. Liu, K. Rajagopal, U.A. Wiedemann, Phys. Rev. Lett. **98**, 182301 (2007)
10. H. Liu, K. Rajagopal, Y. Shi, JHEP **08**, 048 (2008)
11. E. Caceres, M. Natsuume, T. Okamura, JHEP **10**, 011 (2006)
12. S.D. Avramis, K. Sfetsos, D. Zoakos, Phys. Rev. D **75**, 025009 (2007)
13. P. Colangelo, F. Giannuzzi, S. Nicotri, Phys. Rev. D **83**, 035015 (2011)
14. O. DeWolfe, S.S. Gubser, C. Rosen, Phys. Rev. D **83**, 086005 (2011)
15. A. Samberg, Diploma thesis, Heidelberg University (2012)
16. C. Ewerz, L. Lin, A. Samberg, in preparation
17. K. Schade, Ph.D. thesis, Heidelberg University (2012)
18. C. Ewerz, K. Schade, PoS(Confinement X)270 (2013)
19. C. Ewerz, K. Schade, in preparation
20. O. Kaczmarek, F. Karsch, F. Zantow, P. Petreczky, Phys. Rev. D **70**, 074505 (2004)

MIX
Papier aus verantwortungsvollen Quellen
Paper from responsible sources
FSC® C105338

If you have any concerns about our products,
you can contact us on
ProductSafety@springernature.com

In case Publisher is established outside the EU,
the EU authorized representative is:
Springer Nature Customer Service Center GmbH
Europaplatz 3, 69115 Heidelberg, Germany

Printed by Libri Plureos GmbH
in Hamburg, Germany